ISBN 978-1-330-28801-6
PIBN 10014493

A

TREATISE ON ARITHMETIC:

IN

THEORY AND PRACTICE.

A

TREATISE ON ARITHMETIC

IN

Theory and Practice:

FOR THE USE OF

THE IRISH NATIONAL SCHOOLS

DUBLIN;

PUBLISHED BY DIRECTION OF THE
COMMISSIONERS OF NATIONAL EDUCATION IN IRELAND,
At their Office in Marlborough-street.

SOLD BY

W. CURRY, JUN., AND CO., DUBLIN ; R. GROOMBRIDGE AND SONS, LONDON ;
GEORGE PHILIP, LIVERPOOL , FRASER AND CO., EDINBURGH ;
ARMOUR AND RAMSAY, AND DONOGHUE AND MANTZ, MONTREAL, CANADA ;
AND CHUBB AND CO., HALIFAX, NEW BRUNSWICK.

1846.

DUBLIN:

Printed by ALEXANDER THOM, 87, Abbey-street.

PREFACE.

In the present edition a vast number of exercises have been added, that no rule, however trifling, might be left without so many illustrations as should serve to make it sufficiently familiar to the pupil. And when it was feared that the application of any rule to a particular class of cases might not at once suggest itself, some question calculated to remove, or diminish the difficulty has been introduced among the examples.

A considerable space is devoted to the "nature of numbers," and "the principles of notation and numeration;" for the teacher may rest assured, that the facility, and even the success, with which subsequent parts of his instruction will be conveyed to the mind of the learner, depends, in a great degree, upon an adequate acquaintance with them. Hence, to proceed without securing a perfect and practical knowledge of this part of the subject, is to retard, rather than to accelerate improvement.

The pupil, from the very commencement, must be made *perfectly* familiar with the terms and signs which are introduced. Of the great utility of technical language (accurately understood) it is almost superfluous to say anything here: we cannot, however, forbear, upon this occasion, recalling to remembrance what is so admirably and so effectively inculcated in the "Easy Lessons on Reasoning." "Even in the common mechanical arts, something of a technical language is found needful for those who are learning or exercising them. It would be a very great inconvenience, even to a common carpenter, not to have a precise, well understood name for each of the several operations he performs, such as chiselling, sawing, planing, &c., and for the several tools [or instruments] he works with. And if we had not such words as addition, subtraction, multiplication, division, &c., employed in an exactly defined sense, and also fixed rules for conducting these and other arithmetical processes, it would be a tedious and uncertain work to go through even such simple calculations as a child very soon learns to perform with perfect ease. And after all there would be a fresh difficulty in making other persons understand clearly the correctness of the calculations made.

" You are to observe, however, that technical language and rules, if you would make them really useful, must be not only *distinctly understood*, but also learned and *remembered* as familiarly as the alphabet, and employed *constantly*, and with scrupulous *exactness ;* otherwise, technical language will prove an encumbrance instead of an advantage, just as a suit of clothes would be if, instead of putting them on and *wearing* them, you were to carry them about in your hand." Page 11.

What is said of *technical language* is, at least, equally true of the *signs* and *characters* by which we still further facilitate the conveyance of our ideas on such matters as form the subject of the present work. It is much more simple to put down a character which expresses a process, than to write the name, or description of the latter, *in full.* Besides, in glancing over a mathematical investigation, the mind is able, with greater ease, to connect, and understand its different portions when they are briefly expressed by familiar signs, than when they are indicated by words which have nothing particularly calculated to *catch the eye*, and which cannot even be clearly understood without considerable attention. But it must be borne in mind, that, while such a treatise as the present, will seem easy and intelligible enough if the signs, which it contains in almost every page, are as familiar as they should be, it must necessarily appear more or less obscure to those who have not been habituated to the use of them. They are however, so few and so simple that there is no excuse for their not being perfectly understood—particularly by the teacher of arithmetic.

Should peculiar circumstances render a different arrangement of the rules preferable, or make the omission of any of them, for the present at least, advisable, the judicious master will never be at a loss how to act—there *may* be instances in which the shortness of the time, or the limited intelligence of the pupil, will render it necessary to confine his instruction to the more important branches. The teacher should, if possible, make it an inviolable rule to receive no answer unless accompanied by its explanation, and its reason. The references which have been subjoined to the different questions, and which indicate the paragraphs where the answers are *chiefly* to be obtained, and also those references which are scattered through the work, will, be found of considerable assistance ; for, as the most intelligent pupil will occasionally forget something he has learned, he may not at once see that a certain principle is applicable to a particular case, nor even remember where he has seen it explained.

Decimals have been treated of at the same time as integers, because, since both of them follow *precisely* the same laws, when the rules relating to integers are fully understood, there is *nothing new* to be learned on the subject—particularly if what has been said with reference to numeration and notation is carefully borne in mind. Should it, however, in any case, be preferred, what relates to them can be omitted until the learner shall have made some further advance.

The most useful portions of *mental arithmetic* have been introduced into " Practice" and the other rules with which they seemed more immediately connected.

The different rules should be very carefully impressed on the mind of the learner, and when he is found to have been guilty of any inaccuracy, he should be made to *correct him-self* by repeating each part of the appropriate rule, and exemplifying it, until he perceives his error. It should be continually kept in view that, in a work on such a subject as arithmetic, any portion must seem difficult and obscure without a knowledge of what precedes it.

The table of logarithms and article on the subject, also the table of squares and cubes, square roots and cube roots of numbers, which have been introduced at the end of the work, will, it is expected, prove very acceptable to the more dvanced arithmetician.

CONTENTS.

PART I.

SIGNS USED IN THIS TREATISE.

$+$ the sign of addition; as $5+7$, or 5 to be added to 7.

$-$ the sign of subtraction; as $4-3$, or 3 to be subtracted from 4.

$\times$ the sign of multiplication; as 8×9, or 8 to be multiplied by 9.

$\div$ the sign of division; as $18\div6$, or 18 to be divided by 6.

—— the vinculum, which is used to show that all the quantities united by it are to be considered as but one. Thus $\overline{4+3-7}\times6$ means 4 to be added to 3, 7 to be taken from the sum, and 6 to be multiplied into the remainder—the latter is equivalent to the *whole* quantity under the vinculum.

$=$ the sign of equality; as $5+6=11$, or 5 added to 6, is equal to 11.

$\frac{3}{4}>\frac{1}{2}$, and $\frac{2}{3}<\frac{8}{9}$, mean that $\frac{3}{4}$ is greater than $\frac{1}{2}$, and that $\frac{2}{3}$ is less than $\frac{8}{9}$.

$:$ is the sign of ratio or relation; thus $5:6$, means the ratio of 5 to 6, and is read 5 is to 6.

$::$ indicates the equality of ratios; thus, $5:6::7:8$, means that there is the same relation between 5 and 6 as between 7 and 8; and is read 5 is to 6 as 7 is to 8.

$\sqrt{}$ the radical sign. By itself, it is the sign of the *square* root; as $\sqrt{5}$, which is the same as $5^{\frac{1}{2}}$, the square root of 5. $\sqrt[3]{6}$, is the cube root of 3, or $3^{\frac{1}{3}}$. $\sqrt[7]{4}$, is the 7th root of 4, or $4^{\frac{1}{7}}$, &c.

EXAMPLE.—$\sqrt{\overline{8-3+7\times4\div6}}+31\times\sqrt[3]{9}\div10^{\frac{1}{2}}\times5^{2}=$ 641·31, &c. may be read thus: take 3 from 8, add 7 to the difference, multiply the sum by 4, divide the product by 6, take the square root of the quotient and to it add 31, then multiply the sum by the cube root of 9, divide the product by the square root of 10, multiply the quotient by the square of 5, and the product will be equal to 641·31, &c.

These signs are *fully* explained in their proper places.

ARITHMETIC.

PART I.

TABLES.

MULTIPLICATION TABLE.

Twice	3 times	4 times	5 times	6 times	7 times
1 are 2	1 are 3	1 are 4	1 are 5	1 are 6	1 are 7
2 — 4	2 — 6	2 — 8	2 — 10	2 — 12	2 — 14
3 — 6	3 — 9	3 — 12	3 — 15	3 — 18	3 — 21
4 — 8	4 — 12	4 — 16	4 — 20	4 — 24	4 — 28
5 — 10	5 — 15	5 — 20	5 — 25	5 — 30	5 — 35
6 — 12	6 — 18	6 — 24	6 — 30	6 — 36	6 — 42
7 — 14	7 — 21	7 — 28	7 — 35	7 — 42	7 — 49
8 — 16	8 — 24	8 — 32	8 — 40	8 — 48	8 — 56
9 — 18	9 — 27	9 — 36	9 — 45	9 — 54	9 — 63
10 — 20	10 — 30	10 — 40	10 — 50	10 — 60	10 — 70
11 — 22	11 — 33	11 — 44	11 — 55	11 — 66	11 — 77
12 — 24	12 — 36	12 — 48	12 — 60	12 — 72	12 — 84

8 times	9 times	10 times	11 times	12 times
1 are 8	1 are 9	1 are 10	1 are 11	1 are 12
2 — 16	2 — 18	2 — 20	2 — 22	2 — 24
3 — 24	3 — 27	3 — 30	3 — 33	3 — 36
4 — 32	4 — 36	4 — 40	4 — 44	4 — 48
5 — 40	5 — 45	5 — 50	5 — 55	5 — 60
6 — 48	6 — 54	6 — 60	6 — 66	6 — 72
7 — 56	7 — 63	7 — 70	7 — 77	7 — 84
8 — 64	8 — 72	8 — 80	8 — 88	8 — 96
9 — 72	9 — 81	9 — 90	9 — 99	9 — 108
10 — 80	10 — 90	10 — 100	10 — 110	10 — 120
11 — 88	11 — 99	11 — 110	11 — 121	11 — 132
12 — 96	12 —108	12 — 120	12 — 132	12 — 144

It appears from this table, that the multiplication of the same two numbers, in whatever order taken, produces the same result; thus 5 times 6, and 6 times 5 are 30 :—the reason will be explained when we treat of multiplication. There are, therefore, several repetitions, which, although many persons conceive them unnecessary, are not, perhaps, quite unprofitable.   The following is free from such an objection :—

Twice 2 are 4	5 times 7 are 35	10 times 8 are 80
,, 3 — 6	,, 8 — 40	,, 9 — 90
,, 4 — 8	,, 9 — 45	,, 10 — 100
,, 5 — 10		,, 11 — 110
,, 6 — 12	6 times 6 — 36	
,, 7 — 14	,, 7 — 42	11 times 2 — 22
,, 8 — 16	,, 8 — 48	,, 3 — 33
,, 9 — 18	,, 9 — 54	,, 4 — 44
3 times 3 — 9		,, 5 — 55
,, 4 — 12	7 times 7 — 49	,, 6 — 66
,, 5 — 15	,, 8 — 56	,, 7 — 77
,, 6 — 18	,, 9 — 63	,, 8 — 88
,, 7 — 21		,, 9 — 99
,, 8 — 24	8 times 8 — 64	
,, 9 — 27	,, 9 — 72	12 times 2 — 24
4 times 4 — 16	9 times 9 — 81	,, 3 — 36
,, 5 — 20		,, 4 — 48
,, 6 — 24		,, 5 — 60
,, 7 — 28	10 times 2 are 20	,, 6 — 72
,, 8 — 32	,, 3 — 30	,, 7 — 84
,, 9 — 36	,, 4 — 40	,, 8 — 96
	,, 5 — 50	,, 9 — 108
5 times 5 — 25	,, 6 — 60	,, 10 — 120
,, 6 — 30	,, 7 — 70	,, 11 — 132
		,, 12 — 144

" Ten," or " eleven times," in the above, scarcely requires to be committed to memory; since we perceive, that to multiply a number by 10, we have merely to add a cypher to the right hand side of it :—thus, 10 times 8 are 80 ; and to multiply it by 11 we have only to set it down twice :—thus, 11 times 2 are 22.

The following tables are required for reduction, the compound rules, &c., and may be committed to memory as convenience suggests.

TABLE OF MONEY.

A farthing is the smallest coin generally used in this country, it is represented by . . . . $\frac{1}{4}$

Symbols.

Farthings					
2 .	.	.	.	. make 1 halfpenny,	$\frac{1}{2}$
	halfpence				
4 or	2	.	.	1 penny,	*d.*
		pence			
48	24 or	12	.	1 shilling,	*s.*
			shillings		
960	480	240 or	20	1 pound,	£
1,008	504	252 or	21	1 guinea,	*g.*

The symbols of pounds, shillings, and pence, are placed over the numbers which express them. Thus, 3 ,, 14 ,, 6, means, three pounds, fourteen shillings, and sixpence. Sometimes only the symbol for pounds is used, and is placed before the whole quantity; thus, £3 ,, 14 ,, 6. 3 9½ means three shillings and ninepence halfpenny. 2s. 6¾d. means two shillings and sixpence three farthings, &c.

When learning the above and following tables, the pupil should be required, at first, to commit to memory only those portions which are over the thick angular lines; thus, in the one just given :—2 farthings make one halfpenny; 2 halfpence one penny; 12 pence one shilling; 20 shillings one pound; and 21 shillings one guinea.

¼, ½, ¾, really mean the quarter, half, and three quarters of a penny. *d.* is used as a symbol, because it is the first letter of "denarius," the Latin word signifying a penny; *s.* was adopted for a similar reason—"solidus," meaning, in the same language, a shilling; and £ also—"Libra," signifying a pound.

s.	*d.*	
2	6	make one half Crown.
5	0	one Crown.
13	4	one Mark.

AVOIRDUPOISE WEIGHT.

Its name is derived from French—and ultimately from Latin words signifying "to have weight." It is used in weighing heavy articles.

Drams	ounces	pounds	quarters	hundreds		Symbols.
16	.			.	make 1 ounce,	oz.
256 or	16	.	.	.	. 1 pound,	℔.
7,168	448 or	28	.	.	. 1 quarter,	q.
28,672	1,792	112 or	4	.	. 1 hundred,	cwt.
573,440	35,840	2,240	80 or	20	1 ton,	t.

14 lbs., and in some cases 16 lbs., make 1 stone.
20 stones . . . 1 barrel.

TROY WEIGHT.

It is so called from Troyes, a city in France, where it was first employed; it is used in philosophy, in weighing gold, &c.

Grains	pennyweights	ounces		Symbols.
.			. . . .	grs.
24	.	.	. make 1 pennyweight,	dwt.
480 or	20	.	1 ounce, .	oz.
5,760	240 or	12	1 pound, .	℔.

A grain was originally the weight of a grain of corn, taken from the middle of the ear; a pennyweight, that of the silver penny formerly in use.

APOTHECARIES WEIGHT.

In mixing medicines, apothecaries use Troy weight, but subdivide it as follows:—

Grains	scruples	drams	ounces		Symbols.
20	.	. .	.	. make 1 scruple,	℈
60 or	3	.		1 dram,	ʒ
480	24 or	8	.	. 1 ounce,	ℨ
5,760	288	96 or	12	. 1 pound,	℔.

The "Carat," which is equal to four grains, is used in weighing diamonds. The term carat is also applied in estimating the fineness of gold; the latter, when perfectly pure, is said to be " 24 carats fine." If there are 23 parts gold, and one part some other material, the mixture is said to be " 23 carats fine ;" if 22 parts out of the 24 are gold, it is " 22 carats fine," &c. ;—the whole mass is, in all cases, supposed to be divided into 24 parts, of which the number consisting of gold is specified. Our gold coin is 22 carats fine ; pure gold being very soft would too soon wear out. The degree of fineness of gold articles is marked upon them at the Goldsmith's Hall ; thus we generally perceive "18" on the cases of gold watches ; this indicates that they are " 18 carats fine"—the lowest degree of purity which is stamped.

	grs.
A Troy ounce contains .	480
An avoirdupoise ounce .	437½
A Troy pound . .	5,760
An avoirdupoise pound .	7,000

A Troy pound is equal to 372·965 French grammes.

175 Troy pounds are equal to 144 avoirdupoise ; 175 Troy are equal to 192 avoirdupoise ounces.

CLOTH MEASURE.

Inches					
2¼	.	.	.		make 1 nail.
	nails				
9 or	4	.	.	.	1 quarter.
		quarters			
36	16 or	4	.	.	1 yard.
27	12 or	3	.	.	1 Flemish ell.
45	20 or	5	.	.	1 English ell.
54	24 or	6	.	.	1 French ell.

LONG MEASURE.
(It is used to measure Length.)

Lines						
12	.	.			.	make 1 inch.
	inches					
144 or	12	.	.	.		1 foot.
		feet				
432	36 or	3	.	.		1 yard.
			yards			
2,376	198	16½ or	5½			1 English perch.
3,024	252	21 or	7	.		1 Irish perch.
				perches		
95,040	7,920	660	220 or	40	.	1 English furlong.
120,960	10,080	840	280 or	40	.	1 Irish furlong.
					furlongs	
760,320	63,360	5,280	1,760	320 or	8	1 English mile.
967,680	80,640	6,720	2,240	320 or	8	1 Irish mile.

B

Three miles make one league. $69\frac{1}{15}$ English miles make 60 nautical, or geographical miles; which are equal to one degree, or the three hundred and sixtieth part of the circumference of the globe—as measured on the equator.

4 inches make	1 hand (used in measuring horses).
3 inches	1 palm.
3 palms	1 span.
18 inches	1 cubit.
5 feet	1 pace.
6 feet	1 fathom.
120 fathoms	1 cable's length.

100 links, 4 English perches (or poles), 22 yards, 66 feet, or 792 inches, make one chain. Each link, therefore, is equal to $7\frac{92}{100}$ inches. 11 Irish are equal to 14 English miles. The Paris foot is equal to 12·792 English inches; the Roman foot to 11·604; and the French metre to 39·383.

MEASURE OF SURFACES.

A surface is called a square when it has four equal sides and four equal angles. A square inch, therefore, is a surface one inch long and one inch wide; a square foot, a surface one foot long and one foot wide, &c.

Square inches	square feet	sq. yards	sq. perches	sq. roods	sq. acres	
144						make 1 sq. foot.
1,296 or	9					1 square yard.
39,204	272¼ or	30¼				1 sq Eng. perch.
63,504	441 or	49				1 sq. Irish perch.
1,568,160	10,890	1,210 or	40			1 sq. Eng. rood.
2,540,160	17,640	1,960 or	40			1 sq. Irish rood.
6,272,640	43,560	4,840	160 or	4		1 statute acre.
10,160,640	70,560	7,840	160 or	4		1 plantation acre.
4,014,489,600	27,878,400	3,097,600	102,400	2,560 or	640	1 sq. Eng. mile.
6,502,809,600	25,158,400	5,017,600	102,400	2,560 or	640	1 sq. Irish mile.

The English, called also the statute acre, consists of 10 square chains, or 100,000 square links.

The English acre being 4,840 square yards, and the Irish, or plantation acre, 7,840; 196 square English are equal to 121 square Irish acres.

The English square mile being 3,097,600 square yards, and the Irish 5,017,600; 196 English square miles are equal to 121 Irish:—we have seen, however, that 14 English are equal to 11 Irish *linear* miles.

MEASURE OF SOLIDS.

The teacher will explain that a cube is a solid having six equal square surfaces; and will illustrate this by models or examples—the more familiar the better. A cubic inch is a solid, each of whose six sides or faces is a *square inch;* a cubic foot a solid, each of whose six sides is a *square foot,* &c.

Cubic inches				make 1 cubic foot.
1,728	.	.	.	
	cubic feet			
46,656 or	27	.	.	1 cubic yard.

WINE MEASURE.

Gills or naggins						make 1 pint.
4	.			.		
	pints					
8 or	2	.				1 quart.
		quarts				
32	8 or	4	.			1 gallon.
			gallons			
320	80	40 or	10	.		1 anker.
576	144	72	18	.		1 runlet.
1,344	336	168	42	.		1 tierce.
2,016	504	2·2	63	.		1 hogshead
2,688	672	336	84	.	.	1 puncheon.
				hogsheads		
4,032	1,008	504	126 or	2		1 pipe or butt.
					pipes	
8,064	2,016	1,008	252	4 or	2	1 tun.

In some places a gill is equal to half a pint.

Foreign wines, &c., are often sold by measures differing from the above.

ALE MEASURE.

Gallons						make 1 firkin.
8	.	.		.	.	
	firkins					
16 or	2	.		.	.	1 kilderkin.
		kilderkins				
32	4 or	2	.	.		1 barrel.
			barrels			
48	6	3 or	1½	.		1 hogshead.
64	8	4 or	2	.		1 puncheon.
96	12	6 or	3	.		1 butt.

BEER MEASURE.

Gallons	firkins	kilderkins	barrels	
9	.	.	. .	make 1 firkin.
18 or	2	.	. .	1 kilderkin.
36	4 or	2	. .	1 barrel.
54	6	3 or	1½ .	1 hogshead.
72	8	4 or	2 .	1 puncheon.
108	12	6 or	3 .	1 butt.

DRY MEASURE.

(It is used for wheat, and other dry goods.)

Pints	quarts	pottles	gallons	pecks	bushels	coombs	quarters	weys	
4 or	2	.	.	.	.				make 1 pottle.
8	4 or	2	.	.	.				1 gallon.
16	8	4 or	2	.	.	.			1 peck.
64	32	16	8 or	4	.	.			1 bushel.
192	96	48	24	12 or	3	.	.		1 sack.
256	128	64	32	16 or	4	.	.		1 coomb.
576	288	144	72	36 or	9	.	.		1 vat.
512	256	128	64	32	8 or	2	.		1 quarter.
2,048	1,024	512	256	128	32	8 or	4		1 chaldron.
2,560	1,280	640	320	160	40	10 or	5		1 wey.
5,120	2,560	1,280	640	320	80	20	10 or	2	1 last.

The pint dry measure contains about 34⅔ cubic inches; 277¼ cubic inches was made the standard gallon for both liquid and dry goods, by an Act of Parliament which came into operation in 1826.

Coals are now sold by weight; 140 pounds make one bag; 16 bags one ton.

MEASURE OF TIME.

Thirds							Symbols.
60	.	.	.	.	.	.	make 1 second ″
	seconds						
3600, or	60	.	.	.	.	..	1 minute ′
		minutes					
216,000	3600 or	60	.	.	.	.	1 hour h.
			hours				
5,184,000	86,400	1,440 or	24	.	.		1 day d.
				days			
36,288,000	604,800	10,080	168 or	7	.	..	1 week w.
145,152,000	2,419,200	40,320	672 or	28	.	.	1 lunar month
1,892,160,000	31,536,000	525,600	8,760 or	365	.	..	1 common year
1,897,344,000	31,622,400	527,040	8,784 or	366	.	..	1 leap year.
					calendar mon	12	
1,892,160,000	31,536,000	525,600	8,760	365 or	lunar months	13	} 1 year.

The following will exemplify the use of the above symbols:—
The solar year consists of 365 d. 5 h. 48′ 45″ 30‴; read "three hundred and sixty-five days, five hours, forty-eight minutes, forty-five seconds, and thirty thirds.

The number of days in each of the twelve calendar months will be easily remembered by means of the well known lines,

" Thirty days hath September,
April, June, and November,
February twenty-eight alone
And all the rest thirty-one."

The following table will enable us to find how many days there are from any day in one month to any day in another.

		FROM ANY DAY IN										
		Jan.	Feb.	Mar	April	May	June	July	Aug.	Sept.	Oct.	Nov	Dec.
	Jan.	365	334	306	275	245	214	184	153	122	92	61	31
	Feb.	31	365	337	306	276	245	215	184	153	123	92	62
	Mar.	59	28	365	334	304	273	243	212	181	151	120	90
	April	90	59	31	365	335	304	274	243	212	182	151	121
TO ANY DAY IN	May	120	89	61	30	365	334	304	273	242	212	181	151
	June	151	120	92	61	31	365	335	304	273	243	212	182
	July	181	150	122	91	61	30	365	334	303	273	242	212
	Aug.	212	181	153	122	92	61	31	365	334	304	273	243
	Sept.	243	212	184	153	123	92	62	31	365	335	304	274
	Oct.	273	242	214	183	153	122	92	61	30	365	334	304
	Nov.	304	273	245	214	184	153	123	92	61	31	365	335
	Dec.	334	303	275	241	214	183	153	122	91	61	30	365

To find by this table the distance between any two days in two different months :

RULE.—Look along that vertical row of figures at the head of which stands the first of the given months ; and also along the horizontal row which contains the second ; the number of days from any day in the one month to the same day in the other, will be found where these two rows intersect each other.   If the given day in the latter month is earlier than that in the former, find by how much, and *subtract* the amount from the number obtained by the table.   If, on the contrary, it is later, ascertain by how much, and *add* the amount.

When February is included in the given time, and it is a leap year, add one day to the result.

EXAMPLE 1.—How many days are there between the fifteenth of March and the fourth of October ?   Looking down the vertical row of figures, at the head of which March is placed, and at the same time, along the horizontal row at the left hand side of which is October, we perceive in their intersection the number 214 :—so many days, therefore. intervene between the fifteenth of March and the fifteenth of October.   But the fourth of October is eleven days earlier than the fifteenth ; we therefore subtract 11 from 214, and obtain 203, the number required.

EXAMPLE 2.—How many days are there between the third of January and the nineteenth of May ?   Looking as before in the table, we find that 120 days intervene between the third of January and the third of May ; but as the nineteenth is sixteen days later than the third, we add 16 to 120 and obtain 136, the number required.

Since February is in this case included, if it were a leap year, as that month would then contain 29 days, we should add one to the 136, and 137 would be the answer.

During the lapse of time, the calendar became inaccurate : it was corrected by Pope Gregory.   To understand how this became necessary, it must be borne in mind that the Julian Calendar, formerly in use, added one day every fourth year to the month of February ; but this being somewhat too much, the days of the months were thrown out of their proper places, and to such an extent, that each had become ten days too much in advance.   Pope Gregory, to remedy this, ordained that what, according to the Julian style, would have been the 5th of October 1582, should be considered as the 15th; and to prevent the recurrence of such a mistake, he desired that, in place of the last year of every century being, as hitherto, a leap year, only the last year of every fourth century should be deemed such.

The "New Style," as it is called, was not introduced into England until 1752, when the error had become eleven days. The Gregorian Calendar itself is slightly inaccurate.

To find if any given year be a leap year. If not the last year of a century:

RULE.—Divide the number which represents the given year by 4, and if there be no remainder, it is a leap year. If there be a remainder, it expresses how long the given year is after the preceding leap year.

EXAMPLE 1.—1840 was a leap year, because 1840 divided oy 4 leaves no remainder.

EXAMPLE 2.—1722 was the second year after a leap year, because 1722 divided by 4 leaves 2 as remainder.

If the given year be the last of a century:

RULE.—Divide the number expressing the centuries by 4, and if there be no remainder, the given one is a leap year; if there be a remainder, it indicates the number of centuries between the given and preceding last year of a century which was a leap year.

EXAMPLE 1.—1600 was a leap year, because 16, being divided by 4, leaves nothing.

EXAMPLE 2.—1800 was two centuries after that last year of a century which was a leap year, because, divided by 4, it leaves 2.

DIVISION OF THE CIRCLE.

Thirds						Symbols.
60	.	.	.	.	make 1 second	"
		seconds				
3600	or	60	.		1 minute	'
			minutes			
216,000		3,600 or	60		1 degree	○
				degrees		
77,760,000		1,296,000	21,600 or	360	1 circumference	

Every circle is supposed to be divided into the same number of degrees, minutes, &c.; the greater or less, there-

fore, the circle, the greater or less each of these will be. The
following will exemplify the applications of the symbols :—
60° 5′ 4″ 6‴; which means sixty degrees, five minutes, four
seconds, and six thirds.

DEFINITIONS.

1. *Arithmetic* may be considered either as a science
or as an art. As a science, it teaches the properties of
numbers ; as an art, it enables us to apply this knowledge to practical purposes ; the former may be called
theoretical, the latter practical arithmetic.

2. *A Unit*, or as it is also called, *Unity*, is one of the
individuals under consideration, and may include many
units of another kind or denomination ; thus a unit of
the order called "tens" consists of ten simple units. Or
it may consist of one or more parts of a unit of a higher
denomination ; thus five units of the order of "tens" are
five parts of one of the denomination called "hundreds ;"
three units of the denomination called "tenths" are
three parts of a unit, which we shall presently term the
"unit of comparison."

3. *Number* is constituted of two or more units ;
strictly speaking, therefore, unity itself cannot be considered as a number.

4. *Abstract Numbers* are those the properties of
which are contemplated without reference to their application to any particular purpose—as five, seven, &c. ;
abstraction being a process of the mind, by which it separately considers those qualities which cannot in reality
exist by themselves ; thus, for example, when we attend
only to the length of anything, we are said to abstract
from its breadth, thickness, colour, &c., although these
are necessarily found associated with it. There is nothing
inaccurate in this abstraction, since, although length
cannot exist without breadth, thickness, &c., it has properties independent of them. In the same way, five, seven,
&c., can be considered only by an abstraction of the
mind, as not applied to indicate some particular things.

5. *Applicate Numbers* are exactly the reverse of

abstract, being applied to indicate particular objects—as five men, six houses.

6. *The Unit of Comparison.* In every number there is some unit or individual which is used as a standard: this we shall henceforward call the "unit of comparison." It is by no means necessary that it should always be the same; for at one time we may speak of four objects of one species, at another of four objects of another species, at a third, of four dozen, or four scores of objects; in all these cases *four* is the number contemplated, though in each of them the idea conveyed to the mind is different—this difference arising from the different standard of comparison, or unity assumed. In the first case, the "unit of comparison" was a single object; in the second, it was also a single object, but not of the same kind; in the third, it became a dozen; and in the fourth, a score of objects. Increasing the "unit of comparison" evidently increases the quantity indicated by a given number; while decreasing it has a contrary effect. It will be necessary to bear all this carefully in mind.

7. *Odd Numbers.* One, and every succeeding alternate number, are termed *odd*; thus, three, five, seven, &c.

8. *Even Numbers.* Two, and every succeeding alternate number, are said to be *even*; thus, four, six, eight, &c. It is scarcely necessary to remark, that after taking away the odd numbers, all those which remain are even, and after taking away the even, all those which remain are odd.

We shall introduce many other definitions when treating of those matters to which they relate. A clear idea of what is proposed for consideration is of the greatest importance; this must be derived from the definition by which it is explained.

Since nothing assists both the understanding and the memory more than accurately dividing the subject of instruction, we shall take this opportunity of remarking to both teacher and pupil, that we attach much importance to the divisions which in future shall actually be made, or shall be implied by the order in which the different heads will be examined.

SECTION I.

ON NOTATION AND NUMERATION.

1. To avail ourselves of the properties of numbers, we must be able both to form an idea of them ourselves, and to convey this idea to others by spoken and by written language ;—that is, by the voice, and by characters. The expression of number by characters, is called *notation*, the reading of these, *numeration*. Notation, therefore, and numeration, bear the same relation to each other as *writing* and *reading*, and though often confounded, they are in reality perfectly distinct.

2. It is obvious that, for the purposes of Arithmetic, we require the power of designating all possible numbers ; it is equally obvious that we cannot give a different name or character to each, as their variety is boundless. We must, therefore, by some means or another, make a limited system of words and signs suffice to express an unlimited amount of numerical quantities :—with what beautiful simplicity and clearness this is effected, we shall better understand presently.

3. Two modes of attaining such an object present themselves ; the one, that of *combining* words or characters already in use, to indicate new quantities ; the other, that of representing a variety of different quantities by a *single* word or character, the danger of mistake at the same time being prevented. The Romans simplified their system of notation by adopting the principle of *combination ;* but the still greater perfection of ours is due also to the expression of many numbers by the *same* character.

4. It will be useful, and not at all difficult, to explain to the pupil the mode by which, as we may suppose, an idea of considerable numbers was originally acquired, and of which, indeed, although unconsciously, we still avail ourselves ; we shall see, at the same time, how methods of simplifying both numeration and notation were naturally suggested.

Let us suppose no system of numbers to be as yet constructed, and that a heap, for example, of pebbles, is placed before us that we may discover their amount. If this is considerable, we cannot ascertain it by looking at them all together, nor even by separately inspecting them; we must, therefore, have recourse to that contrivance which the mind always uses when it desires to grasp what, taken as a whole, is too great for its powers. If we examine an extensive landscape, as the eye cannot take it all in at one view, we look successively at its different portions, and form our judgment upon them in detail. We must act similarly with reference to large numbers; since we cannot comprehend them at a single glance, we must divide them into a sufficient number of parts, and, examining these in succession, acquire an indirect, but accurate idea of the entire. This process becomes by habit so rapid, that it seems, if carelessly observed, but one act, though it is made up of many: it is indispensable, whenever we desire to have a *clear idea* of numbers—which is not, however, every time they are mentioned.

5. Had we, then, to form for ourselves a numerical system, we would naturally divide the individuals to be reckoned into equal groups, each group consisting of some number quite within the limit of our comprehension; if the groups were few, our object would be attained without any further effort, since we should have acquired an accurate knowledge of the number of groups, and of the number of individuals in each group, and therefore a satisfactory, although indirect estimate of the whole.

We ought to remark, that different persons have very different limits to their perfect comprehension of number; the intelligent can conceive with ease a comparatively large one; there are savages so rude as to be incapable of forming an idea of one that is extremely small.

6. Let us call the *number* of individuals that we choose to constitute a group, the *ratio;* it is evident that the larger the ratio, the smaller the number of groups, and the smaller the ratio, the larger the number of groups— but the smaller the number of groups the better.

7. If the groups into which we have divided the objects to be reckoned exceed in amount that number of which we have a perfect idea, we must continue the process, and considering the groups themselves as individuals, must form with them new groups of a higher order.   We must thus proceed until the number of our highest group is sufficiently small.

8. The *ratio* used for groups of the second and higher orders, would naturally, but not necessarily, be the same as that adopted for the lowest; that is, if seven individuals constitute a group of the first order, we would probably make seven groups of the first order constitute a group of the second also; and so on.

9. It might, and very likely would happen, that we should not have so many objects as would exactly form a certain number of groups of the highest order—some of the next lower might be left.   The same might occur in forming one or more of the other groups.   We might, for example, in reckoning a heap of pebbles, have two groups of the fourth order, three of the third, none of the second, five of the first, and seven individuals or " units of comparison."

10. If we had made each of the first order of groups consist of ten pebbles, and each of the second order consist of ten of the first, each group of the third of ten of the second, and so on with the rest, we had selected the *decimal* system, or that which is not only used at present, but which was adopted by the Hebrews, Greeks, Romans, &c.   It is remarkable that the language of every civilized nation gives names to the different groups of this, but not to those of any other numerical system; its very general diffusion, even among rude and barbarous people, has most probably arisen from the habit of counting on the fingers, which is not altogether abandoned, even by us.

11. It was not indispensable that we should have used the same *ratio* for the groups of all the different orders; we might, for example, have made four pebbles form a group of the first order, twelve groups of the first order a group of the second, and twenty groups of the second a group of the third order :—in such a

case we had adopted a system exactly like that to be found in the table of money (page 3), in which four farthings make a group of the order *pence,* twelve pence a group of the order *shillings,* twenty shillings a group of the order *pounds.* While it must be admitted that the use of the same system for applicate, as for abstract numbers, would greatly simplify our arithmetical processes—as will be very evident hereafter, a glance at the tables given already, and those set down in treating of exchange, will show that a great variety of systems have actually been constructed.

12. When we use the same *ratio* for the groups of all the orders, we term it a *common* ratio. There appears to have been no particular reason why *ten* should have been selected as a " common ratio" in the system of numbers ordinarily used, except that it was suggested, as already remarked, by the mode of counting on the fingers ; and that it is neither so low as unnecessarily to increase the number of orders of groups, nor so high as to exceed the conception of any one for whom the system was intended.

13. A system in which ten is the " common ratio" is called *decimal,* from " decem," which in Latin signifies ten :—ours is, therefore, a "decimal system" of numbers. If the common ratio were sixty, it would be a *sexagesimal* system ; such a one was formerly used, and is still retained—as will be perceived by the tables already given for the measurement of arcs and angles, and of time. A *quinary* system would have five for its " common ratio ;" a *duodecimal,* twelve ; a *vigesimal,* twenty, &c.

14. A little reflection will show that it was useless to give different names and characters to any numbers except to those which are less than that which constitutes the lowest group, and to the *different orders* of groups ; because all possible numbers must consist of individuals, or of groups, or of both individuals and groups :—in neither case would it be required to specify more than the number of individuals, and the number of each species of group, none of which numbers—as is evident—can be greater than the common ratio. This

is just what we have done in our numerical system,
except that we have formed the names of some of the
groups by combination of those already used; thus,
" tens of thousands," the group next higher than thousands, is designated by a combination of words already
applied to express other groups—which tends yet further
to simplification.

15. ARABIC SYSTEM OF NOTATION :—

		Names.	Characters.
Units of Comparison,	One . .	1
	Two . .	2
	Three . .	3 .
	Four . .	4
	Five . .	5
	Six . .	6
	Seven . .	7
	Eight . .	8
	Nine . .	9
First group, or units of the second order, .	Ten . .	10
Second group, or units of the third order, .	Hundred . .	100
Third group, or units of the fourth order, .	Thousand .	1,000
Fourth group, or units of the fifth order, .	Ten thousand .	10.000
Fifth group, or units of the sixth order, .	Hundred thousand	100,000
Sixth group, or units of the seventh order,	Million . .	1,000,000

16. The characters which express the nine first numbers are the only ones used; they are called *digits*, from
the custom of counting them on the fingers, already
noticed—" digitus" meaning in Latin a finger; they are
also called *significant figures*, to distinguish them from
the cypher, or 0, which is used merely to give the digits
their proper *position* with reference to the *decimal point*.
The pupil will distinctly remember that the place where
the "units of comparison" are to be found is that immediately to the left hand of this point, which, if not expressed, is supposed to stand to the right hand side of
all the digits—thus, in 468·76 the 8 expresses " units
of comparison," being to the left of the decimal point;
in 49 the 9 expresses " units of comparison," the decimal point being understood to the right of it.

17. We find by the table just given, that after the
nine first numbers, the same digit is constantly repeated,
its position with reference to the decimal point being,
however, changed:—that is, to indicate each succeeding
group it is moved, by means of a cypher, one place
farther to the left. Any of the digits may be used to

express its respective number of any of the groups :—
thus 8 would be eight "units of comparison ;" 80,
eight groups of the first order, or eight "tens" of
simple units ; 800, eight groups of the second, or units
of the third order ; and so on. We might use any of
the digits with the different groups; thus, for example,
5 for groups of the third order, 3 for those of the second,
7 for those of the first, and 8 for the "units of comparison ;" then the whole set down in full would be 5000,
300, 70, 8, or for brevity sake, 5378—for we never use
the cypher when we can supply its place by a significant
figure, and it is evident that in 5378 the 378 keeps the
5 four places from the decimal point (understood), just as
well as cyphers would have done ; also the 78 keeps the
3 in the third, and the 8 keeps the 7 in the second place.

18. It is important to remember that each digit has
two values, an *absolute* and a *relative ;* the absolute
value is the number of units it expresses, whatever these
units may be, and is unchangeable ; thus 6 always
means six, sometimes, indeed, six tens, at other times
six hundred, &c. The relative value depends on the
order of units indicated, and on the nature of the "unit
of comparison."

19. What has been said on this very important sub-
jcet, is intended principally for the teacher, though an
ordinary amount of industry and intelligence will be
quite sufficient for the purpose of explaining it, even to
a child, particularly if each point is illustrated by an
appropriate example ; the pupil may be made, for instance, to arrange a number of pebbles in groups, sometimes of one, sometimes of another, and sometimes of
several orders, and then be desired to express them by
figures--the "unit of comparison" being occasionally
changed from individuals, suppose to tens, or hundreds, or
to scores, or dozens, &c. Indeed the pupils *must* be well
acquainted with these introductory matters, otherwise
they will contract the habit of answering without any
very definite ideas of many things they will be called
upon to explain, and which they should be expected
perfectly to understand. Any trouble bestowed by the
teacher at this period will be well repaid by the ease

and rapidity with which the scholar will afterwards advance; to be assured of this, he has only to recollect that most of his future reasonings will be derived from, and his explanations grounded on the very principles we have endeavoured to unfold. It may be taken as an important truth, that what a child learns without understanding, he will acquire with disgust, and will soon cease to remember; for it is with children as with persons of more advanced years, when we appeal successfully to their understanding, the pride and pleasure they feel in the attainment of knowledge, cause the labour and the weariness which it costs to be undervalued, or forgotten.

20. Pebbles will answer well for examples; indeed, their use in computing has given rise to the term *calculation*, "calculus" being, in Latin, a pebble: but while the teacher illustrates what he says by groups of particular objects, he must take care to notice that his remarks would be equally true of any others. He must also point out the difference between a group and its equivalent unit, which, from their perfect equality, are generally confounded. Thus he may show, that a penny, while equal to, is not identical with four farthings. This seemingly unimportant remark will be better appreciated hereafter; at the same time, without inaccuracy of result, we may, if we please, consider any group *either* as a unit of the order to which it belongs, *or* so many of the next lower as are equivalent.

21. *Roman Notation.*—Our ordinary numerical characters have not been always, nor every where used to express numbers; the letters of the alphabet naturally presented themselves for the purpose, as being already familiar, and, accordingly, were very generally adopted— for example, by the Hebrews, Greeks, Romans, &c., each, of course, using their own alphabet. The pupil should be acquainted with the Roman notation on account of its beautiful simplicity, and its being still employed in inscriptions, &c.: it is found in the following table:—

ROMAN NOTATION.

	Characters.	Numbers Expressed.
	I.	One.
	II.	Two.
	III.	Three.
Anticipated change	IIII. or IV.	Four.
Change	V.	Five.
	VI.	Six.
	VII.	Seven.
	VIII.	Eight.
Anticipated change	IX.	Nine.
Change	X.	Ten.
	XI.	Eleven.
	XII.	Twelve.
	XIII.	Thirteen.
	XIV.	Fourteen.
	XV.	Fifteen.
	XVI.	Sixteen.
	XVII.	Seventeen.
	XVIII.	Eighteen.
	XIX.	Nineteen.
	XX.	Twenty.
	XXX., &c.	Thirty, &c.
Anticipated change	XL.	Forty.
Change	L.	Fifty.
	LX., &c.	Sixty, &c.
Anticipated change	XC.	Ninety.
Change	C.	One hundred.
	CC., &c.	Two hundred, &c.
Anticipated change	CD.	Four hundred.
Change	D. or IↃ.	Five hundred, &c.
Anticipated change	CM.	Nine hundred.
Change	M. or CIↃ.	One thousand, &c.
	V̄. or IↃↃ.	Five thousand, &c.
	X̄. or CCIↃↃ.	Ten thousand, &c.
	IↃↃↃ	Fifty thousand, &c.
	CCCIↃↃↃ.	One hundred thousand,&c.

22. Thus we find that the Romans used very few characters—fewer, indeed, than we do, although our system is still more simple and effective, from our applying the principle of " position," unknown to them.

They expressed all numbers by the following symbols, or combinations of them : I. V. X. L. C. D. or IↃ. M., or CLↃ. In constructing their system, they evidently had a quinary in view; that is, as we have said, one in which five would be the *common ratio ;* for we find that they changed their character, not only at ten, ten times ten, &c., but also at five, ten times five, &c. :—a purely decimal system would suggest a change only at ten, ten times ten, &c. ; a purely quinary, only at five, five times five, &c. As far as notation was concerned, what they adopted was neither a decimal nor a quinary system, nor even a combination of both; they appear to have supposed *two* primary groups, one of five, the other of ten " units of comparison ;" and to have formed all the other groups from these, by using ten as the *common ratio* of each resulting series.

23. They anticipated a change of character ; one unit before it would naturally occur—that is, not one " unit of comparison," but one of the units under consideration. In this point of view, four is one unit before five ; forty, one unit before fifty—tens being now the units under consideration ; four hundred, one unit before five hundred—hundreds having become the units contemplated.

24. When a lower character is placed before a higher its value is to be subtracted from, when placed after it, to be added to the value of the higher ; thus, IV. means one less than five, or four ; VI., one more than five, or six.

25. To express a number by the Roman method of notation :—

RULE.—Find the highest number within the given one, that is expressed by a single character, or the " anticipation " of one [21] ; set down that character, or anticipation—as the case may be, and take its value from the given number. Find what highest number less than the remainder is expressed by a single character, or " anticipation ;" put that character or " anticipation " to the right hand of what is already written, and take its value from the last remainder : proceed thus until nothing is left.

EXAMPLE.—Set down the present year, eighteen hundred and fourty-four, in Roman characters. One thousand, expressed by M., is the highest number within the given one, indicated by one character, or by an anticipation ; we put down

M,

and take one thousand from the given number, which leaves

eight hundred and fourty-four. Five hundred, D, is the highest number within the last remainder (eight hundred and forty-four) expressed by one character, or an " anticipation ;" we set down D to the right hand of M,

MD,

and take its value from eight hundred and forty-four, which leaves three hundred and forty-four. In this the highest number expressed by a single character, or an " anticipation," is one hundred, indicated by C ; which we set down ; and for the same reason two other Cs.

MDCCC.

This leaves only forty-four, the highest number within which, expressed by a single character, or an " anticipation," is forty, XL—an anticipation ; we set this down also,

MDCCCXL.

Four, expressed by IV., still remains ; which, being also added, the whole is as follows :—

MDCCCXLIV.

26. *Position.*—The same character may have different values, according to the *place* it holds with reference to the decimal point, or, perhaps, more strictly, to the " unit of comparison." This is the principle of *position.*

27. The places occupied by the units of the different orders, according to the Arabic, or ordinary notation [15], may be described as follows :—units of comparison, one place to the left of the decimal point, expressed, or understood ; tens, two places ; hundreds, three places, &c. The pupil should be made so familiar with these, as to be able, at once, to name the " place" of any order of units, or the " units" of any place.

28. When, therefore, we are desired to write any number, we have merely to put down the digits expressing the amounts of the different units in their proper places, according to the order to which each belongs. If, in the given number, there is any order of which there are no units to be expressed, a cypher must be set down in the place belonging to it ; the object of which is, to keep the significant figures in their own *positions.* A cypher produces no effect when it is not between significant figures and the decimal point ; thus 0536, 536·0, and 536 would mean the same thing—the

second is, however, the correct form. 536 and 5360 are different ; in the latter case the cypher affects the value, because it alters the *position* of the digits.

EXAMPLE.—Let it be required to set down six hundred and two. The six must be in the third, and the two in the first place ; for this purpose we are to put a cypher between the 6 and 2—thus, 602 : without the cypher, the six would be in the second place—thus, 62 ; and would mean not six hundreds, but six tens.

29. In numerating, we begin with the digits of the highest order and proceed downwards, stating the number which belongs to each order.

To facilitate notation and numeration, it is usual to divide the places occupied by the different orders of units into periods ; for a certain distance the English and French methods of division agree ; the English billion is, however, a thousand times greater than the French. This discrepancy is not of much importance, since we are rarely obliged to use so high a number,—we shall prefer the French method. To give some idea of the amount of a billion, it is only necessary to remark, that according to the English method of notation, there has not been one billion of seconds since the birth of Christ. Indeed, to reckon even a million, counting on an average three per second for eight hours a day, would require nearly 12 days. The following are the two methods.

ENGLISH METHOD.

Trillions.	Billions.	Millions.	Units.
000·000	000·000	000·000	000·000

FRENCH METHOD.

Billions.			Millions.			Thousands.			Units.		
Hundreds.	Tens.	Units.	Hund.	Tens.	Units.	Hund.	Tens.	Units.	Hund.	Tens.	Units.
0	0	0	0	0	0	0	0	0	0	0	0

30. *Use of Periods.*—Let it be required to read off the following number, 576934. We put the first point to the left of the hundreds' place, and find that there are *exactly* two periods—576,934 ; this does not always occur, as the highest period is often imperfect, consisting only of one or two digits. Dividing the number thus

into parts, shows at once that 5 is in the third place of the second period, and of course in the sixth place to the left hand of the decimal point (understood) ; and, therefore, that it expresses hundreds of thousands. The 7 being in the fifth place, indicates tens of thousands ; the 6 in the fourth, thousands ; the 9 in the third, hundreds ; the 3 in the second, tens ; and the 4 in the first, units (of "comparison"). The whole, therefore, is five hundreds of thousands, seven tens of thousands, six thousands, nine hundreds, three tens, and four units,— or more briefly, five hundred and seventy-six thousand, nine hundred and thirty-four.

31. To prevent the separating point, or that which divides into periods, from being mistaken for the decimal point, the former should be a comma (,)—the latter a full stop (·) Without this distinction, two numbers which are very different might be confounded : thus, 498·763 and 498,763,—one of which is a thousand times greater than the other. After a while, we may dispense with the separating point, though it is convenient to use it with considerable numbers, as they are then read with greater ease.

32. It will facilitate the reading of large numbers not separated into periods, if we begin with the units of comparison, and proceed onwards to the left, saying at the first digit " units," at the second " tens," at the third " hundreds," &c., marking in our mind the denomination of the highest digit, or that at which we *stop*. We then commence with the highest, express its number and denomination, and proceed in the same way with each, until we come to the last to the right hand.

EXAMPLE.—Let it be required to read off 6402. Looking at the 2 (or pointing to it), we say "units;" at the 0, "tens;" at the 4, " hundreds ;" and at the 6, " thousands." The latter, therefore, being six thousands, the next digit is four hundreds, &c. Consequently, six thousands, four hundreds, no tens, and two units ; or, briefly, six thousand four hundred and two, is the reading of the given number.

33. Periods may be used to facilitate notation. The pupil will first write down a number of periods of cyphers,

to represent the places to be occupied by the various orders of units. He will then put the digits expressing the different denominations of the given number, under, or instead of those cyphers which are in corresponding positions, with reference to the decimal point—beginning with the highest.

EXAMPLE.—Write down three thousand six hundred and fifty-four. The highest denomination being thousands, will occupy the fourth place to the left of the decimal point. It will be enough, therefore, to put down four cyphers, and under them the corresponding digits—that expressing the thousands under the fourth cypher, the hundreds under the third, the tens under the second, and the units under the first ; thus

$$0,000$$
$$3,654$$

A cypher is to be placed under any denomination in which there is no significant figure.

EXAMPLE.—Set down five hundred and seven thousand, and sixty-three.

$$000,000$$
$$507,063$$

After a little practice the periods of cyphers will become unnecessary, and the number may be rapidly put down at once.

34. The units of comparison are, as we have said, always found in the first place to the left of the decimal point ; the digits to the left hand progressively increase in a tenfold degree—those occupying the first place to the left of the units of comparison being ten times greater than the units of comparison ; those occupying the second place, ten times greater than those which occupy the first, and one hundred times greater than the units of comparison themselves ; and so on. Moving a digit one place to the left multiplies it by ten, that is, makes it ten times greater ; moving it two places multiplies it by one hundred, or makes it one hundred times greater ; and so of the rest. If all the digits of a quantity be moved one, two, &c., places to the left, the whole is increased ten, one hundred, &c., times—as the case may be. On the other hand, moving

a digit, or a quantity one place to the right, divides it by ten, that is, makes it ten times smaller than before ; moving it two places, divides it by one hundred, or makes it one hundred times smaller, &c.

35. We possess this power of easily increasing, or diminishing any number in a tenfold, &c. degree, whether the digits are all at the right, or all at the left of the decimal point ; or partly at the right, and partly at the left. Though we have not hitherto considered quantities to the *left* of the decimal point, their relative value will be very easily understood from what we have already said. For the pupil is now aware that in the decimal system the quantities increase in a tenfold degree to the left, and decrease in the same degree to the right ; but there is nothing to prevent this decrease to the right from proceeding beyond the units of comparison, and the decimal point ;—on the contrary, from the very nature of notation, we ought to put quantities ten times less than units of comparison one place to the right of them, just as we put those which are ten times less than hundreds, &c., one place to the right of hundreds, &c. We accordingly do this, and so continue the notation, not only upwards, but downwards, calling quantities to the left of the decimal point *integers*, because none of them is less than a *whole* " unit of comparison ;" and those to the right of it *decimals*. When there are decimals in a given number, the decimal point is actually expressed, and is always found at the right hand side of the units of comparison.

36. The quantities equally distant from the unit of comparison bear a very close relation to each other, which is indicated even by the similarity of their names ; those which are one place to the *left* of the units of comparison are called " tens," being each identical with, *or* equivalent to ten units of comparison ; those which are one place to the *right* of the units of comparison are called " tenths," each being the tenth part of, that is, ten times as small as a unit of comparison ; quantities two places to the *left* of the units of comparison are called " hundreds," being one hundred times greater ; and those two places to the *right*, " hundredths," being one

hundred times less than the units of comparison; and so of all the others to the right and left. This will be more evident on inspecting the following table :—

Ascending Series, or Integers.		Descending Series, or Decimals.
One Unit . . 1	1·	One Unit.
Ten . . 10	·1	Tenth.
Hundred . ` 100	·01	Hundredth.
Thousand . 1,000	·001,	Thousandth.
Ten thousand . 10,000	·000,1	Ten-thousandth.
Hundred thousand 100,000	·000,01	Hundred-thousandth.
&c.		&c.

We have seen that when we divide integers into periods [29], the first separating point must be put to the right of the thousands; in dividing decimals, the first point must be put to the right of the thousandths.

37. Care must be taken not to confound what we now call "decimals," with what we shall hereafter designate "decimal fractions;" for they express equal, but not identically the same quantities—the decimals being what shall be termed the "quotients" of the corresponding decimal fractions. This remark is made here to anticipate any inaccurate idea on the subject, in those who already know something of Arithmetic.

38. There is no reason for treating integers and decimals by different rules, and at different times, since they follow precisely the same laws, and constitute parts of the very same series of numbers. Besides, any quantity may, as far as the decimal point is concerned, be expressed in different ways; for this purpose we have merely to change the unit of comparison. Thus, let it be required to set down a number indicating five hundred and seventy-four men. If the "unit of comparison" be *one man*, the quantity would stand as follows, 574. If a band of *ten men*, it would become 57·4—for, as each man would then constitute only the tenth part of the "unit of comparison," four men would be only four-tenths, or 0·4; and, since ten men would form but one unit, seventy men would be merely seven units of comparison, or 7; &c. Again, if it were a band of one *hundred men*, the number must be written 5·74; and lastly, if a band of a *thousand men*, it would be 0·574.

Should the "unit" be a band of a dozen, or a score men, the change would be still more complicated; as, not only the position of the decimal point, but the very digits also, would be altered.

39. It is not necessary to remark, that moving the decimal point so many places to the left, or the digits an equal number of places to the right, amount to the same thing.

Sometimes, in changing the decimal point, one or more cyphers are to be added; thus, when we move 42·6 three places to the left, it becomes 42600; when we move 27 five places to the right, it is ·00027, &c.

40. It follows, from what we have said, that a decimal, though less than what constitutes the unit of comparison, may itself consist of not only one, but several individuals. Of course it will often be necessary to indicate the "unit of comparison,"—as 3 scores, 5 dozen, 6 men, 7 companies, 8 regiments, &c.; but its nature does not affect the abstract properties of numbers; for it is true to say that seven and five, when added together, make twelve, whatever the unit of comparison may be:— provided, however, that the *same* standard be applied to both; thus 7 men and 5 men are 12 men; but 7 men and 5 horses are neither 12 men nor 12 horses; 7 men and 5 dozen men are neither 12 men nor 12 dozen men. When, therefore, numbers are compared, &c., they must have the same unit of comparison, or—without altering their value—they must be reduced to those which have. Thus we may consider 5 *tens* of men to become 50 *individual* men—the unit of comparison being altered from *ten men* to *one man*, without the value of the quantity being changed. This principle must be kept in mind from the very commencement, but its utility will become more obvious hereafter.

EXAMPLES IN NUMERATION AND NOTATION.

Notation.

		Ans.
1. Put down one hundred and four	. .	104
2. One thousand two hundred and forty	.	1,240
3. Twenty thousand, three hundred and forty-five		20,345

c

		Ans.
4.	Two hundred and thirty-four thousand, five hundred and sixty-seven . . .	234,567
5.	Three hundred and twenty-nine thousand, seven hundred and seventy-nine . .	329,779
6.	Seven hundred and nine thousand, eight hundred and twelve	709,812
7.	Twelve hundred and forty-seven thousand, four hundred and fifty-seven . .	1,247,457
8.	One million, three hundred and ninety-seven thousand, four hundred and seventy-five	1,397,475
9.	Put down fifty-four, seven-tenths . .	54·7
10.	Ninety-one, five hundredths . .	91·05
11.	Two, three-tenths, four thousandths, and four hundred-thousandths . . .	2·30404
12.	Nine thousandths, and three hundred-thousandths	0·00903
13.	Make 437 ten thousand times greater .	4,370,000
14.	Make 2·7 one hundred times greater .	270
15.	Make 0 056 ten times greater . .	0·56
16.	Make 430 ten times less . . .	43
17.	Make 2·75 one thousand times less . .	0·00275

Numeration.

1. read 132		7. read 8540326	
2. — 407		8. — 5210007	
3. — 2760		9. — 6030405	
4. — 5060		10. — 56·0075	
5. — 37654		11. — 3·000006	
6. — 8700002		12. — 0·0040007	

13. Sound travels at the rate of about 1142 feet in a second; light moves about 195,000 miles in the same time.

14. The sun is estimated to be 886,149 miles in diameter; its size is 1,377,613 times greater than that of the earth.

15. The diameter of the planet mercury is 3,108 miles, and his distance from the sun 36,814,721 miles.

16. The diameter of Venus is 7,498 miles, and her distance from the sun 68,791,752 miles.

17. The diameter of the earth is about 7,964 miles; it is 95,000,000 miles from the sun, and travels round the latter at the rate of upwards of 68,000 miles an hour.

18. The diameter of the moon is 2,144 miles, and her distance from the earth 236,847 miles.

19. The diameter of Mars is 4,218 miles, and his distance from the sun 144,907,630 miles.

20. The diameter of Jupiter is 89,069 miles, and his distance from the sun 494,499,108 miles.

21. The diameter of Saturn is 78,730 miles, and his distance from the sun 907,089,032 miles.

22. The length of a pendulum which would vibrate seconds at the equator, is 39·011,684 inches; in the latitude of 45 degrees, it is 39·116,820 inches; and in the latitude of 90 degrees, 39·221,956 inches.

23. It has been calculated that the distance from the earth to the nearest fixed star is 40,000 times the diameter of the earth's orbit, or annual path in the heavens; that is, about 7,600,000,000,000 miles. Now suppose a cannon ball to fly from the earth to this star, with a uniform velocity equal to that with which it first leaves the mouth of the gun — say 2,500 feet in a second—it would take nearly 1,000 years to reach its destination.

24. A piece of gold equal in bulk to an ounce of water, would weigh 19·258 ounces; a piece of iron of exactly the same size, 7·788 ounces; of copper, 8·788 ounces; of lead, 11·3 2 ounces; and of silver, 10·474 ounces.

Note.—The examples in notation may be made to answer for numeration; and the reverse.

QUESTIONS IN NOTATION AND NUMERATION.

[The references at the end of the questions show in what paragraphs of the preceding section the respective answers are principally to be found.]

1. What is notation? [1].
2. What is numeration? [1].
3. How are we able to express an infinite variety of numbers by a few names and characters? [3].
4. How may we suppose ideas of numbers to have been originally acquired? [4, &c.].
5. What is meant by the *common ratio* of a system of numbers? [12].
6. Is any particular number better adapted than another for the common ratio? [12].
7. Are there systems of numbers without a *common* ratio? [11].
8. What is meant by quinary, decimal, duodecimal, vigesimal, and sexagesimal systems? [13].
9. Explain the Arabic system of notation? [15].
10. What are digits? [16];
11. How are they made to express all numbers? [17]

12. What is meant by their absolute and rélative values? [18].

13. Are a digit of a higher, and the equivalent number of units of a lower order precisely the same thing? [20].

14. Have the characters we use, always and every where been employed to express numbers? [21].

15. Explain the Roman method of notation? [22, &c.].

16. What is the decimal point, and the position of the different orders of units with reference to it? [26 and 27].

17. When and how do cyphers affect significant figures? [28].

18. What is the difference between the English and French methods of dividing numbers into periods? [29].

19. What is the difference between integers and decimals? [35].

20. What is meant by the ascending and descending series of numbers; and how are they related to each other? [36].

21. Show that in expressing the same quantity, we must place the decimal point differently, according to the unit of comparison we adopt? [38].

22. What effect is produced on a digit, or a quantity by removing it a number of *places* to the right, or left, or similarly removing the decimal point? [34 and 39].

SECTION II.

THE SIMPLE RULES.

SIMPLE ADDITION.

1. If numbers are changed by any arithmetical process, they are either increased or diminished; if increased, the effect belongs to *Addition;* if diminished, to *Subtraction.* Hence all the rules of Arithmetic are ultimately resolvable into either of these, or combinations of both.

2. When *any* number of quantities, either *different,* or *repetitions* of the same, are united together so as to form but one, we term the process, simply, " Addition." When the quantities to be added are the *same,* but we may have *as many* of them as we please, it is called " Multiplication;" when they are not only the *same,* but their number is indicated by *one of them,* the process belongs to " Involution." That is, addition restricts us neither as to the kind, nor the number of the quantities to be added; multiplication restricts us as to the kind, but not the number; involution restricts us both as to the kind and number:—all, however, are really comprehended under the same rule—*addition.*

3. *Simple Addition* is the addition of abstract numbers; or of applicate numbers, containing but one denomination.

The quantities to be added are called the *addends;* the result of the addition is termed the *sum.*

4. The process of addition is expressed by +, called the plus, or positive sign; thus $8+6$, read 8 plus 6, means, that 6 is to be added to 8. When no sign is prefixed, the positive is understood.

The equality of two quantities is indicated by =; thus $9+7=16$, means that the sum of 9 and 7 is equal to 16.

Quantities connected by the sign of addition, or that of equality, may be read in any order; thus if $7+9=16$, it is true, also, that $9+7=16$, and that $16=7+9$, or $9+7$.

5. Sometimes a single horizontal line, called a *vinculum*, from the Latin word signifying a bond or tie, is placed over several numbers; and shows that all the quantities under it are to be considered, and treated as but one; thus in $\overline{4+7}=11$, $\overline{4+7}$ is supposed to form but a single term. However, a vinculum is of little consequence in addition, since putting it over, or removing it from an additive quantity—that is, one which has the sign of addition prefixed, or understood—does not in any way alter its value. Sometimes a parenthesis ( ) is used in place of the vinculum; thus $\overline{5+6}$ and $(5+6)$ mean the same thing.

6. The pupil should be made *perfectly* familiar with these symbols, and others which we shall introduce as we proceed; or, so far from being, as they ought, a great advantage, they will serve only to embarass him. There can be no doubt that the expression of quantities by characters, and not by words written in full, tends to brevity and clearness; the same is equally true of the processes which are to be performed—the more concisely they are indicated the better.

7. Arithmetical rules are, naturally, divided into two parts; the one relates to the setting down of the quantities, the other to the operations to be described. We shall generally distinguish these by a line.

To add Numbers.

RULE.—I. Set down the addends under each other, so that digits of the same order may stand in the same vertical column—units, for instance, under units, tens under tens, &c.

II. Draw a line to separate the addends from their sum.

III. Add the units of the same denomination together, beginning at the right hand side.

IV. If the sum of any column be less than ten, set it down under that column; but if it be greater, for every ten it contains, carry one to the next column, and put down only what remains after deducting the tens: if nothing remains, put down a cypher.

V. Set down the sum of the last column in full.

8. EXAMPLE.—Find the sum of 542 + 375 + 984—

$$
\begin{array}{r}
542\ \big) \\
375\ \big\}\ \text{addends.} \\
984\ \big) \\
\hline
1901\ \text{sum.}
\end{array}
$$

We have placed 2, 5, and 4, which belong to the order "units," in one column; 4, 7, and 8, which are "tens," in another; and 5, 3, and 9, which are "hundreds," in another.

4 and 5 units are 9 units, and 2 are 11 units—equivalent to one ten and one unit; we add, or as it is called, "carry" the ten to the other tens found in the next column, and set down the unit, in the units' place of the "sum."

The pupil, having learned notation, can easily find how many tens there are in a given number; since all the digits that express it, except *one* to the right hand side, will indicate the number of "tens" it contains; thus in 14 there are 1 ten, and 4 units; in 32, 3 tens, and 2 units; in 143, 14 tens, and 3 units, &c.

The ten obtained from the sum of the units, along with 8, 7, and 4 tens, makes 20 tens; this, by the method just mentioned, is found to consist of 2 tens (of tens), that is, two of the next denomination, or hundreds, to be carried, and no units (of tens) to be set down. We "carry," 2 to the hundreds, and write down a cypher in the tens' place of the "sum."

The two hundreds to be "carried," added to 9, 3, and 5, hundreds, make 19 hundreds; which are equal to 1 ten (of hundreds); or one of the next denomination, and 9 units (of hundreds); the former we "carry" to the tens of hundreds, or thousands, and the latter we set down in the hundreds' place of the "sum."

As there are no thousands in the next column—that is, nothing to which we can "carry" the thousand obtained by adding the hundreds, we put it down in the thousands' place of the "sum;" in other words, we set down the sum of the last column in full.

9. REASON OF I. (the first part of the rule).—We put units of the same denomination in the same vertical column,

that we may easily find those quantities which are to be added together; and that the value of each digit may be more clear from its being of the same denomination as those which are under, and over it.

REASON OF II.—We use the separating line to prevent the sum from being mistaken for an addend.

REASON OF III.—We obtain a correct result only by adding units of the same denomination together [Sec. I. 40]:—hundreds, for instance, added to tens, would give neither hundreds nor tens as their sum.

We begin at the right hand side to avoid the necessity of more than one addition; for, beginning at the left, the process would be as follows—

$$
\begin{array}{r}
542 \\
375 \\
984 \\
\hline
1,700 \\
190 \\
11 \\
\hline
1,000 \\
800 \\
100 \\
1 \\
\hline
1,901
\end{array}
$$

The first column to the *left* produces, by addition, 17 hundred, or 1 thousand and 7 hundred; the next column 19 tens, or 1 hundred and 9 tens; and the next 11 units, or 1 ten and 1 unit. But these quantities are still to be added:—beginning again, therefore, at the *left* hand side, we obtain 1000, 800, 100, and 1, as the respective sums. These being added, give 1,901 as the *total* sum. Beginning at the right hand rendered the successive additions unnecessary.

REASON OF IV.—Our object is to obtain the sum, expressed in the highest orders, since these, only, enable us to represent any quantity with the lowest numbers; we therefore consider ten of one denomination as a unit of the next, and add it to those of the next which we already have.

After taking the "tens" from the sums of the different columns, we must set down the remainders, since they are parts of the *entire* sum; and they are to be put under the columns that produced them, since they have not ceased to belong to the denominations in these columns.

REASON OF V.—It follows, that the sum of the last column is to be set down in full; for (in the above example, for instance,) there is nothing to be added to the tens (of hundreds) it contains.

10. *Proof of Addition.*—Cut off the upper addend, by a separating line; and add the sum of the quantities

under, to what is above this line. If all the additions have been correctly performed, the latter sum will be equal to the result obtained by the rule : thus—

$$
\begin{array}{r}
5,673 \\
\hline
4,632 \\
8,697 \\
2,543 \\
\hline
21,545 \\
\end{array}
$$ sum of all the addends.

15,872 sum of all the addends, but one.
 5,673 upper addend.

21,545 same as sum to be proved.

This mode of proof depends on the fact, that the whole is equal to the sum of its parts, in whatever order they are taken ; but it is liable to the objection, that any error committed in the first addition, is not unlikely to be repeated in the second, and the two errors would then conceal each other.

To prove addition, therefore, it is better to go through the process again, beginning at the top, and proceeding downwards. From the principle on which the last mode of proof is founded, the result of both additions—the direct and reversed—ought to be the same.

It should be remembered that these, and other proofs of the same kind, afford merely a high degree of probability, since it is not in any case quite certain, that two errors calculated to conceal each other, have not been committed.

11. *To add Quantities containing Decimals.*—From what has been said on the subject of notation (Sec. I. 35), it appears that decimals, or quantities to the right hand side of the decimal point, are merely the continuation, downwards, of a series of numbers, all of which follow the same laws; and that the decimal point is intended, not to show that there is a difference in the nature of quantities at opposite sides of it, but to mark where the "unit of comparison" is placed. Hence the rule for addition, already given, applies at whatever side all, or any of the digits in the addends may be found. It is necessary to remember that the decimal point in the sum, should stand precisely under the decimal points of the addends; since the digits of the sum must be, from the very nature of the process [9], of exactly the same values, respectively, as the digits of the addends under

c 2

which they are; and if set down as they should be, their denominations are ascertained, not only by their position with reference to their own decimal point, but also by their position with reference to the digits of the addends above them.

EXAMPLE.

$$
\begin{array}{r}
263\cdot785 \\
460\cdot502 \\
637\cdot008 \\
526\cdot3 \\
\hline
1887\cdot595
\end{array}
$$

It is not necessary to fill up the columns, by adding cyphers to the last addend; for it is sufficiently plain that we are not to notice any of its digits, until we come to the *third* column.

12. It follows from the nature of notation [Sec. I. 40], that however we may alter the decimal points of the addends—provided they are all in the same vertical column—the digits of the sum will continue unchanged; thus in the following :—

4785	478·5	47·85	·4785	·004785
3257	325·7	32·57	·3257	·003257
6546	654·6	65·46	·6546	·006546
14588	1458·8	145·88	1 4588	·014588

EXERCISES.

(Add the following numbers.)

	Addition.			Multiplication.			Involution.	
(1)	(2)	(3)	(4)	(5)	(6)	(7)	(8)	(9)
4	8	3	6	4	9	3	4	5
5	4	9	6	4	9	3	4	5
3	7	7	6	4	9	3	4	5
6	6	6	6	4	9	—	4	5
7	2	5	6	4	9	—	—	5

(10)	(11)	(12)	(13)	(14)	(15)
6763	3707	2367	6978	5767	7647
2341	2465	3246	3767	4579	1239
5279	5678	1239	1236	1236	3789

(16)	(17)	(18)	(19)	(20)	(21)
5673	3767	3001	5147	34567	73456
1237	4567	2783	3745	47891	45678
2345	1234	4567	6789	41234	91234

(22)	(23)	(24)	(25)	(26)	(27)
76789	34567	78789	34676	73412	36707
46767	89123	01007	78767	70760	46770
12476	45678	34657	45679	47076	36767

(28)	(29)	(30)	(31)	(32)	(33)
45697	76767	23456	45678	23745	87964
37676	45677	78912	91234	67891	32785
36767	76988	34567	56789	23456	64127

(34)	(35)	(36)	(37)	(38)	(39)
30071	45676	37645	47656	76767	45676
45667	37412	67456	12345	12345	34567
12345	37373	12345	67891	37676	12345
47676	45674	67891	10707	71267	67891

(40)	(41)	(42)	(43)	(44)	(45)
71234	19123	93456	45678	45679	76756
12498	67345	13767	34567	34567	34567
91379	67777	37124	12345	12345	12345
92456	88899	12456	99999	76767	67891

(46)	(47)	(48)	(49)	(50)	(51)
37676	78967	34567	47676	67678	57667
12677	12345	12345	12345	12345	34567
88991	73767	77766	67671	67912	23456
23478	12671	67345	10070	46767	76799

(52)	(53)	(54)	(55)	(56)	(57)
76769	57567	767346	473894	376767	576
12345	19807	476734	767367	123764	4589
76775	34076	467007	412345	345678	87
45666	13707	123456	671234	912345	84028

(58)	(59)	(60)	(61)	(62)	(63)
74564	5676	76746	67674	42·37	0·87
7674	1567	71207	75670	56·84	5·273
376	63	100	36	27·93	8·127
6	6767	56	77	62·41	25·63

(64)	(65)	(66)	(67)
03·785	85·772	·00007	5471·3
20·766	6034·82	·06236	563·47
00·253	57·8563	·0572	21·502
10·004	712·52	·21	0·00007

(68)	(69)	(70)	(71)
81·0235	0·0007	8456·5	576·34
576·03	5000·	·37	4000·005
4712·5	427·	8456·302	213·5
6·53712	37·12	·007	2753·

72. £7654 + £50121 + £100 + £76767 + £675
    =£135317.

73. £10 + £7676 + £97674 + £676 + £9017
    =£115053.

74. £971 + £400 + £97476 + £30 + £7000 + £76734
    =£182611.

75. 10000 + 76567 + 10 + 76734 + 6763 + 6767 + 1
    =176842.

76. 1 + 2 + 7676 + 100 + 9 + 7767 + 67 = 15622.

77. 76 + 9970 + 33 + 9977 + 100 + 67647 + 676760
    =764563.

78. $\cdot 75 + \cdot 6 + \cdot 756 + \cdot 7254 + \cdot 345 + \cdot 5 + \cdot 005 + \cdot 07$
$= 3 \cdot 7514.$

79. $\cdot 4 + 74 \cdot 47 + 37 \cdot 007 + 75 \cdot 05 + 747 \cdot 077 = 934 \cdot 004.$

80. $56 \cdot 05 + 4 \cdot 75 + \cdot 007 + 36 \cdot 14 + 4 \cdot 672 = 101 \cdot 619.$

81. $\cdot 76 + \cdot 0076 + 76 + \cdot 5 + 5 + \cdot 05. = 82 \cdot 3176.$

82. $\cdot 5 + \cdot 05 + \cdot 005 + 5 + 50 + 500 = 555 \cdot 555.$

83. $\cdot 367 + 56 \cdot 7 + 762 + 97 \cdot 6 + 471 = 1387 \cdot 667.$

84. $1 + \cdot 1 + 10 + \cdot 01 + 160 + \cdot 001 = 171 \cdot 111.$

85. $3 \cdot 76 + 44 \cdot 3 + 476 \cdot 1 + 5 \cdot 5 = 529 \cdot 66.$

86. $36 \cdot 77 + 4 \cdot 42 + 1 \cdot 1001 + \cdot 6 = 42 \cdot 8901 \cdot$

87. A merchant owes to A. £1500; to B. £408; to C. £1310; to D. £50; and to E. £1900; what is the sum of all his debts? *Ans.* £5168.

88. A merchant has received the following sums:— £200, £315, £317, £10, £172, £513 and £9; what is the amount of all? *Ans.* £1536

89. A merchant bought 7 casks of merchandize. No. 1 weighed 310 ℔; No. 2, 420 ℔; No. 3, 338 ℔; No. 4, 335 ℔; No. 5, 400 ℔; No. 6, 412 ℔; and No. 7, 429 ℔: what is the weight of the entire?

Ans. 2644 ℔.

90. What is the total weight of 9 casks of goods:— Nos. 1, 2, and 3, weighed each 350 ℔; Nos. 4 and 5, each 331 ℔; No. 6, 310 ℔; Nos. 7, 8, and 9, each 342 ℔? *Ans.* 3048 ℔.

91. A merchant paid the following sums:—£5000, £2040, £1320, £1100, and £9070; how much was the amount of all the payments? *Ans.* £18530.

92. A linen draper sold 10 pieces of cloth, the first contained 34 yards; the second, third, fourth, and fifth, each 36 yards; the sixth, seventh, and eighth, each 33 yards; and the ninth and tenth each 35 yards; how many yards were there in all? *Ans.* 347.

93. A cashier received six bags of money, the first held £1034; the second, £1025; the third, £2008; the fourth, £7013; the fifth, £5075; and the sixth, £89: how much was the whole sum? *Ans.* £16244.

94. A vintner buys 6 pipes of brandy, containing as follows:—126, 118, 125, 121, 127, and 119 gallons; how many gallons in the whole? *Ans.* 736 gals.

95. What is the total weight of 7 casks, No. 1, con-

taining, 960 ℔ ; No. 2, 725 ℔ ; No. 3, 830 ℔ ; No. 4, 798 ℔ ; No. 5, 697 ℔ ; No. 6, 569 ℔ ; and No. 7, 987 ℔ ?   *Ans.* 5566 ℔.

96. A merchant bought 3 tons of butter, at £90 per ton ; and 7 tons of tallow, at £40 per ton ; how much is the price of both butter and tallow ?   *Ans.* £550.

97. If a ton of merchandize cost £39, what will 20 tons come to ?   *Ans.* £780.

98. How much are five hundred and seventy-three ; eight hundred and ninety-seven ; five thousand six hundred and eighty-two ; two thousand seven hundred and twenty-one ; fifty-six thousand seven hundred and seventy-one ?   *Ans.* 66644.

99. Add eight hundred and fifty-six thousand, nine hundred and thirty-three ; one million nine hundred and seventy-six thousand, eight hundred and fifty-nine ; two hundred and three millions, eight hundred and ninety-five thousand, seven hundred and fifty-two.
         Ans. 206729544.

100. Add three millions and seventy-one thousand ; four millions and eighty-six thousand ; two millions and fifty-one thousand ; one million ; twenty-five millions and six ; seventeen millions and one ; ten millions and two ; twelve millions and twenty-three ; four hundred and seventy-two thousand, nine hundred and twenty-three ; one hundred and forty-three thousand ; one hundred and forty-three millions.   *Ans.* 217823955.

101. Add one hundred and thirty-three thousand ; seven hundred and seventy thousand ; thirty-seven thousand ; eight hundred and forty-seven thousand ; thirty-three thousand ; eight hundred and seventy-six thousand ; four hundred and ninety-one thousand. *Ans.* 3187000.

102. Add together one hundred and sixty-seven thousand ; three hundred and sixty-seven thousand ; nine hundred and six thousand ; two hundred and forty-seven thousand ; ten thousand ; seven hundred thousand ; nine hundred and seventy-six thousand ; one hundred and ninety-five thousand ; ninety-seven thousand.
         Ans. 3665000.

103. Add three ten-thousandths ; forty-four, five tenths ; five hundredths ; six thousandths, eight ten-thou-

sandths; four thousand and forty-one; twenty-two, one tenth; one ten-thousandth. *Ans.* 4107·6572.

104. Add one thousand; one ten-thousandth; five hundredths; fourteen hundred and forty; two tenths, three ten-thousandths; five, four tenths, four thousandths.

Ans. 2445·6544.

105. The circulation of promissory notes for the four weeks ending February 3, 1844, was as follows :—Bank of England, about £21,228,000; private banks of England and Wales, £4,980,000; Joint Stock Banks of England and Wales, £3,446,000; all the banks of Scotland, £2,791,000; Bank of Ireland, £3,581,000; all the other banks of Ireland, £2,429,000 : what was the total circulation ? *Ans.* £38,455,000.

106. Chronologers have stated that the creation of the world occurred 4004 years before Christ; the deluge, 2348; the call of Abraham, 1921; the departure of the Israelites, from Egypt, 1491; the foundation of Solomon's temple, 1012; the end of the captivity, 536. This being the year 1844, how long is it since each of these events ? *Ans.* From the creation, 5848 years; from the deluge, 4192; from the call of Abraham, 3765; from the departure of the Israelites, 3335; from the foundation of the temple, 2856; and from the end of the captivity, 2380.

107. The deluge, according to this calculation, occurred 1656 years after the creation; the call of Abraham 427 after the deluge; the departure of the Israelites, 430 after the call of Abraham; the foundation of the temple, 479 after the departure of the Israelites; and the end of the captivity, 476 after the foundation of the temple. How many years from the first to the last ? *Ans.* 3468 years.

108. Adam lived 930 years; Seth, 912; Enos, 905; Cainan, 910; Mahalaleel, 895; Jared, 962; Enoch, 365; Methuselah, 969; Lamech, 777; Noah, 950; Shem, 600; Arphaxad, 438; Salah, 433; Heber, 464; Peleg, 239; Reu, 239; Serug, 230; Nahor, 148; Terah, 205; Abraham, 175; Isaac, 180; Jacob, 147. What is the sum of all their ages? *Ans.* 12073 years.

13. The pupil should not be allowed to leave addition,

until he can, with great rapidity, continually add any of the nine digits to a given quantity; thus, beginning with 9, to add 6, he should say :—9, 15, 21, 27, 33, &c., without hesitation, or further mention of the numbers. For instance, he should not be allowed to proceed thus : 9 and 6 are 15; 15 and 6 are 21; &c.; nor even 9 and 6 are 15; and 6 are 21; &c.   He should be able, ultimately, to add the following—

$$
\begin{array}{r}
5638 \\
4756 \\
9342 \\
\hline
19736
\end{array}
$$

in this manner:—2, 8 ... 16 (the sum of the column; of which 1 is to be carried, and 6 to be set down); 5, 10 ... 13; 4, 11 ... 17; 10, 14 ... 19.

QUESTIONS TO BE ANSWERED BY THE PUPIL.

1. To how many rules may all those of arithmetic be ·educed ? [1].

2. What is addition ? [3].

3. What are the names of the quantities used in addition ? [3].

4. What are the signs of addition, and equality ? [4].

5. What is the vinculum; and what are its effects on additive quantities ? [5].

6. What is the rule for addition ? [7].

7. What are the reasons for its different parts ? [9].

8. Does this rule apply, at whatever side of the decimal point all, or any of the quantities to be added are found ? [11].

9. How is addition proved ? [10].

10. What is the reason of this proof ? [10].

SIMPLE SUBTRACTION.

14. *Simple* subtraction is confined to abstract numbers, and applicate which consist of but one denomination.

Subtraction enables us to take one number called the *subtrahend,* from another called the *minuend.* If anything is left, it is called the *excess ;* in commercial concerns, it is termed the remainder; and in the mathematical sciences, the *difference.*

15. Subtraction is indicated by —, called the minus, or negative sign. Thus $5-4=1$, read five minus four equal to one, indicates that if 4 is subtracted from 5, unity is left.

Quantities connected by the negative sign cannot be taken, indifferently, in any order; because, for example, $5-4$ is not the same as $4-5$. In the former case the positive quantity is the greater, and 1 (which means $+1[4]$) is left; in the latter, the negative quantity is the greater, and -1, or one to be subtracted, still remains. To illustrate yet further the use and nature of the signs, let us suppose that we *have* five pounds, and *owe* four ;—the five pounds we *have* will be represented by 5, and our *debt* by -4; taking the 4 from the 5, we shall have 1 pound $(+1)$ remaining. Next let us suppose that we *have* only four pounds and *owe* five; if we take the 5 from the 4—that is, if we pay as far as we can—a debt of one pound, represented by -1, will still remain;—consequently $5-4=1$; but $4-5=-1$.

16. A vinculum placed over a subtractive quantity, or one having the negative sign prefixed, alters its value, unless we change all the signs but the first :— thus $5-3+2$, and $5-\overline{3+2}$, are not the same thing; for, $5-3+2=4$; but $5-\overline{3+2}$ $(3+2$ being considered now as but one quantity$)=0$; for $3+2=5$;—therefore $-\overline{3+2}$ is the same as $5-5$, which leaves nothing ; or, in other words, it is equal to 0. If, however, we change all the signs, except the first, the value of the quantity is

not altered by the vinculum ;—thus $5-3+2=4$; and $5-\overline{3-2}$, also, is equal to 4.

Again, $27-4+7-3=27.$

$\qquad 27-\overline{4+7-3}=19.$

But $\qquad 27-\overline{4-7+3}$ (changing all the signs of the original quantities, but the first) $\Big\} =27.$

The following example will show how the vinculum affects numbers, according as we make it include an additive or a subtractive quantity :—

$48+7-3-8+7-2=49.$

$48+\overline{7-3}-8+7-2=49$; what is under the vinculum being additive, it is not necessary to change any signs.

$48+7-\overline{3+8-7+2}=49$; it is now necessary to change all the signs *under* the vinculum.

$48+7-\overline{8-7+2}=49$; it is necessary in this case, also, to change the signs.

$48+7-3-8+\overline{7-2}=49$; it is not necessary in this case.

In the above, we have sometimes put an additive, and sometimes a subtractive quantity, under the vinculum ; in the former case, we were obliged to change the signs of all the terms connected by the vinculum, except the first—that is, to change all the signs *under* the vinculum ; in the latter, to preserve the original value of the quantity, it was not necessary to change any sign.

To Subtract Numbers.

17. RULE.—I. Place the digits of the subtrahend under those of the same denomination in the minuend—units under units, tens under tens, &c.

II. Put a line under the subtrahend, to separate it from the remainder.

III. Subtract each digit of the subtrahend from the one over it in the minuend, beginning at the right hand side.

IV. If any order of the minuend be smaller than the quantity to be subtracted from it, increase it by ten ; and either consider the next order of the minuend as lessened by unity, or the next order of the subtrahend as increased by it.

V. After subtracting any denomination of the sub-

trahend from the corresponding part of the minuend,
set down what is left, if any thing, in the place which
belongs to the same denomination of the "remainder."

VI. But if there is nothing left, put down a cypher—
provided any digit of the "remainder" will be more distant from the decimal point, and at the same side of it.

18. EXAMPLE 1.—Subtract 427 from 792.

792 minuend.
427 subtrahend.
―――
365 remainder, difference, or excess.

We cannot take 7 units from 2 units; but "borrowing," as
it is called, *one* of the 9 tens in the minuend, and considering it as *ten* units, we add it to the 2 units, and then have
12 units; taking 7 from 12 units, 5 are left:—we put 5 in
the units' place of the "remainder." We may consider the
9 tens of the minuend (one having been taken away, or
borrowed) as 8 tens; or, which is the same thing, may
suppose the 9 tens to remain as they were, but the 2 tens
of the subtrahend to have become 3; then, 2 tons from 8
tens, or 3 tens from 9 tens, and 6 tens are left:—we put 6
in the tens' place of the "remainder." 4 hundreds, of the
subtrahend, taken from the 7 hundreds of the minuend,
leave 3 hundreds—which we put in the hundreds' place of
the "remainder."

EXAMPLE 2.—Take 564 from 768.

768
564
―――
204

When 6 tens are taken from 6 tens, nothing is left; we
therefore put a cypher in the tens' place of the "remainder."

EXAMPLE 3.—Take 537 from 594.

594
537
―――
57

When 5 hundreds are taken from 5 hundreds, nothing
remains; but we do not here set down a cypher, since no
significant figure in the remainder is at the same side of,
and farther from the decimal point, than the place which
would be occupied by this cypher.

19. REASON or I.—We put digits of the same denominations in the same vertical column, that the different parts

of the subtrahend may be near those of the minuend from which they are to be taken; we are then sure that the *corresponding* portions of the subtrahend and minuend may be easily found. By this arrangement, also, we remove any doubt as to the denominations to which the digits of the subtrahend belong—their values being rendered more certain, by their position with reference to the digits of the minuend.

REASON OF II.—The separating line, though convenient, is not of such importance as in addition [9]; since the "remainder" can hardly be mistaken for another quantity.

REASON OF III.—When the numbers are considerable, the subtraction cannot be effected at once, from the limited powers of the mind; we therefore divide the given quantities into parts; and it is clear that the sum of the differences of the corresponding parts, is equal to the difference between the sums of the parts:—thus, $578 - 327$ is evidently equal to $500 - 300 + 70 - 20 + 8 - 7$, as can be shown to the child by pebbles, &c. We begin at the right hand side, because it may be necessary to alter some of the digits of the minuend, so as to make it possible to subtract from them the corresponding ones of the subtrahend; but, unless we begin at the right hand side, we cannot know what alterations may be required.

REASON OF IV.—If any digit of the minuend be smaller than the corresponding digit of the subtrahend, we can proceed in either of two ways. First, we may increase that denomination of the minuend which is too small, by borrowing *one* from the next higher, (considered as *ten* of the lower denomination, or that which is to be increased,) and adding it to those of the lower, already in the minuend. In this case we alter the form, but not the value of the minuend; which, in the example given above, would become—

Hundreds.	tens.	units.		
7	8	12	=	792, the minuend.
4	2	7	=	427, the subtrahend.
3	6	5	=	365, the difference.

Or, secondly, we may add equal quantities to both minuend and subtrahend, which will not alter the difference; then we would have

Hundreds.	tens.	units.		
7	9	2 + 10	= 792 + 10,	the minuend + 10.
4	2 + 1	7	= 427 + 10,	the subtrahend + 10.
3	6	5	= 365 + 0,	the same difference.

In this mode of proceeding we do not use the given minuend and subtrahend, but others which produce the same remainder.

REASON OF V.—The remainders obtained by subtracting, successively, the different denominations of the subtrahend from those which correspond in the minuend are the *parts* of the *total* remainder. They are to be set down under the denominations which produced them, since they belong to these denominations.

REASON OF VI.—Unless there is a significant figure at the same side of the decimal point, and more distant from it than the cypher, the latter—not being between the decimal point and a significant figure—will be useless [Sec. I. 28], and may therefore be omitted.

20. *Proof of Subtraction.*—Add together the remainder and subtrahend; and the sum should be equal to the minuend. For, the remainder expresses by how much the subtrahend is smaller than the minuend; adding, therefore, the remainder to the subtrahend, should make it equal to the minuend; thus

$$
\begin{array}{ll}
8754 & \text{minuend.} \\
5839 & \text{subtrahend.} \\
\hline
2915 & \text{difference.}
\end{array}
$$

Sum of difference and subtrahend, 8754=minuend.

Or; subtract the remainder from the minuend, and what is left should be equal to the subtrahend. For the remainder is the excess of the minuend above the subtrahend; therefore, taking away this excess, should leave both equal; thus

8634 minuend.   PROOF: 8634 minuend.
7985 subtrahend.   649 remainder.
———   ———
649 remainder.   New remainder, 7985=subtrahend.

In practice, it is sufficient to set down the quantities once; thus

8634 minuend.
7985 subtrahend.
———
649 remainder.

Difference between remainder and minuend, 7985=subtrahend.

21. *To Subtract, when the quantities contain Decimals.*—The rule just given is applicable, at whatever side of the decimal point all or any of the digits may be found;—this follows, as in addition [11], from the very nature of notation. It is necessary to put the decimal point of the remainder under the decimal points of the minuend and subtrahend; otherwise the digits of the remainder will not, as they ought, have the same value as the digits from which they have been derived.

EXAMPLE.—Subtract 427·85 from 563·04.

$$563·04$$
$$427·85$$
$$\overline{135\ 19·}$$

Since the digit to the right of the decimal point in the remainder, indicates what is left after the subtraction of the tenths, it expresses so many tenths; and since the digit to the left of the decimal point indicates what remains after the subtraction of the units, it expresses so many units;— all this is shown by the position of the decimal point.

22. It follows, from the principles of notation [Sec. I. 40], that however we may alter the decimal points of the minuend and subtrahend, as long as they stand in the same vertical column, the digits of the difference are not changed; thus, in the following examples, the same digits are found in all the remainders:—

4362	436·2	43·62	·4362	·0004362
3547	354·7	35·47	·3547	·0003547
815	81·5	8·15	·0815	·0000815

EXERCISES IN SUBTRACTION.

	(1)	(2)	(3)	(4)	(5)	(6)
From	1969	7432	9076	8146	3176	76377
Take	1408	6711	4567	4377	2907	45761

	(7)	(8)	(9)	(10)	(11)	(12)
From	86167	67777	71234	900076	376704	745674
Take	61376	46699	43412	899934	297610	376789

	(13)	(14)	(15)	(16)	(17)	(18)
From	67001	9733376	567674	473676	6310756	376576
Take	35690	4124767	476476	321799	3767016	240940

	(19)	(20)	(21)	(22)	(23)	(24)
From	345676	234100	4367676	345673	70101076	67360000
Take	1799	990	256569	124799	37691734	31237777

	(25)	(26)	(27)	(28)	(29)	(30)
From	1970000	7010707	67345001	1674561	14767674	4007070
Take	1361111	3441216	47134777	1123640	7476909	3713916

	(31)	(32)	(33)	(34)	(35)
From	7045676	37670070	70000000	70040500	50070007
Take	3077037	26716645	9999999	56767767	41234016

	(36)	(37)	(38)	(39)	(40)
From	11000000	3000001	8000800	8000000	4040053
Take	9919919	2199077	377776	62358	220202

	(41)	(42)	(43)	(44)	(45)
From	85·73	865·4	594·763	47·630	52·137
Take	42·16	73·2	85·600	0·078	20·005

	(46)	(47)	(48)	(49)	(50)
From	0·00063	874·32	57·004	47632·	400·327
Take	0·00048	5·63705	2·3	0·845003	0·0006

51. 745676—567456=178220.
52. 566789—75674=501115.
53. 941000—5007=935993.
54. 97001—50077=46924.
55. 76734—977=75757.
56. 56400—100=56300.
57. 700000—99=699901.
58. 5700—500=5200.
59. 9777—89=9688.
60. 76000—1=75999.
61. 90017—3=90014.

62. 97777—4=97773.
63. 60000—1=59999.
64. 75477—76=75401.
65. 7·97—1·05=6·92.
66. 1·75—·074=1·676.
67. 97·07—4·769=92 301.
68. 7·05—4·776=2·274.
69. 10·761—9·001=1·76.
70. 12·10009—7·121=4 97909.
71. 176·1—·007=176·093.
72. 15·06—7·863=7·197.

73. What number, added to 9709, will make it 10901 ?
Ans. 1192.

74. A vintner bought 20 pipes of brandy, containing 2459 gallons, and sold 14 pipes, containing 1680 gallons ; how many pipes and gallons had he remaining ?
Ans. 6 pipes and 779 gallons.

75. A merchant bought 564 hides, weighing 16800 ℔, and sold of them 260 hides, weighing 7809 ℔ ; how many hides had he unsold, and what was their weight ?
Ans. 304 hides, weighing 8991 ℔.

76. A gentleman who had 1756 acres of land, gives 250 acres to his eldest, and 230 to his second son ; how many acres did he retain in his possession ? *Ans.* 1276.

77. A merchant owes to A. £800 ; to B. £90 ; to C. £750 ; to D. £600. To meet these debts, he has but £971 ; how much is he deficient ? *Ans.* £1269.

78. Paris is about 225 English miles distant from London ; Rome, 950 ; Madrid, 860 ; Vienna, 820 ; Copenhagen, 610 ; Geneva, 460 ; Moscow, 1660 ; Gibraltar, 1160 ; and Constantinople, 1600. How much more distant is Constantinople than Paris ; Rome than Madrid ; and Vienna than Copenhagen. And how much less distant is Geneva than Moscow ; and Paris than Madrid ? *Ans.* Constantinople is 1375 miles more distant than Paris ; Rome, 90 more than Madrid ; and Vienna, 210 more than Copenhagen. Geneva is 1200 miles less distant than Moscow ; and Paris, 635 less than Madrid.

79. How much was the Jewish greater than the English mile; allowing the former to have been 1·3817 miles English ? *Ans.* 0·3817.

80. How much is the English greater than the Roman mile; allowing the latter to have been 0·915719 of a mile English ? *Ans.* 0·084281.

81. What is the value of $6-3+15-4$? *Ans.* 14.

82. Of $43+\overline{7-3-14}$? *Ans.* 33.

83. Of $47\cdot6-\overline{2+1-24+16}-\cdot34$? *Ans.* 52·94.

84. What is the difference between $15+13-6-81+62$, and $15+13-\overline{6-81+62}$? *Ans.* 38.

23. Before leaving this rule, the pupil should be able

to take any of the nine digits continually from a given number, without stopping or hesitating. Thus, subtracting 7 from 94, he should say, 94, 87, 80, &c.; and should proceed, for instance, with the following example

$$\begin{array}{r} 5376 \\ 4298 \\ \hline 1078 \end{array}$$

in this manner:—8, 16...8 (the difference, to be set down); 10, 17...7; 3, 3...0; 4, 5...1.

QUESTIONS TO BE ANSWERED BY THE PUPIL.

1. What is subtraction? [14].
2. What are the names of the terms used in subtraction? [14].
3. What is the sign of subtraction? [15].
4. How is the vinculum used, with a subtractive quantity? [16].
5. What is the rule for subtraction? [17].
6. What are the reasons of its different parts? [19].
7. Does it apply, when there are decimals? [21].
8. How is subtraction proved, and why? [20].
9. Exemplify a brief mode of performing subtraction? [23].

SIMPLE MULTIPLICATION

24. Simple multiplication is confined to abstract numbers, and applicate which contain but one denomination.

Multiplication enables us to add a quantity, called the *multiplicand,* a number of times indicated by the *multiplier.* The multiplicand, therefore, is the number multiplied; the multiplier is that *by* which we multiply: the result of the multiplication is called the *product.* It follows, that what, in addition, would be called an " addend," in multiplication, is termed the " multiplicand;" and what, in the former, would be called the " sum," in the latter, is designated the " product." The quantities which, when multiplied together, give the

D

product, are called also *factors*, and, when they are integers, *submultiples*. There may be more than two factors; in that case, the multiplicand, multiplier, or both, will consist of more than one of them. Thus, if 5, 6, and 7, be the factors, either 5 times 6 may be considered as the multiplicand, and 7 as the multiplier—or 5 as the multiplicand, and 6 times 7 as the multiplier.

25. Quantities not formed by the continued addition of any number, but unity—that is, which are not the products of any two numbers, unless unity is taken as one of them—are called *prime* numbers: all others are termed *composite*. Thus 3 and 5 are prime, but 9 and 14 are composite numbers; because, only *three*, multiplied by *one*, will produce " three," and only *five*, multiplied by *one*, will produce " five,"—but, *three* multiplied by *three* will produce " nine," and *seven* multiplied by *two* will produce " fourteen."

26. Any quantity contained in another, some number of times, expressed by an *integer*—or, in other words, that can be subtracted from it without leaving a remainder—is said to be a *measure*, or *aliquot part* of that other. Thus 5 is a measure of 15, because it is contained in it three times *exactly*—or can be subtracted from it a number of times, expressed by 3, an integer, without leaving a remainder; but 5 is not a measure of 14, because, taking it as often as possible from 14, 4 will still be left;—thus, $15-5=10$, $10-5=5$, $5-5=0$, but $14-5=9$, and $9-5=4$. Measure, submultiple, and aliquot part, are synonymous.

27. The *common measure* of two or more quantities is a number that will measure each of them: it is a measure *common* to them. Numbers which have no *common* measure but unity, are said to be *prime to each other*; all others are *composite to each other*. Thus 7 and 5 are *prime* to each other, for unity alone will measure both; 9 and 12 are *composite* to each other, because 3 will measure either. It is evident that two *prime* numbers must be prime to *each other*; thus 3 and 7; for 3 cannot measure seven, nor 7 three, and—except unity—there is no other number that will measure either of them.

Two numbers may be composite to *each other*, and yet *one* of them may be a *prime* number; thus 5 and 25 are both measured by 5, still the former is *prime*.

Two numbers may be *composite*, and yet prime to *each other;* thus 9 and 14 are both *composite* numbers, yet they have no *common* measure but unity.

28. The *greatest common measure* of two or more numbers, is the *greatest* number which is their common measure; thus 30 and 60 are measured by 5, 10, 15, and 30; therefore each of these is their *common* measure;—but 30 is their *greatest* common measure. When a product is formed by factors which are integers, it is measured by each of them.

29. One number is the *multiple* of another, if it contain the latter a number of times expressed by an integer. Thus 27 is a multiple of 9, because it contains it a number of times expressed by 3, an integer. Any quantity is the multiple of its measure, and the measure of its multiple.

30. The *common multiple* of two or more quantities, is a number that is the multiple of each, by an integer;— thus 40 is the common multiple of 8 and 5; since it is a multiple of 8 by 5, an integer, and of 5 by 8, an integer.

The *least common multiple* of two or more quantities, is the *least* number which is their common multiple;— thus 30 is a *common* multiple of 3 and 5; but 15 is their *least* common multiple; for no number smaller than 15 contains each of them exactly.

31. The *equimultiples* of two or more numbers, are their products, when multiplied by the *same* number;— thus 27, 12, and 18, are equimultiples of 9, 4, and 6; because, multiplying 9 by *three*, gives 27, multiplying 4 by *three*, gives 12, and multipying 6 by *three*, gives 18.

32. Multiplication greatly abbreviates the process of addition;—for example, to add 68965 to itself 7000 times by "addition," would be a work of great labour, and consume much time; but by "multiplication," as we shall find presently, it can be done with ease, in less than a minute.

33. At first it may seem inaccurate, to have stated [2] that multiplication is a species of addition; since we an know the product of two quantities without having

recourse to that rule, if they are found in the multiplication table. But it must not be forgotten that the multiplication table is actually the result of additions, long since made; without its assistance, to multiply so simple a number as 4 by so small a one as five, we should be obliged to proceed as follows,

$$\begin{array}{r} 4 \\ 4 \\ 4 \\ 4 \\ 4 \\ \hline 20 \end{array}$$

performing the addition, as with any other addends.

The multiplication table is due to Pythagoras, a celebrated Greek philosopher, who was born 590 years before Christ.

34. We express multiplication by $\times$; thus $5\times7=35$, means that 5 multiplied by 7 are equal to 35, or that the product of 5 *and* 7, or of 5 *by* 7, is equal to 35.

When a quantity under the vinculum is to be multiplied by any number, each of its parts must be multiplied—for, to multiply the whole, we must multiply *each* of its parts :—thus, $3\times\overline{7+8-3}=3\times7+3\times8-3\times3$; and $\overline{4+5}\times\overline{8+3-6}$, means that each of the terms under the *latter* vinculum, is to be multiplied by each of those under the *former*.

35. Quantities connected by the sign of multiplication may be read in any order; thus $5\times6=6\times5$. This will be evident from the following illustration, by which it appears that the very same number may be considered either as 5×6, or 6×5, according to the view we take of it :—

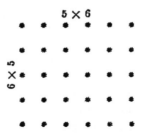

Quantities connected by the sign of multiplication,

are multiplied if we multiply *one* of the factors; thus $6 \times 7 \times 3$ multiplied by $4 = 6 \times 7$ multiplied by 3×4.

36. To prepare him for multiplication, the pupil should be made, on seeing any two digits, to name their product, without mentioning the digits themselves. Thus, a large number having been set down, he may begin with the product of the first and second digits; and then proceed with that of the second and third, &c. Taking

$$5\,8\,7\,6\,3\,4\,9\,2\,5\,8\,6\,7$$

for an example, he should say:—40 (the product of 5 and 8); 56 (the product of 8 and 7); 42; 18; &c., as rapidly as he could read 5, 8, 7, &c.

To Multiply Numbers.

37. When neither multiplicand, nor multiplier exceeds 12—

RULE.—Find the product of the given numbers by the multiplication table, page 1.

The pupil should be perfectly familiar with this table.

EXAMPLE.—What is the product of 5 and 7? The multiplication table shows that $5 \times 7 = 35$, (5 times 7 are 35).

38. This rule is applicable, whatever may be the *relative* values of the multiplicand and multiplier; that is [Sec. I. 18 and 40], whatever may be the *kind* of units expressed—provided their *absolute* values do not exceed 12. Thus, for instance, 1200×90, would come under it, as well as 12×9; also $\cdot 0009 \times 0 \cdot 8$, as well as 9×8. We shall reserve what is to be said of the management of cyphers, and decimals for the next rule; it will be equally true, however, in all cases of multiplication.

39. When the multiplicand does, but the multiplier does not exceed 12—

RULE.—I. Place the multiplier under that denomination of the multiplicand to which it belongs.

II. Put a line under the multiplier, to separate it from the product.

III. Multiply each denomination of the multiplicand by the multiplier—beginning at the right hand side.

IV. If the product of the multiplier and any digit of the multiplicand is less than ten, set it down under that digit; but if it be greater, for every ten it contains carry one to the next product, and put down only what remains, after deducting the tens; if nothing remains, put down a cypher.

V. Set down the last product in full.

40. EXAMPLE 1.—What is the product of 897351×4 ?

897351 multiplicand.
4 multiplier.
_____
3589404 product.

4 times one unit are 4 units; since 4 is less than *ten*, it gives nothing to be " carried," we, therefore, set it down in the units' place of the product. 4 times 5 are twenty (tens); which are equal to 2 tens of tens, or hundreds to be carried, and *no* units of tens to be set down in the tens' place of the product—in which, therefore, we put a cypher. 4 times 3 are 12 (hundreds), which, with the 2 hundreds to be carried from the tens, make 14 hundreds; these are equal to one thousand to be carried, and 4 to be set down in the thousands' place of the product. 4 times 7 are 28 (thousands), and 1 thousand to be carried, are 29 thousands; or 2 to be carried to the next product, and 9 to be set down. 4 times 9 are 36, and 2 are 38; or 3 to be carried, and 8 to be set down. 4 times 8 are 32, and 3 to be carried are 35; which is to be set down, since there is nothing in the next denomination of the multiplicand.

EXAMPLE 2.—Multiply 80073 by 2.

80073
2
_____
160146

Twice 3 units are 6 units; 6 being less than *ten*, gives nothing to be carried, hence we put it down in the units' place of the quotient. Twice 7 tens are 14 tens; or 1 hundred to be carried, and 4 tens to be set down. As there are no hundreds in the multiplicand, we can have none in the product, except that which is derived from the multiplication of the tens; we accordingly put the 1, to be carried, in the hundreds' place of the product. Since there are no thousands in the multiplicand, nor any to be carried, we put a cypher in that denomination of the product, to keep any significant figures that follow, in their proper places.

· 41. REASON or I.—The multiplier is to be placed under that denomination of the multiplicand to which it belongs; since there is then no doubt of its value. Sometimes it is necessary to add cyphers in putting down the multiplier; thus,

EXAMPLE 1.—478 multiplied by 2 hundred—
478 multiplicand.
200 multiplier.

EXAMPLE 2.—539 multiplied by 3 ten-thousandths—
539· multiplicand.
0·0003 multiplier.

REASON or II.—It is similar to that given for the separating line in subtraction [19].

REASON OF III.—When the multiplicand exceeds a certain amount, the powers of the mind are too limited to allow us to multiply it at once; we therefore multiply its parts, in succession, and add the results as we proceed. It is clear that the sum of the products of the parts by the multiplier, is equal to the product of the sum of the parts by the same multiplier:—thus, 537×8 is evidently equal to $500 \times 8 + 30 \times 8 + 7 \times 8$. For multiplying all the parts, is multiplying the whole; since the whole is equal to the sum of all its parts.

We begin at the right hand side to avoid the necessity of *afterwards* adding together the subordinate products. Thus, taking the example given above; were we to begin at the left hand, the proccess would be—

$$897351$$
$$4$$

$$3200000 = 800000 \times 4$$
$$360000 = 90000 \times 4$$
$$28000 = 7000 \times 4$$
$$1200 = 300 \times 4$$
$$200 = 50 \times 4$$
$$4 = 1 \times 4$$

3589404 = sum of products.

REASON OF IV.—It is the same as that of the fourth part of the rule for addition [9]; the *product* of the multiplier and any denomination of the multiplicand, being equivalent to the *sum* of a column in addition. It is easy to change the given example to an exercise in addition; for 897351×4, is the same thing as

$$897351$$
$$897351$$
$$897351$$
$$897351$$

$$3589404$$

REASON OF V.—It follows, that the last product is to be set down in full; for the tens it contains will not be increased: they may, therefore, be set down at once.

This rule includes all cases in which the *absolute* value of the digits in the multiplier does not exceed 12. Their relative value is not material; for it is as easy to multiply by 2 thousands as by 2 units.

42. To *prove* multiplication, when the multiplier does not exceed 12. Multiply the multiplicand by the multiplier, minus one; and add the multiplicand to the product. The sum should be the same as the product of the multiplicand and multiplier.

EXAMPLE.—Multiply 6432 by 7, and prove the result.
6432 multiplicand.
6=7 (the multiplier) −1

6432	38592 multiplicand × 6.
7 (=6+1)	6432 multiplicand × 1.
45024 =	45024 multiplicand multiplied by $\overline{6+1}=7$.

We have multiplied by 6, and by 1, and added the results; but six times the multiplicand, plus once the multiplicand, is equal to seven times the multiplicand. What we obtain from the two processes should be the same, for we have merely used two methods of doing one thing.

EXERCISES FOR THE PUPIL.

	(1)	(2)	(3)	(4)
Multiply	76762	67456	78976	57346
By	2	2	6	5

	(5)	(6)	(7)	(8)
Multiply	763452	456769	354709	456789
By	6	7	8	9

	(9)	(10)	(11)	(12)
Multiply	866342	738579	4763875	8429763
By	11	12	11	12

43. *To Multiply when the Quantities contain Cyphers, or Decimals.*—The rules already given are applicable : those which follow are consequences of them.

When there are cyphers at the *end* of the multiplicand (cyphers in the *middle* of it, have been already noticed [40])—

RULE.—Multiply as if there were none, and add to the product as many cyphers as have been neglected. For

The greater the quantity multiplied, the greater ought to be the product.

EXAMPLE.—Multiply 56000 by 4.

$$
\begin{array}{r}
56000 \\
\text{-} \quad 4 \\
\hline
224000
\end{array}
$$

4 times 6 units in the fourth place from the decimal point, are evidently 24 units in the *same* place ;—that is, 2 in the *fifth* place, to be carried, and 4 in the *fourth*, to be set down. That we may leave no doubt of the 4 being in the fourth place of the product, we put three cyphers to the right hand. 4 times 5 are 20, and the 2 to be carried, make 22.

44. If the multiplier contains cyphers—

Rule.—Múltiply as if there were none, and add to the product as many cyphers as have been neglected.

The greater the multiplier, the greater the number of times the multiplicand is added to itself; and, therefore, the greater the product.

EXAMPLE.—Multiply 567 by 200.

$$
\begin{array}{r}
567 \\
200 \\
\hline
113400
\end{array}
$$

From what we have said [35], it follows that 200×7 is the same as 7×200; but 7 times 2 hundred are 14 hundred; and, consequently, 200 times 7 are 14 hundred ;—that is, 1 in the *fourth* place, to be carried, and 4 in the *third*, to be set down. We add two cyphers, to show that the 4 *is* in the third place.

45. If both multiplicand and multiplier contain cyphers—

RULE.—Multiply as if there were none in either, and add to the product as ·many cyphers as are found in both.

Each of the quantities to be multiplied adds cyphers to the product [43 and 44].

EXAMPLE.—Multiply 46000 by 800.

$$\begin{array}{r} 46000 \\ 800 \\ \hline 36800000 \end{array}$$

8 times 6 thousand would be 48 thousand; but 8 *hundred* times six thousand ought to produce a number 100 times greater—or 48 hundred thousand ;—that is, 4 in the *seventh* place from the decimal point, to be carried, and 8 in the *sixth* place, to be set down. But, 5 cyphers are required, to keep the 8 in the sixth place. After ascertaining the position of the first digit in the product—from what the pupil already knows—there can be no difficulty with the other digits.

46. When there are decimal places in the multiplicand—

RULE.—Multiply as if there were none, and remove the product (by means of the decimal point) so many places to the right as there have been decimals neglected.

The smaller the quantity multiplied, the less the product.

EXAMPLE.—Multiply 5·67 by 4.

$$\begin{array}{r} 5\cdot67 \\ 4 \\ \hline 22\cdot68 \end{array}$$

4 times 7 hundredths are 28 hundredths ;—or 2 tenths, to be carried, and 8 hundredths—or 8 in the *second* place, to the right of the decimal point, to be set down. 4 times 6 tenths are 24 tenths, which, with the 2 tenths to be carried, make 26 tenths ;—or 2 units to be carried, and 6 tenths to be set down. To show that the 6 represents tenths, we put the decimal point to the left of it. 4 times 5 units are 20 units, which, with the 2 to be carried, make 22 units.

47. When there are decimals in the multiplier—

RULE.—Multiply as if there were none, and remove the product so many places to the right as there are decimals in the multiplier.

The smaller the quantity by which we multiply, the less must be the result.

EXAMPLE.—Multiply 563 by ·07

$$563$$
$$0·07$$
$$\overline{39·41}$$

3 multiplied by 7 hundredths, is the same [35] as 7 hundredths multiplied by 3 ; which is equal to 21 hundredths ;— or 2 tenths to be carried, and 1 hundredth—or 1 in the *second* place to the right of the decimal point, to be set down. Of course the 4, derived from the next product, must be *one* place from the decimal point, &c.

48. When there are decimals in both multiplicand and multiplier—

RULE.—Multiply as if there were none, and move the product so many places to the right as there are decimals in both.

In this case the product is diminished, by the smallness of *both* multiplicand and multiplier

EXAMPLE 1.—Multiply 56·3 by ·08·

$$56·3$$
$$·08$$
$$\overline{4·504}$$

8 times 3 tenths are 2·4 [46] ; consequently a quantity one hundred times less than 8—or ·08, multiplied by three-tenths, will give a quantity one hundred times less than 2·4— or ·024 ; that is, 4 in the *third* place from the decimal point, to be set down, and 2 in the *second* place, to be carried.

EXAMPLE 2.—Multiply 5·63 by 0·00005.

$$5·63$$
$$0·00005$$
$$\overline{0·0002815}$$

49. When there are decimals in the multiplicand, and cyphers in the multiplier ; or the contrary—

RULE.—Multiply as if there were neither cyphers nor decimals ; then, if the decimals exceed the cyphers, move the product so many places to the *right* as will be equal to the excess ; but if the cyphers exceed the decimals, move it so many places to the *left* as will be equal to the excess.

The cyphers move the product to the left, the decimals to the right ; the effect of both together, therefore, will be equal to the difference of their separate effects.

EXAMPLE. 1.—Multiply 4600 by ·06·

  4600

     0·06   2 cyphers and 2 decimals; **excess =0.**

   276

EXAMPLE 2.—Multiply 47·63 by 300.

   47·63

   300   2 decimals and 2 cyphers; excess =0.

  14289

EXAMPLE 3.—Multiply 85·2 by 7000.

   85·2

  7000   1 decimal and 3 cyphers; excess =2 cyphers.

 596400

EXAMPLE 4.—Multiply 578·36 by 20.

  578·36

  20   2 decimals and 1 cypher; excess=1 decimal.

11567·2

EXERCISES FOR THE PUPIL.

	(13)	(14)	(15)	(16)
Multiply	48960	75460	678000	57800
By	5	9	8	6

	(17)	(18)	(19)	(20)
Multiply	7463	770967	147005	56976748
By	80	900	4000	30000

	(21)	(22)	(23)	(24)
Multiply	743560	534900	50000	86000
By	800	30·000	300	5000

	(25)	(26)	(27)	(28)
Multiply	52736	8·7563	·21375	0·0007
By	2	4	6	8

	(29)	(30)	(31)	(32)
Multiply	56341	85637	72168	2176·38
By	0·0003	0·005	0·0007	0·06

	(33)	(34)	(35)	(36)
Multiply	875·432	78000	51·721	32
By	0·04	0·3	6000·	0·00007
				·00224

In the last example we are obliged to add cyphers to the product, to make up the required number of decimal places.

50. When both multiplicand and multiplier exceed 12—

RULE.—I. Place the digits of the multiplier under those denominations of the multiplicand to which they belong.

II. Put a line under the multiplier, to separate it from the product.

III. Multiply the multiplicand, and *each* part of the multiplier (by the preceding rule [39]), beginning with the digit at the right hand, and taking care to move the product of the multiplicand and each *successive* digit of the multiplier, so many places more to the left, than the preceding product, as the digit of the multiplier which produces it is more to the left than the significant figure by which we have *last* multiplied.

IV. Add together all the products; and their sum will be the product of the multiplicand and multiplier.

51. EXAMPLE.—Multiply 5634 by 8073.

$$
\begin{array}{r}
5634 \\
8073 \\
\hline
\end{array}
$$

16902 = product by 3.
39438 = product by 70.
45072 = product by 8000.

45483282 = product by 8073.

The product of the multiplicand by 3, requires no explanation.

7 tens times 4, or [35] 4 times 7 tens are 28 tens :—2 hundreds, to be carried, and 8 tens (8 in the *second* place from the decimal point) to be set down, &c. 8000 times 4, or 4 times 8000, are 32 thousand :—or 3 tens of thousands to be carried, and 2 thousands (2 in the *fourth* place) to be set down, &c. It is unnecessary to add cyphers, to show the values of the first digits of the different products ; as they are sufficiently indicated by the digits above. The products by 3, by 70, and by 8000, are added together in the ordinary way.

52. REASONS or I. and II.—They are the same as those given for corresponding parts of the preceding rule [41].

REASON or III.—We are obliged to multiply *successively* by the parts of the multiplier ; since we cannot multiply by the *whole at once.*

REASON or IV.—The sum of the products of the multiplicand by the parts of the multiplier, is evidently equal to the product of the multiplicand by the whole multiplier ; for, in the example just given, $5634 \times 8073 = 5634 \times \overline{8000 + 70 + 3} =$ [34] $5634 \times 8000 + 5634 \times 70 + 5634 \times 3$. Besides [35], we may consider the multiplicand as multiplier, and the multiplier as multiplicand ; then, observing the rule would be the same thing as multiplying the new multiplier into the different parts of the new multiplicand ; which, we have already seen [41], is the same as multiplying the whole multiplicand by the multiplier. The example, just given, would become 8073×5634.

> 8073 new multiplicand.
> 5634 new multiplier.

We are to multiply 3, the first digit of the multiplicand, by 5634, the multiplier ; then to multiply 7 (tens), the second digit of the multiplicand, by the multiplier ; &c. When the multiplier was small, we could add the different products as we proceeded ; but we now require a *separate* addition,—which, however, does not affect the nature, nor the reasons of the process.

53. To *prove* multiplication, when the multiplier exceeds 12—

RULE.—Multiply the multiplier by the multiplicand ; and the product ought to be the same as that of the multiplicand by the multiplier [35]. It is evident, that we could not avail ourselves of this mode of proof, in the last rule [42] ; as it would have supposed the pupil o be then able to multiply by a quantity greater than 12.

54. We may prove multiplication by what is called " casting out the nines."

RULE.—Cast the nines from the sum of the digits of the multiplicand and multiplier; multiply the remainders, and cast the nines from the product :—what is now left should be the same as what is obtained, by casting the nines, out of the sum of the digits of the product of the multiplicand by the multiplier.

EXAMPLE 1.—Let the quantities multiplied be 9426 and 3785.

Taking the nines from 9426, we get 3 as remainder.
And from 3785, we get 5.

$$
\begin{array}{l}
47130 \\
75408 \\
65982 \\
28278
\end{array}
\qquad
\begin{array}{l}
3 \times 5 = 15, \text{ from which 9} \\
\text{being taken,} \\
6 \text{ are left.}
\end{array}
$$

Taking the nines from 35677410, 6 are left.

The remainders being equal, we are to presume the multiplication is correct. The same result, however, would have been obtained, even if we had misplaced digits, added or omitted cyphers, or fallen into errors which had counteracted each other :—with ordinary care, however, none of these is likely to occur.

EXAMPLE 2.—Let the numbers be 76542 and 8436.

Taking the nines from 76542, the remainder is 6.
Taking them from 8436, it is 3.

$$
\begin{array}{l}
459252 \\
229626 \\
306168 \\
612336
\end{array}
\qquad
\begin{array}{l}
6 \times 3 = 18, \text{ the} \\
\text{remainder from which is 0.}
\end{array}
$$

Taking the nines from 645708312 also, the remainder is 0.

The remainders being the same, the multiplication may be considered right.

EXAMPLE 3.—Let the numbers be 463 and 54.

From 463, the remainder is 4.
From 54, it is 0.

$$
\begin{array}{l}
1852 \\
2315
\end{array}
\qquad
4 \times 0 = 0 \cdot \text{from which the remainder is 0.}
$$

From 25002 the remainder is 0.

The remainder being in each case 0, we are to suppose that the multiplication is correctly performed.

This proof applies whatever be the position of the decimal point in either of the given numbers.

55. To understand this rule, it must be known that "a number, from which 9 is taken as often as possible, will leave the same remainder as will be obtained if 9 be taken as often as possible from the sum of its digits."

Since the pupil is not supposed, as yet, to have learned *division*, he cannot use that rule for the purpose of casting out the nines ; — nevertheless, he can easily effect this object.

Let the given number be 563. The sum of its digits is $5+6+3$, while the number itself is $500+60+3$.

First, to take 9 as often as possible from the *sum of its digits.* 5 and 6 are 11; from which, 9 being taken, 2 are left. 2 and 3 are 5, which, not containing 9, is to be set down as the *remainder.*

Next, to take 9 as often as possible from the *number itself.* $563 = 500 + 60 + 3 = 5 \times 100 + 6 \times 10 + 3 = 5 \times \overline{99+1} + 6 \times \overline{9+1} + 3,$ = (if we remove the vinculum [34]), $5 \times 99 + 5 + 6 \times 9 + 6 + 3$. But any number of nines, will be found to be the product of the same number of ones by 9 :—thus $999 = 111 \times 9$; $99 = 11 \times 9$; and $9 = 1 \times 9$. Hence 5×99 expresses a certain number of nines—being $\overline{5 \times 11} \times 9$; it may, therefore, be cast out; and for a similar reason, 6×9; after which, there will then be left $5+6+3$—from which the nines are still to be rejected; but, as this is the *sum of the digits*, we must, in casting the nines out of it, obtain the *same* remainder as before. Consequently "we get the same remainder whether we cast the nines out of the number itself, or out of the sum of its digits."

Neither the above, nor the following reasoning can offer any difficulty to the pupil who has made himself as familiar with the use of the signs as he ought :— they will both, on the contrary, serve to show how much simplicity, is derived from the use of characters expressing, not only quantities, but processes ; for, by means of such characters, a long series of argunentation may be seen, as it were, at a single glance.

56. "Casting the nines from the factors, multiplying the resulting remainders, and casting the nines from this product,

will leave the same remainder, as if the nines were cast from the product of the factors,"—provided the multiplication has been rightly performed.

To show this, set down the quantities, and take away the nines, as before. Let the factors be 573×464.

Casting the nines from $5 + 7 + 3$ (which we have just seen is the same as casting the nines from 573), we obtain 6 as *remainder*. Casting the nines from $4 + 6 + 4$, we get 5 as *remainder*. Multiplying 6 and 5 we obtain 30 as product; which, being equal to $3 \times 10 = 3 \times \overline{9 + 1} = 3 \times 9 + 3$, will, when the nines are taken away, give 3 as *remainder*.

We can show that 3 will be the remainder, also, if we cast the nines from the product of the factors;—which is effected by setting down this product; and taking, in succession, quantities that are equal to it—as follows,

$$573 \times 464 \text{ (the product of the factors)} =$$
$$\overline{5 \times 100 + 7 \times 10 + 3} \times \overline{4 \times 100 + 6 \times 10 + 4} =$$

$$\overline{5 \times \overline{99 + 1} + 7 \times \overline{9 + 1} + 3} \times \overline{4 \times \overline{99 + 1} + 6 \times \overline{9 + 1} + 4} =$$
$$\overline{5 \times 99 + 5 + 7 \times 9 + 7 + 3} \times \overline{4 \times 99 + 4 + 6 \times 9 + 6 + 4}.$$

5×99, as we have seen [55], expresses a number of nines; it will continue to do so, when multiplied by all the quantities under the second vinculum, and is, therefore, to be cast out; and, for the same reason, 7×9. 4×99 expresses a number of nines; it will continue to do so when multiplied by the quantities under the first vinculum, and is, therefore, to be cast out; and, for the same reason, 6×9. There will then be left, only $\overline{5 + 7 + 3} \times \overline{4 + 6 + 4}$,—from which the nines are still to be cast out, the *remainders* to be multiplied together, and the nines to be cast from their product;—but we have done all this already, and obtained 3, as the remainder.

EXERCISES FOR THE PUPIL.

	(37)	(38)	(39)	(40)
Multiply	765	732	997	767
By	765	456	345	347
Products				

	(41)	(42)	(43)	(44)
Multiply	657	456	767	745
By	789	791,	789	741
Products				

57. If there are cyphers, or decimals in the multiplicand, multiplier, or both; the same rules apply as when the multiplier does not exceed 12 [43, &c.].

EXAMPLES.

(1)	(2)	(3)	(4)	(5)	(6)
4600	2784	32·68	7856	87·96	482000
57	620	26·	0·32	220·	0·37
262200	1726080	849·68	2513·92	19351·2	178340

Contractions in Multiplication.

58. When it is not necessary to have as many decimal places in the product, as are in both multiplicand and multiplier—

RULE.—Reverse the multiplier, putting its units' *place* under the *place* of that denomination in the multiplicand, which is the lowest of the required product.

Multiply by each digit of the multiplier, beginning with the denomination over it in the multiplicand; but adding what would have been obtained, on multiplying the preceding digit of the multiplicand—unity, if the number obtained would be between 5 and 15; 2, if between 15 and 25; 3, if between 25 and 35; &c.

Let the lowest denominations of the products, arising from the different digits of the multiplicand, stand in the same vertical column.

Add up all the products for the total product; from which cut off the required number of decimal places.

59. EXAMPLE 1.—Multiply 5·6784 by 9·7324, so as to have four decimals in the product.

Short Method.	Ordinary Method.
56784	5·6784
42379	9·7324
511056	22|7136
39749	113|568
1703	1703|52
113	39748|8
22	511056|
55·2643	55·2644|6016

9 in the multiplier, expresses units; it is therefore put under the *fourth* decimal place of the multiplicand—that being the place of the lowest decimal required in the product.

In multiplying by each succeeding digit of the multiplier, we neglect an additional digit of the multiplicand; because, as the multiplier decreases, the number multiplied must increase—to keep the lowest denomination of the different products, the same as the lowest denomination required in the *total* product. In the example given, 7 (the second digit of the multiplier) multiplied by 8 (the second digit of the multiplicand), will evidently produce the same denomination as 9 (one denomination higher than the 7), multiplied by 4 (one denomination lower than the 8). Were we to multiply the lowest denomination of the multiplicand by 7, we should get [46] a result in the *fifth* place to the right of the decimal point; which is a denomination supposed to be, in the present instance, too inconsiderable for notice—since we are to have only *four* decimals in the product. But we add unity for every ten that would arise, from the multiplication of an additional digit of the multiplicand; since every such *ten* constitutes *one*, in the lowest denomination of the required product. When the multiplication of an additional digit of the multiplicand would give *more* than 5, and *less* than 15; it is nearer to the truth, to suppose we have 10, than either 0, or 20; and therefore it is more correct to add 1, than either 0, or 2. When it would give more than 15, and less than 25, it is nearer to the truth to suppose we have 20, than either 10, or 30; and, therefore it is more correct to add 2, than 1, or 3; &c. We may consider 5 *either* as 0, or 10; 15 *either* as 10, or 20; &c.

On inspecting the results obtained by the abridged, and ordinary methods, the difference is perceived to be inconsiderable. When greater accuracy is desired, we should proceed, as if we intended to have more decimals in the product, and afterwards reject those which are unnecessary.

EXAMPLE 2.—Multiply 8·76532 by ·5764, so as to have 3 decimal places.

$$
\begin{array}{r}
8{\cdot}76532 \\
4675 \\
\hline
4383 \\
613 \\
52 \\
3 \\
\hline
5{\cdot}051
\end{array}
$$

There are no units in the multiplier; but, as the rule directs, we put its units' *place* under the third decimal place of the multiplicand. In multiplying by 4, since there is no digit over it in the multiplicand, we merely set down what would have resulted from multiplying the preceding denomination of the multiplicand.

EXAMPLE 3.—Multiply ·4737 by ·6731 so as to have 6 decimal places in the product.

```
        ·47370
         1376
        ──────
        284220
         33159
          1421
            47
        ──────
       ·318847
```

We have put the units' *place* of the multiplier under the sixth decimal *place* of the multiplicand, adding a cypher, or supposing it to be added.

EXAMPLE 4.—Multiply 84·6732 by ·0056, so as to have four decimal places.

```
       84·6732
            65
        ──────
          4234
           508
         ─────
        ·4742
```

EXAMPLE 5.—Multiply ·23257 by ·243, so as to have four decimal places.

```
        23257
          342
         ─────
          465
           93
            7
         ─────
        ·0565
```

We are obliged to place a cypher in the product, to make up the required number of decimals.

60. To multiply by a Composite Number—
RULE.—Multiply, successively, by its factors.

EXAMPLE.—Multiply 732 by 96. 96=8×12; therefore 732×96=732×8×12. [35].

$$732$$
$$8$$
_____

5856, product by 8.
$$12$$
_____

70272, product by 8×12, or 96.

If we multiply by 8 only, we multiply by a quantity 12 times too small; and, therefore, the product will be 12 times less than it should. We rectify this, by making the product 12 times greater—that is, we multiply it by 12.

61. When the multiplier is not exactly a Composite Number—

RULE.—Multiply by the factors of the nearest composite; and add to, or subtract from the last product, so many times the multiplicand, as the assumed composite is less or greater than the given multiplier.

EXAMPLE 1.—Multiply 927 by 87.

$87 = 7 \times 12 + 3$; therefore $927 \times 87 = 927 \times \overline{7 \times 12 + 3} = 927 \times 7 \times 12 + 927 \times 3$. [34].

$$927$$
$$7$$
_____

$6489 = 927 \times 7$.
$$12$$
_____

$77868 = 927 \times 7 \times 12$.
$2781 = 927 \times 3$.
_____

$80649 = 927 \times 7 \times 12 + 927 \times 3$, or 927×87.

If we multiply only by 84 (7 × 12), we take the number to be multiplied 3 times less than we ought; this is rectified, by adding 3 times the multiplicand.

EXAMPLE 2.—Multiply 432 by 79. $79 = 81 - 2 = 9 \times 9 - 2$; therefore $432 \times 79 = 432 \times \overline{9 \times 9 - 2} = 432 \times 9 \times 9 - 432 \times 2$.

$$432$$
$$9$$
_____

$3888 = 432 \times 9$.
$$9$$
_____

$34992 = 432 \times 9 \times 9$.
$864 = 432 \times 2$.
_____

$34128 = 432 \times 9 \times 9 - 432 \times 2$, or 432×79.

In multiplying by 81, the composite number, we have taken the number to be multiplied twice too often; but the inaccuracy is rectified by subtracting twice the multiplicand from the product.

62. This method is particularly convenient, when the multiplier consists of *nines*.

To Multiply by any Number of Nines,—

RULE.—Remove the decimal point of the multiplicand so many places to the right (by adding cyphers if necessary) as there are nines in the multiplier; and subtract the multiplicand from the result.

EXAMPLE.—Multiply 7347 by 999.
$$7347 \times 999 = 7347000 - 7347 = 7339653.$$

We, in such a case, merely multiply by the next higher convenient composite number, and subtract the multiplicand so many times as we have taken it too often; thus, in the example just given—
$$7347 \times 999 = 7347 \times \overline{1000 - 1} = 7347000 - 7347 = 7339653.$$

63. We may sometimes abridge multiplication by considering a part or parts of the multiplier as produced by multiplication of one or more other parts.

EXAMPLE.—Multiply 57839268 by 62421648. The multiplier may be divided as follows:—6, 24, 216, and 48.

$$6 = 6$$
$$24 = 6 \times 4$$
$$216 = 24 \times 9$$
$$48 = 24 \times 2$$

57839268, multiplicand.
62421648, multiplier.

347035608 ⋮ ⋮ ⋮	product by 6 (60000000).
1388142432 ⋮ ⋮	product by 24 (2400000).
12493281888 ⋮	product by 216 (21600).
2776284864	product by 48.

3610422427673664 product by 62421648.

The product by 6 when multiplied by 4 will give the product by 24; the product by 24, multiplied by 9, will give the product by 216—and, multiplied by 2, the product by 48.

64. There can be no difficulty in finding the places of the first digits of the different products. For when there are neither cyphers nor decimals in the multiplicand— and *during* multiplication, we may suppose that there are neither [48, &c.]—the lowest denomination of each pro-

duct, will be the same as the lowest denomination of the multiplier that produced it ;—thus 12 units multiplied by 4 units will give 48 units ; 14 units multiplied by 4 tens will give 56 tens ; 124 units multiplied by 35 units will be 4340 units, &c. ; and, therefore, the beginning of each product—if a significant figure—must stand under the lowest digit of the multiplier from which it arises. When the process is finished, cyphers or decimals, if necessary, may be added, according to the rules already given.

The vertical dotted lines show that the places of the lowest digits of the respective multipliers, or those parts into which the whole multiplier has been divided, and the lowest digits of their resulting products are—as they ought to be—of the same denomination.

48 being of the denomination units, when multiplied into 8 units, will produce units ; the first digit, therefore, of the product by 48 is in the units' place. 216, being of the denomination hundreds when multiplied into units will give hundreds ; hence the first digit of the product by 216 will be in the hundreds' place, &c. The parts into which the multiplier is divided are, in reality,

$$\left.\begin{array}{r}60000000 \\ 2400000 \\ 21600 \\ 48\end{array}\right\} = 62421648, \text{ the whole multiplier.}$$

We shall give other contractions in multiplication hereafter, at the proper time.

EXERCISES.

45. $745 \times 456 = 339720.$
46. $476 \times 767 = 365092.$
47. $345 \times 579 = 199755.$
48. $476 \times 479 = 228004.$
49. $897 \times 979 = 878163.$
50. $4\cdot59 \times 705 = 3235\cdot95.$
51. $767 \times 407 = 312169.$
52. $\cdot457 \times \cdot606 = \cdot276942.$
53. $700 \times 810 = 567000.$
54. $670 \times 910 = 609700.$
55. $910 \times 870 = 791700.$
56. $5001\cdot4 \times 70 = 350098.$
57. $64\cdot001 \times 40 = 2560\cdot04.$
58. $91009 \times 79 = 7189711.$
59. $40170 \times 80 = 3213600.$

60. $707 \times 604 = 427028.$
61. $777 \times \cdot407 = 316\cdot239$
62. $7407 \times 4404 = 32620428.$
63. $5767 \times 1307 = 7537469.$
64. $67\cdot74 \times \cdot1706 = 11\cdot556444.$
65. $4567 \times 2002 = 9143134.$
66. $7\cdot767 \times 301\cdot2 = 2339\ 4204.$
67. $9600 \times 7100 = 68160000.$
68. $7800 \times 9100 = 70980000.$
69. $6700 \times 6700 = 44890000.$
70. $5000 \times 7600 = 38000000.$
71. $70\ 814 \times 901\cdot07 = 63808\cdot37098.$
72. $97001 \times 76706 = 7440558706.$
73. $93400 \times 67407 = 6295813800.$
74. $\cdot56007 \times 45070 = 25242\cdot35490$

75. How many shillings in £1395; a pound being 20 shillings?   *Ans.* 27900.

76. In 2480 pence how many farthings; four farthings being a penny?   *Ans.* 9920.

77. If 17 oranges cost a shilling, how many can be had for 87 shillings?   *Ans.* 1479.

78. How much will 245 tons of butter cost at £25 a ton?   *Ans.* 6125.

79. If a pound of any thing cost 4 pence, how much will 112 pounds cost?   *Ans.* 448 pence.

80. How many pence in 100 pieces of coin, each of which is worth 57 pence?   *Ans.* 5700 pence.

81. How many gallons in 264 hogsheads, each containing 63 gallons?   *Ans.* 16632.

82. If the interest of £1 be £0·05, how much will be the interest of £376?   *Ans.* £18·8.

83. If one article cost £0·75, what will 973 such cost?   *Ans.* £729·75.

84. It has been computed that the gold, silver, and brass expended in building the temple of Solomon at Jerusalem, amounted in value to £6904822500 of our money; how many pence are there in this sum, one pound containing 240?   *Ans.* 1657157400000.

85. The following are the lengths of a degree of the meridian, in the following places: 60480·2 fathoms in Peru; 60486·6 in India; 60759·4 in France; 60836·6 in England; and 60952·4 in Lapland. 6 feet being a fathom, how many feet in each of the above? *Ans.* 362881·2 in Peru; 362919·6 in India; 364556·4 in France; 365019·6 in England; and 365714·4 in Lapland.

86. The width of the Menai bridge between the points of suspension is 560 feet; and the weight between these two points 489 tons. 12 inches being a foot, and 2240 pounds a ton, how many inches in the former, and pounds in the latter?
                Ans. 6720 inches, and 1095360 pounds.

87. There are two minims to a semibreve; two crotchets to a minim; two quavers to a crotchet; two semiquavers to a quaver: and two demi-semiquavers to a semiquaver: how many demi-semiquavers are equal to seven semibreves?   *Ans.* 224.

88. 32,000 seeds have been counted in a single poppy; how many would be found in 297 of these? *Ans.* 9504000.

89. 9,344,000 eggs have been found in a single cod fish; how many would there be in 35 such?

Ans. 327040000.

65. When the pupil is familiar with multiplication, in working, for instance, the following example,

897351, multiplicand.
4, multiplier.

3589404, product.

He should say :—4 (the product of 4 and 1), 20 (the product of 4 and 5), 14 (the product of 4 and 3 plus 2, to be carried), 29, 38, 35; at the same time putting down the units, and carrying the tens of each.

QUESTIONS TO BE ANSWERED BY THE PUPIL.

1. What is multiplication ? [24].
2. What are the multiplicand, multiplier, and product? [24].
3. What are factors, and submultiples ? [24].
4. What is the difference between prime and composite numbers [25]; and between those which are prime and those which are composite to *each other*? [27].
5. What is the measure, aliquot part, or submultiple of a quantity ? [26].
6. What is a multiple ? [29].
7. What is a *common* measure ? [27].
8. What is meant by the *greatest* common measure? [28].
9. What is a *common* multiple ? [30].
10. What is meant by the *least* common multiple ? [30].
11. What are *equi*multiples ? [31].
12. Does the use of the multiplication table prevent multiplication from being a species of addition ? [33].
13. Who first constructed this table ? [33].
14. What is the sign used for multiplication ? [34].
15. How are quantities under the vinculum affected by the sign of multiplication ? [34].
16. Show that quantities connected by the sign of multiplication may be read in any order ? [35].

E

17. What is the rule for multiplication, when neither multiplicand nor multiplier exceeds 12 ? [37].

18. What is the rule, when only the multiplicand exceeds 12 ? [39].

19. What is the rule when both multiplicand and multiplier exceed 12 ? [50].

20. What are the rules when the multiplicand, multiplier, or both, contain cyphers, or decimals ? [43, &c.]: and what are the reasons of these, and the preceding rules ? [41, 43, &c., 52].

21. How is multiplication proved ? [42 and 53].

22. Explain the method of proving multiplication, by "casting out the nines [54];" and show that we can cast the nines out of any number, without supposing a knowledge of *division*. [55].

23. How do we multiply so as to have a required number of decimal places ? [58].

24. How do we multiply by a composite number [60]; or by one that is a little more, or less than a composite number ? [61].

25. How may we multiply by any number of nines ? [62].

26. How is multiplication very briefly performed ? [65].

SIMPLE DIVISION.

66. Simple Division is the division of abstract numbers, or of those which are applicate, but contain only one denomination.

Division enables us to find out how often one number, called the *divisor*, is *contained in*, or can *be taken from* another, termed the *dividend ;*—the number expressing *how often* is called the *quotient*. Division also enables us to tell, if a quantity be divided into a certain number of equal parts, what will be the amount of each.

When the divisor is not contained in the dividend any number of times exactly, a quantity, called the *remainder*, is left after the division.

67. It will help us to understand how greatly division abbreviates subtraction, if we consider how long a process would be required to discover—by actually sub-

tracting it—how often 7 is contained in 8563495724, while, as we shall find, the same thing can be effected by *division*, in less than a minute.

68. Division is expressed by ÷, placed between the dividend and divisor; or by putting the divisor under the dividend, with a separating line between :—thus $6 \div 3 = 2$, or $\frac{6}{3} = 2$ (read 6 divided by 3 is equal to 2) means, that if 6 is divided by 3, the quotient will be 2.

69. When a quantity under the vinculum is to be divided, we must, on removing the vinculum, put the divisor under each of the terms connected by the sign of addition, or subtraction, otherwise the value of what was to be divided will be changed;—thus $\overline{5 + 6 - 7} \div 3 = \frac{5}{3} + \frac{6}{3} - \frac{7}{3}$; for we do not divide the whole unless we divide *all* its parts.

The line placed between the dividend and divisor occasionally assumes the place of a vinculum; and therefore, when the quantity to be divided is subtractive, it will sometimes be necessary to change the signs—as already directed [16] :—thus $\frac{6}{2} + \frac{13 - 3}{2} = \frac{6 + 13 - 3}{2}$;

but $\frac{27}{3} - \frac{15 - 6 + 9}{3} = \frac{27 - 15 + 6 - 9}{3}$. For when, as in these cases, *all* the terms are put under the vinculum, the effect—as far as the subtractive signs are concerned—is the same as if the vinculum were removed altogether; and then the signs should be changed *back again* to what they must be considered to have been *before* the vinculum was affixed [16].

When quantities connected by the sign of multiplication are to be divided, dividing any one of the factors, will be the same as dividing the product; thus, $5 \times 10 \times 25 \div 5 = \frac{5}{5} \times 10 \times 25$; for each is equal to 250.

To Divide Quantities.

70. When the divisor does not exceed 12, nor the dividend 12 times the divisor.

RULE.—I. Find by the multiplication table the greatest number which, multiplied by the divisor, will give a product that does not exceed the dividend: this will be the quotient required.

II. Subtract from the dividend the product of this number and the divisor; setting down the remainder, if any, with the divisor under it, and a line between them.

EXAMPLE.—Find how often 6 is contained in 58; or, in other words, what is the quotient of 58 divided by 6.

We learn from the multiplication table that 10 times 6 are 60. But 60 is greater than 58; the latter, therefore, does not contain 6 10 times. We find, by the same table, that 9 times 6 are 54, which is less than 58:—consequently 6 is contained 9, but not 10 times in 58; hence 9 is the quotient; and 4—the difference between 9 times 6 and the given number—is the *remainder*.

The total quotient is $9 + \frac{4}{6}$, or $9\frac{4}{6}$; that is, $\frac{58}{6} = 9\frac{4}{6}$.

If we desire to carry the division farther, we can effect it by a method to be explained presently.

71. REASON OF I.—Our object is to find the greatest number of times the divisor can be taken from the dividend; that is, the greatest multiple of 6 which will not exceed the number to be divided. The multiplication table shows the products of any two numbers, neither of which exceeds 12; and therefore it enables us to obtain the product we require; this must not exceed the dividend, nor, being subtracted from it, leave a number equal to, or greater than, the divisor. It is hardly necessary to remark, that the divisor would not have been subtracted as often as possible from the dividend if a number equal to or greater than it were left; nor would the quotient answer the question, *how often* the divisor could be taken from the dividend.

REASON OF II.—We subtract the product of the divisor and quotient from the dividend, to learn, if there be any remainder, what it is. When there is a remainder, we in reality suppose the dividend divided into two parts; one of these is equal to the product of the divisor and quotient—and this we actually divide; the other is the difference between that product and the given dividend—this we express, by the notation already explained, as still *to be* divided. In the example given, $\frac{58}{6} = \frac{54 + 4}{6} = \frac{54}{6} + \frac{4}{6} = 9 + \frac{4}{6}$.

72. When the divisor does not exceed 12, but **the** dividend exceeds 12 times the divisor—

RULE.--I. Set down the dividend with a line under it to separate it from the future quotient : and put the divisor to the left hand side of the dividend, with a line between them.

II. Divide the divisor into all the denominations of the dividend, beginning with the highest.

III. Put the resulting quotients under those denominations of the dividend which produced them.

IV. If there be a remainder, after subtracting the product of the divisor and any denomination of the quotient from the corresponding denomination of the dividend, consider it ten times as many of the next lower denomination, and add to it the next digit of the dividend.

V. If any denomination of the dividend (the preceding remainder, when there is one, included) does not contain the divisor, consider it ten times as many of the next lower, and add to it the next digit of the dividend—putting a cypher in the quotient, under the digit of the dividend thus reduced to a lower denomination, unless there are no significant figures in the quotient at the same side of, and farther removed from the decimal point.

VI. If there be a remainder, after dividing the "units of comparison," set it down—as already directed [70]—with the divisor under it, and a separating line between them ; or, writing the decimal point in the quotient, proceed with the division, and consider each remainder ten times as many of the next lower denomination ; proceed thus until there is no remainder, or until it is so trifling that it may be neglected without inconvenience.

73. EXAMPLE.—What is the quotient of 64456÷7 ?

Divisor 7)64456 dividend.
9208 quotient.

6 tens of thousands do not contain 7, even *once* ten thousand times ; for ten thousand times 7 are 70 thousand, which is greater than 60 thousand ; there is, therefore, no digit to be put in the ten-thousands' place of the quotient—we do not, however, put a cypher in that place, since no digit

of the quotient can be further removed from the decimal point than this cypher; for it would, in such a case, produce no effect [Sec. I. 28]. Considering the 6 tens of thousands as 60 thousands, and adding to these the 4 thousands already in the dividend, we have 64 thousands. 7 will "go" into (that is, 7 can be taken from) 64 thousand, 9 thousand times; for 7 times 9 thousand are 63 thousand—which is less than 64 thousand, and therefore is not too large; it does not leave a remainder equal to the divisor—and therefore it is not too small:—9 is to be set down in the thousands' place of the quotient; and the 4 already in the dividend being added to one thousand (the difference between 64 and 63 thousand) considered as ten times so many hundreds, we have 14 hundreds. 7 will go 2 hundred times into 14 hundreds, and leave no remainder; for 7 times 2 hundreds are exactly 14 hundreds:—2 is, therefore, to be put in the hundreds' place of the quotient, and there is nothing to be carried. 7 will not go into 5 tens, even *once* ten times; since 10 times 7 are 7 tens, which is more than 5 tens. But considering the 5 tens as 50 units, and adding to them the other 6 units of the dividend, we have 56 units. 7 will go into 56, 8 times, leaving no remainder. As the 5 tens gave no digit in the tens' place of the quotient, and there are significant figures farther removed from the decimal point than this denomination of the dividend, we have been obliged to use a cypher. The division being finished, and no remainder left, the required quotient is found to be 9208 exactly; that is, $\frac{64456}{7} = 9208$

74. EXAMPLE 2.—What is the quotient of 73268, divided by 6?

$$6) \underline{73268}$$
$$12211\tfrac{2}{6}$$

We may set down the 2 units, which remain after the units of the quotient are found, as represented; or we may proceed with the division as follows—

$$6) \underline{73268}$$
$$12211 \cdot 333, \&c.$$

Considering the 2 units, left from the units of the dividend, as 20 tenths, we perceive that 6 will go into them three tenths times, and leave 2 tenths—since 3 tenths times 6 (=6 times 3 tenths [35]) are 18 tenths:—we put 3 in the tenths' place of the quotient, and consider the 2 tenths remaining, as 20 hundredths. For similar reasons, 6 will go into 20 hundredths 3 hundredths times, and leave 2 hundredths. Considering these 2 hundredths as 20 thousandths, they will give 3 thousandths as quotient, and 2 thousandths as remainder, &c. The same remainder, constantly recurring, will evidently produce the same digit in the successive denominations of the quotient; we may, therefore, at once put down in the quotient as many threes as will leave the final remainder so small, that it may be neglected.

75. EXAMPLE 3.—Divide 47365 by 12.

$$12)47365$$
$$\overline{3947{\cdot}08},\ \&c.$$

In this example, the one unit left (after obtaining the 7 in the quotient) even when considered as 10 tenths, does not contain 12 :—there is, therefore, nothing to be set down in the tenths' place of the quotient—except a cypher, to keep the following digits in their proper places. The 10 tenths are by consequence to be considered as 100 hundredths ; 12 will go into 100 hundredths 8 hundredths times, &c.

This may be applied to the last rule [70], when we desire to continue the division.

EXAMPLE.—Divide 8 by 5.

$$8 \div 5 = 1\tfrac{3}{5},\ or\ 1{\cdot}37,\ \&c.$$

76. When the pupil fully understands the real denominations of the dividend and quotient, he may proceed, for example, with the following

$$5)46325$$

in this manner :—5 will not go into 4. 5 into 46, 9 times and 1 over (the 46 being of the denomination to which 6 belongs [thousands], the first digit of the quotient is to be put under the 6—that is, under the denomination which produced it). 5 into 13, twice and 3 over. 5 into 32, 6 times and 2 over. 5 into 25, 5 times and no remainder.

When the divisor does not exceed 12, the process is called *short* division.

77. REASON OF I.—In this arrangement of the quantities— which is merely a matter of convenience—the values of the digits of the quotient are ascertained, both by their position with reference to the digits of the dividend, and to their own decimal point. The separating lines prevent the dividend, divisor, or quotient from being in any way mistaken.

REASON OF II.—We divide the divisor successively into all the parts of the dividend, because we cannot divide it at once into the whole :—the sum of the numbers of times it can be subtracted from these parts is evidently equal to the number

of times it can be subtracted from their sum. Thus, if 5 goes
into 500, 100 times, into 50, 10 times, and into 5, once; it
will go into $500+50+5$ ($=555$), $100+10+1$ ($=111$) times.

The pupil perceives by the examples given above, that, in
dividing the divisor successively into the parts of the dividend,
each, or any of these parts does not necessarily consist of one
or more digits of the dividend. Thus, in finding, for example,
the quotient $64456 \div 7$, we are not obliged to consider the parts
as 60000, 4000, 400, 50, and 6:—on the contrary, to render the
dividend suited to the process of division, we alter its form,
while, at the same time, we leave its value unchanged; it becomes

Thousands.		Hundreds.		Tens.		Units.	
63	+	14	+	0	+	56	($=64456$).

Each part being divided by 7, the different portions of the
dividend, with their respective quotients, will be,

	Thousands.	Hundreds.	Tens.	Units.		
7	63	14	0	56	=	64456.
	9	2	0	8	=	9208.

We begin at the left hand side, because what remains of the
higher denomination, may still give a quotient in a lower; and
the question is, *how often* the divisor will go into the
dividend—its different denominations being taken in *any* convenient way. We cannot know how many of the higher we
shall have to add to the lower denominations, unless we begin
with the higher.

REASON OF III.—Each digit of the quotient is put under
that denomination of the dividend which produced it, because
it belongs to that denomination; for it expresses what number
of times (indicated by a digit of that denomination) the divisor
can be taken from the corresponding part of the dividend:—
thus the tens of the quotient express how many *tens* of times
the divisor can be taken from the tens of the dividend; the
hundreds of the quotient, how many hundreds of times it can
be taken from the hundreds, &c.

REASON OF IV.—Since what is left belongs to the *total* remainder, it must be added to it; but unless considered as of a
lower denomination, it will give nothing further in the quotient.

REASON OF V.—We are to look upon the remainder as of
the highest denomination capable of giving a quotient; and
though it may not contain the divisor a number of times expressed by a digit of one denomination, it may contain it some
number of times expressed by one that is lower.

The true remainder, after subtracting each product, is the
whole remainder of the dividend; but we "bring down" only
so much of it as is necessary for our present object. Thus, in
looking for a digit in the hundreds' place of the quotient, it
will not be necessary to take into account the tens, or units
of the dividend; since they cannot add to the number of *hundreds* of times the divisor may be taken from the dividend.

A cypher must be added [Sec. I. 28], when it is required, to give significant figures their proper value—which is never the case, except it comes between them and the decimal point.

REASON or VI.—We may continue the process of division, if we please, as long as it is possible to obtain quotients of *any* denomination. Quotients will be produced although there are no longer any significant figures in the dividend, to which we can add the successive remainders.

78. The smaller the divisor the larger the quotient—for, the smaller the parts of a given quantity, the greater their number will be; but 0 is the least possible divisor, and therefore any quantity divided by 0 will give the largest possible quotient—which is *infinity*. Hence, though any quantity multiplied by 0 is equal to 0, any number divided by 0 is equal to an infinite number.

It appears strange, but yet it is true, that $\dfrac{5}{0} = \dfrac{1}{0}$; for each is equal to the *greatest* possible number, and one, therefore, cannot be greater than another—the apparent contradiction arises from our being unable to form a true conception of an *infinite* quantity. It is necessary to bear in mind also that 0, in this case, indicates a quantity infinitely small, rather than absolutely nothing.

79. *To prove Division.*—Multiply the quotient by the divisor; the product should be equal to the dividend, minus the remainder, if there is one.

For, the dividend, exclusive of the remainder, contains the divisor a number of times indicated by the quotient; if, therefore, the divisor, is taken that number of times, a quantity equal to the dividend, minus the remainder, will be produced. It follows, that adding the remainder to the product of the divisor and quotient should give the dividend.

EXAMPLE 1.—Prove that $\dfrac{6832}{4} = 1708$.

4)6832
‾1708‾

PROOF. 1708, quotient.
        4, divisor.

6832, product of divisor and quotient, equal to the dividend.

EXAMPLE 2.—Prove that $\dfrac{85643}{7} = 12234\,\dfrac{5}{7}$.

PROOF.   PROOF.
12234   or   12234
    7   7

85638 = dividend minus 5, the remainder. 85638 + 5 = dividend.

E 2

EXERCISES.

(1)	(2)	(3)	(4)
2)78345	8)91234	- 3)67859	9)71234

(5)	(6)	(7)	(8)
4)96707	10)134567	5)767456	11)37067

(9)	(10)	(11)	(12)
6)970763	12)876967	7)891023	9)763457

80. *When the dividend, divisor, or both contain cyphers or decimals.*—The rules already given are applicable : those which follow are consequences of them.

When the dividend contains cyphers—

RULE.—Divide as if there were none, and remove the quotient so many places to the left as there have been cyphers neglected.

The greater the dividend, the greater ought to be the quotient; since it expresses the number of times the divisor can be subtracted from the dividend. Hence, if 8 will go into 56 7 times, it will go into 5600 (a quantity 100 times greater than 56) 100 times more than 7 times—or 700 times.

EXAMPLE 1.—What is the quotient of $568000 \div 4$?

$$\frac{568}{4} = 142 ; \text{ therefore } \frac{568000}{4} = 142000.$$

EXAMPLE 2.—What is the quotient of $4060000 \div 5$?

$$\frac{406}{5} = 81\cdot2 ; \text{ therefore } \frac{4060000}{5} = 812000 \quad [\text{Sec. I. 39.}].$$

81. When the divisor contains cyphers—

RULE.—Divide as if there were none, and move the quotient so many places to the right as there are cyphers in the divisor.

The greater the divisor, the smaller the number of times it can be subtracted from the dividend. If, for example, 6 can be taken from a quantity any number of times, 100 times 6 can be taken from it 100 times less often.

EXAMPLE.—What is the quotient of $\frac{56}{800}$?

$$\frac{56}{8} = 7 ; \text{ therefore } \frac{56}{800} = \cdot07.$$

82. If both dividend and divisor contain cyphers—

RULE.—Divide as if there were none, and move the quotient a number of places equal to the difference between the numbers of cyphers in the two given quantities :—if the cyphers in the dividend exceed those in the divisor, move to the left; if the cyphers in the divisor exceed those in the dividend, move to the right.

We have seen that the effect of cyphers in the dividend is to move the quotient to the left and of cyphers in the divisor, to move it to the right; when, therefore, both causes act together, their effect must be equal to the difference between their separate effects.

EXAMPLES.

(1)	(2)	(3)	(4)	(5)	(6)
7)63	7)6300	70)63	70)6300	700)630	700)6300
9	900	0·9	90	0·9	9

In the sixth example, the *difference* between the numbers of cyphers being $= 0$, the quotient is moved neither to the right nor the left.

83. If there are decimals in the dividend—

RULE.—Divide as if there were none, and move the quotient so many places to the right as there are decimals.

The smaller the dividend, the less the quotient.

EXAMPLE.—What is the quotient of ·048 ÷ 8 ?

$$\frac{48}{8} = 6, \text{ therefore } \frac{·048}{8} = ·006.$$

84. If there are decimals in the divisor—

RULE.—Divide as if there were none, and move the quotient so many places to the left as there are decimals.

The smaller the divisor, the greater the quotient.

EXAMPLE.—What is the quotient of 54 ÷ ·006 ?

$$\frac{54}{·6} = 9, \text{ therefore } \frac{54}{·006} = 9000.$$

85. If there are decimals in both dividend and divisor—

RULE.—Divide as if there were none, and move the quotient a number of places equal to the difference between the numbers of decimals in the two given quantities :—if the decimals in the dividend exceed those in the divisor, move to the right ; if the decimals in the divisor exceed those in the dividend, move to the left.

We have seen that decimals in the dividend move the quotient to the right, and that decimals in the divisor move it to the left; when, therefore, both causes act together, the effect must be equal to the difference between their separate effects.

EXAMPLES.

(1)	(2)	(3)	(4)	(5)	(6)
5)45	5)·45	·05)45	·5)·045	·005) 450	·05)·45
9	·09	900	·09	90000	9·00

86. If there are cyphers in the dividend, and decimals in the divisor—

RULE.—Divide as if there were neither, and move the quotient a number of places to the left, equal to the number of both cyphers and decimals.

Both the cyphers in the dividend, and the decimals in the divisor increase the quotient.

EXAMPLE.—What is the quotient of $270 \div ·03$?

$$\frac{27}{3} = 9, \text{therefore, } 270 \div ·03 = 9000.$$

87. If there are decimals in the dividend, and cyphers in the divisor—

RULE.—Divide as if there were neither, and move the quotient a number of places to the right equal to the number of both cyphers and decimals.

Both the decimals in the dividend, and the cyphers in the divisor diminish the quotient.

EXAMPLE.—What is the quotient of $·18 \div 20$?

$$\frac{18}{2} = 9, \text{ therefore } \frac{·18}{20} = ·009.$$

The rules which relate to the management of cyphers and decimals, in multiplication and in division—though numerous—will be very easily remembered, if the pupil merely considers what *ought* to be the effect of either.

EXERCISES.

(13)	(14)	(15)	(16)	(17)
8)10000	11)16000	3)70170	6)68530	20)36526

(18)	(19)	(20)	(21)
3000)47865	40)56020	80)75686	12)63·075

(22)	23)	(24)	(25)
10)·08756	·07)54268	·09)57·368	·0005)60300

(26)	(27)	(28)	(29)
700)·03576	·008)57·362	400)63700	110)97·634

88. When the divisor exceeds 12—

The process used is called *long* division ; that is, we perform the multiplications, subtractions, &c., in full, and not, as before, merely in the mind. This will be understood better, by applying the method of long division to an example in which—the divisor not being greater than 12—it is unnecessary.

```
Short Division :      the same by    Long Division.
  8)5763472                          8)5763472(720434
    720434                             56

                                       16
                                       16

                                          34
                                          32

                                            27
                                            24

                                              32
                                              32
```

In the second method, we multiply the divisor by the different parts of the quotient, and in each case *set down* the product, *subtract* it from the corresponding portion of the dividend, *write* the remainder, and *bring down* the required digits of the dividend. All this *must* be done when the divisor becomes large, or the memory would be too heavily burdened.

89. RULE—I. Put the divisor to the left of the dividend, with a separating line.

II. Mark off, by a separating line, a place for the quotient, to the right of the dividend.

III. Find the smallest number of digits at the left hand side of the dividend, which expresses a quantity not less than the divisor.

IV. Put under these, and subtract from them, the greatest multiple of the divisor which they contain; and set down, underneath, the remainder, if there is any. The digit by which we have multiplied the divisor is to be placed in the quotient.

V. To the remainder just mentioned add, or, as it is said, "bring down" so many of the next digits (or cyphers, as the case may be) of the dividend, as are required to make a quantity not less than the divisor; and for every digit or cypher of the dividend thus brought down, *except one*, add a cypher after the digit last placed in the quotient.

VI. Find out, and set down in the quotient, the *number* of times the divisor is contained in this quantity; and then subtract from the latter the product of the divisor and the digit of the quotient just set down. Proceed with the resulting remainder, and with all that succeed, as with the last.

VII. If there is a remainder, after the *units* of the dividend have been "brought down" and divided, either place it into the quotient with the divisor under it, and a separating line between them [70]; or, putting the decimal point in the quotient—and adding to the remainder as many cyphers as will make it at least equal to the divisor, and to the quotient as many cyphers *minus one* as there have been cyphers added to the remainder—proceed with the division.

90. EXAMPLE 1.—Divide 78325826 by 82.

```
82)78325826(955193
   738

   452
   410

   425
   410

   158
    82

   762
   738

   246
   246
```

82 will not go into 7; nor into 78; but it will go 9 times into 783 :—9 is to be put in the quotient.

The values of the higher denominations in the quotient will be sufficiently marked by the digits which succeed them—it will, however, sometimes be proper to ascertain, if the pupil, as he proceeds, is acquainted with the orders of units to which they belong.

9 times 82 are 738, which, being put under 783, and subtracted from it, leaves 45 as remainder; since this is less than the divisor, the digit put into the quotient is—as it ought to be [71]—the largest possible. 2, the next digit of the dividend, being brought down, we have 452, into which 83 goes 5 times ;—5 being put in the quotient, we subtract 5 times the divisor from 452, which leaves 42 as remainder. 42, with 5, the next digit of the dividend, makes 425, into which 82 goes 5 times, leaving 15 as remainder ;—we put another 5 in the quotient. The last remainder, 15, with 8 the next digit of the dividend, makes 158, into which 82 goes once, leaving 76 as remainder ;—1 is to be put in the quotient. 2, the next digit of the dividend, along with 76, makes 762, into which the divisor goes 9 times, and leaves 24 as remainder ;—9 is to be put in the quotient. The next digit being brought down, we have 246, into which 82 goes 3 times exactly ;—3 is to be put in the quotient. This 3 indicates 3 units, as the last digit brought down expressed units.

Therefore $\frac{78325826}{82} = 955193$.

EXAMPLE 2.—Divide 6421284 by 642.

642)6421284(10002
 642
 ————
 1284
 1284

642 goes once into 642, and leaves no remainder. Bringing down the next digit of the dividend gives no digit in
the quotient, in which, therefore, we put a cypher after the
1. The next digit of the dividend, in the same way, gives
no digit in the quotient, in which, consequently, we put
another cypher; and, for similar reasons, another in bringing
down the next; but the next digit makes the quantity
brought down 1284, which contains the divisor twice, and
gives no remainder :—we put 2 in the quotient.

91. When there is a remainder, we may continue the
division, adding decimal places to the quotient, as follows—

EXAMPLE 3.—Divide 796347 by 847.

847)796347(940·19, &c.
 7623
 ————
 3404
 3388
 ————
 1670
  847
 ————
 8230
 7623

92. The learner, after a little practice, will guess
pretty accurately what, in each case, should be the next
digit of the quotient. He has only to multiply in his mind
the last digit of the divisor, adding to the product what
he would probably have to carry from the multiplication of the second last :—if this sum can be taken from
the corresponding part of what is to be the minuend,
leaving little, or nothing, the assumed number is likely
to answer for the next digit of the quotient.

93. REASON OF I.—This arrangement is merely a matter of
convenience; some put the divisor to the right of the dividend,
and immediately over the quotient—believing that it is more
convenient to have two quantities which are to be multiplied
together as near to each other as possible. Thus, in dividing
6425 by 54—

$$6425 \Big(\frac{54}{118, \&c.}$$
$$54$$
$$\overline{102}$$
$$54$$
$$\overline{485}$$
$$432$$
$$\overline{53, \&c.}$$

REASON OF II.—This, also, is only a matter of convenience.

REASON OR III.—A smaller part of the dividend would give *no* digit in the quotient, and a larger would give more than *one*.

REASON OR IV.—Since the numbers to be multiplied, and the products to be subtracted, are considerable, it is not so convenient as in short division, to perform the multiplications and subtractions mentally. The rule directs us to set down each multiplier in the quotient, because the latter is the sum of the multipliers.

REASON OF V.—One digit of the dividend brought down would make the quantity to be divided *one* denomination lower than the preceding, and the resulting digit of the quotient also *one* denomination lower. But if we are obliged to bring down two digits, the quantity to be divided is *two* denominations lower, and consequently the resulting digit of the quotient is *two* denominations lower than the preceding—which, from the principles of notation [Sec. I. 28], is expressed by using a cypher. In the same way, bringing down three figures of the dividend reduces the denomination three places, and makes the new digit of the quotient three denominations lower than the last—two cyphers must then be used. The same reasoning holds for any number of characters, whether significant or otherwise, brought down to any remainder.

REASON OF VI.—We subtract the products of the different parts of the quotient and the divisor (these different parts of the quotient being put down successively according as they are found), that we may discover what the remainder is from which we are to expect the next portion of the quotient. From what we have already said [77], it is evident that, if there are no decimals in the divisor, the quotient figure will always be of the same denomination as the lowest in the quantity from which we subtract the product of it and the divisor.

REASON OF VII.—The reason of this is the same as what was given for the sixth part of the preceding rule [77].

It is proper to put a dot over each digit of the dividend, as we bring it down; this will prevent our forgetting any one, or bringing it down twice.

94. When there are cyphers, decimals, or both, the rules already given [80, &c.] are applicable.

95. *To prove the Division.*—Multiply the quotient by the divisor ; the product should be equal to the dividend, minus the remainder, if there is any [79].

To prove it by the method of " casting out the nines".—

RULE.—Cast the nines out of the divisor, and the quotient ; multiply the remainders, and cast the nines from their product :—that which is now left ought to be the same as what is obtained by casting the niues out of the dividend minus the remainder obtained from the process of division.

EXAMPLE.—Prove that $\dfrac{63776}{54} = 1181\frac{2}{54}$.

Considered as a question in multiplication, this becomes $1181 \times 54 = 63776 - 2 = 63774$. To try if this be true,

Casting the nines from 1181, the remainder is 2. $\Big\}$ $2 \times 0 = 0$
„   „   from   54,   „   is 0.
Casting the nines from 63774, the remainder is . . 0

The two remainders are equal, both being 0 ; hence the multiplication is to be presumed right, and, consequently, the process of division which supposes it.

The division involves an example of multiplication ; since the product of the divisor and quotient ought to be equal to the dividend minus the remainder [79]. Hence, in proving the multiplication (supposed), as already explained [54], we indirectly prove the division.

EXERCISES.

(30)	(31)	(32)	(33)
24)7654	15)6783	16)5674	17)4675
318$\frac{22}{24}$	452$\frac{3}{15}$	354$\frac{10}{16}$	275

(34)	(35)	(36)	(37)
18)7831	19)5977	21)6783	22)9767
435$\frac{1}{18}$	314$\frac{11}{19}$	323	443$\frac{21}{22}$

(38)	(39)	(40)
23)767500	390)5807	1460)6767600
33369$\frac{13}{23}$	14·8897	4635·3425

(41)	(42)	(43)
253)77676700	67·1)·1342	·153)·829749
303424·6094	·002	5·4232

(44)	(45)	(46)
54·25)123·70536	14·35)269·0625	·0037) 555
2·2803	18·75	150000

In example 40—and some of those which follow—after obtaining as many decimal places in the quotient as are deemed necessary, it will be more accurate to consider the remainder as equal to the divisor (since it is *more* than one half of it), and add unity to the last digit of the quotient.

CONTRACTIONS IN DIVISION.

96. We may abbreviate the process of division when there are many decimals, by cutting off a digit to the right hand of the divisor, at each new digit of the quotient; remembering to carry what would have been obtained by the multiplication of the figure neglected— unity if this multiplication would have produced more than 5, or less than 15; 2 if more than 15, or less than 25, &c. [59].

EXAMPLE.—Divide 754·337385 by 61·347.

Ordinary Method.	Contracted Method.
61·347)754·33|7385(12·296	61·347)754·337385(12·296
61347|	61347
14086|7	14086
12269|4	12269
1817|33	1817
1226|94	1227
590|398	590
552|123	552
38|2755	38
36|8082	37
1|46730	1

According as the denominations of the quotient become small, their products by the lower denomination of the divisor become inconsiderable, and may be neglected, and, conse_quently, the portions of the dividend from which they would have been subtracted. What should have been *carried* from the multiplication of the digit neglected—since it belongs to a higher denomination than what is neglected, should still be retained [59].

97. We may avail ourselves, in division, of contrivances very similar to those used in multiplication [60].

To divide by a composite number—

RULE.—Divide successively by its factors.

EXAMPLE.—Divide 98 by 49. $49 = 7 \times 7$.

$$7)98$$
$$\overline{7)14}$$
$$2 = 98 \div 7 \times 7, \text{ or } 49.$$

Dividing only by 7 we divide by a quantity 7 times too small, for we are to divide by 7 times 7; the result is, therefore, 7 times too great :—this is corrected if we divide again by 7.

98. If the divisor is not a composite number, we cannot, as in multiplication, abbreviate the process, except it is a quantity which is but little less than a number expressed by unity and one or more cyphers. When this is the case—

RULE.—Divide by the nearest higher number expressed by unity and one or more cyphers; add to remainder so many times the quotient as the assumed exceeds the given divisor, and divide the sum by the preceding divisor. Proceed thus, adding to the remainder in each case so many times the foregoing quotient as the assumed exceeds the given divisor until the exact, or a sufficiently near approximation to the exact quotient is obtained—the *last* divisor must be the given, and not the assumed one. The last remainder will be the true one; and the sum of all the quotients will be the true quotient.

EXAMPLE.—Divide 987663425 by 998.

$$987663_\wedge 425 = 987663425 \div 1000.$$
$$1975_\wedge 751 = \overline{987663 \times 2 + 425} \div 1000.$$
$$4_\wedge 701 = \overline{1975 \times 2 + 751} \div 1000.$$
$$0{\cdot}7_\wedge 090 = \overline{4 \times 2 + 701} \div 1000.$$
$$0{\cdot}01_\wedge 040 = \overline{{\cdot}7 \times 2 + 9} \div 1000.$$
$$0{\cdot}000_\wedge 420 = \overline{{\cdot}01 \times 2 + {\cdot}4} \div 1000.$$
$$0{\cdot}0004_\wedge 0208 = \overline{{\cdot}01 \times 2 + {\cdot}4} \div 998.$$

that is, the last quotient is 0·0004, and ·0208 is the last remainder.

all the quotients are $\left\{ \begin{array}{l} 987663 \\ 1975 \\ 4 \\ 0{\cdot}7 \\ 0{\cdot}01 \\ 0{\cdot}0004 \end{array} \right.$

The true quotient is 989642·7104, or the *sum* of the quotients.
And the true remainder 0·0208, or the *last* remainder.

Unless we add twice the preceding quotient to each successive remainder, we shall have subtracted from the dividend, or the part of it just divided, 1000, and not 998 times the quotient—in which case the remainder would be too small to the amount of twice the quotient.—We have used ($_\wedge$) to separate the quotients from the remainders.

There can be no difficulty when the learner, by this process, comes to the decimals of the quotient. Thus in the third line, 4701 gives, when divided by 1000, 4 units as quotient, and 701 units still *to be* divided—that is, 701 as remainder. 4·701 would express 4701 *actually* divided by 1000. A number occupying four places, all to the left of the decimal point, when divided by 1000, gives units as quotient; but if, as in 709·0 (in the next line), one is a decimal place, the quotient must be of a lower denomination than before—that is, of the order tenths; and in 010·40 (next line), since *two* out of the four places are decimals, the quotient must be hundredths, &c.

In adding the necessary quantities, we must carefully bear in mind to what denominations the quotient multiplied, and the remainder to which the product is to be added, belong.

47. $56789 \div 741 = 76\frac{443}{741}$.
48. $478907 \div 971 = 493\frac{704}{971}$.
49. $977076 \div 47600 = 20\frac{24076}{47600}$.
50. $567897 \div 842 = 674\frac{389}{842}$.
51. $7867674 \div 9712 = 810\frac{3954}{9712}$.
52. $3070700 \div 457000 = 6\cdot 7193$.
53. $6765158 \div 7894 = 857$.
54. $67470 \div 3900 = 17\cdot 3$.
55. $69000 \div 47600 = 1\cdot 4496$.
56. $76767 \div 40700 = 1\cdot 8862$.
57. $6114592 \div 764324 = 8$.
58. $9676744 \div 910076 = 10\cdot 6329$.
59. $740070000 \div 741000 = 998\cdot 7449\cdot$
60. $9410607111 \div 45673 = 206043\cdot 1132$.
61. $454076000 \div 400100 = 1134\cdot 9063\cdot$
62. $7376476767 \div 345670 = 21339\cdot 649\cdot$
63. $47\cdot 5782975 \div 26\cdot 175 = 1\cdot 8177$.
64. $47\cdot 655 \div 4\cdot 5 = 10\cdot 59$.
65. $756\cdot 98 \div 76\cdot 73612 = 9\cdot 866$.
66. $75\cdot 3470 \div \cdot 3829 = 196\cdot 7798$.
67. $0\cdot 1 \div 7\cdot 6345 = 0\cdot 0000131$.
68. $5378 \div 0\cdot 00096 = 5602083\cdot 33$, &c.

69. If £7500 were to be divided between 5 persons, how much ought each person to receive? *Ans.* £1500.

70. Divide 7560 acres of land between 15 persons.
Ans. Each will have 504 acres.

71. Divide £2880 between 60 persons.
Ans. Each will receive £48.

72. What is the ninth of £972? *Ans.* £108.

73. What is each man's part if £972 be divided among 108 men? *Ans.* £9.

74. Divide a legacy of £8526 between 294 persons.
Ans. Each will have £29.

75. Divide 340480 ounces of bread between 1792 persons. *Ans.* Each person's share will be 190 ounces.

76. There are said to be seven bells at Pekin, each of which weighs 120,000 pounds; if they were melted up, how many such as great Tom of Lincoln, weighing 9894 pounds, or as the great bell of St. Paul's, in London, weighing 8400 pounds, could be made from them? *Ans.* 84 like great Tom of Lincoln, with 8904 pounds left; and 100 like the great bell of St. Paul's.

77. Mexico produced from the year 1790 to 1830 a

quantity of gold which was worth £6,436,443, or 6,178,985,280 farthings. How many dollars, at 207 farthings each, are in that sum? *Ans.* 29,850170 nearly.

78. A single pound of cotton has been spun into a thread 76 miles in length, and a pound of wool into a thread 95 miles long; how many pounds of each would be required for threads 5854 miles in length? *Ans.* 77·0263 pounds of cotton, and 61·621 pounds of wool.

79. The earth travels round its orbit, a space equal to 567,019,740 miles, in about 365 days, 8765 hours, 525948 minutes, 31556925 seconds, and 1893415530 thirds; supposing its motion uniform, how much would it travel per day, hour, minute, second, and third? *Ans.* About 1553480 miles a day, 64691 an hour, 1078 a minute, 18 a second, and 0·3 a third.

80. All the iron produced in Great Britain in the year 1740 was 17,000 tons from 59 furnaces; and in 1827, 690,000 from 284. What may be considered as the produce of each furnace in 1740, one with another; and of each in 1827. *Ans.* 288·1356 in 1740; and 2429·5775 in 1827.

81. In 1834, 16,000 steam engines in Great Britain saved the labour of 450,000 horses, or 2 millions and a half of men; to how many horses, and how many men, may each steam engine be supposed equivalent, one with another? *Ans.* About 28 horses; and 156 men.

99. Before the pupil leaves division, he should be able to carry on the process as follows :—

EXAMPLE.—Divide 84380848 by 87532.

$$87532)84380848(964$$
$$\underline{560204}$$

$$350128$$

He will say (at first aloud) 4 (the digit of the dividend to be brought down). 18 (9 times 2); 0 (the remainder after subtracting the right hand digit of 18 from 8 in the dividend). 28 (9 times 3 + the 1 to be carried from the 18); 2 (the remainder after subtracting the right hand digit of 28 from 0, or rather 10 in the dividend). 48 (9 times 5 + the 2 to be carried from 28, and 1 to compensate for what we borrowed when we considered 0 in the dividend as 10); 0 (the remainder when we subtract the right hand digit of 48 from 8 in the dividend). 67 (9 times 7 + the 4 to be carried from the 48); 6 (the remainder after subtracting the right hand digit of 67 from 3, or rather 13 in the dividend). 79 (9 times 8 + the 6 to be carried from the 67 + the 1, for what we borrowed to make 3 in the dividend become 13); 5 (the remainder after subtracting 79 from 84 in the dividend).

As the parts in the parentheses are merely explanatory, and not to be repeated, the whole process would be,
    First part, 4. 18 ; 0. 28 ; 2. 48 ; 0. 67 ; 6. 79 ; 5.
    Second part, 8. 12 ; 2. 19 ; 1. 32 ; 0. 45 ; 5. 53 ; 3
    Third part, 8 ; 0. 12 ; 0. 21 ; 0. 30 ; 0. 35 ; 0.
The remainders in this case being cyphers, are omitted.

All this will be very easy to the pupil who has practised what has been recommended [13, 23, and 65]. The chief exercise of the memory will consist in recollecting to add to the products of the different parts of the divisor by the digit of the quotient under consideration, what is to be carried from the preceding product, and unity besides—when the preceding digit of the dividend has been increased by 10; then to subtract the right hand digit of this sum from the proper digit of the dividend (increased by 10 if necessary).

QUESTIONS FOR THE PUPIL.

1. What is division ? [66].

2. What are the dividend, divisor, quotient, and remainder ? [66].

3. What is the sign of division ? [68].

4. How are quantities under the vinculum, or united by the sign of multiplication, divided ? [69].

5. What is the rule when the divisor does not exceed 12, nor the dividend 12 times the divisor ? [70].

6. Give the rule, and the reasons of its different parts, when the divisor does not exceed 12, but the dividend is more than 12 times the divisor ? [72 and 77].

7. How is division proved ? [79 and 95].

8. What are the rules when the dividend, divisor, or both contain cyphers or decimals ? [80].

9. What is the rule, and what are the reasons of its different parts, when the divisor exceeds 12? [89 and 93].

10. What is to be done with the remainder? [72 and 89].

11. How is division proved by casting out the nines? [95].

12. How may division be abbreviated, when there are decimals? [96].

13. How is division performed, when the divisor is a composite number? [97].

14. How is the division performed, when the divisor is but little less than a number which may be expressed by unity and cyphers? [98].

15. Exemplify a very brief mode of performing division. [99]

THE GREATEST COMMON MEASURE OF NUMBERS.

100. To find the greatest common measure of two quantities—

RULE.—Divide the larger by the smaller; then the divisor by the remainder; next the preceding divisor by the new remainder:—continue this process until nothing remains, and the last divisor will be the greatest common measure. If this be unity, the given numbers are prime to *each other*.

EXAMPLE.—Find the greatest common measure of 3252 and 4248.

$$
\begin{array}{l}
3252)4248(1 \\
\quad 3252 \\
\quad \overline{\quad\quad} \\
\quad\quad 996)3252(3 \\
\quad\quad\;\; 2988 \\
\quad\quad\;\; \overline{\quad\quad} \\
\quad\quad\quad\; 264)996(3 \\
\quad\quad\quad\;\; 792 \\
\quad\quad\quad\;\; \overline{\quad\quad} \\
\quad\quad\quad\quad\; 204)264(1 \\
\quad\quad\quad\quad\;\; 204 \\
\quad\quad\quad\quad\;\; \overline{\quad\quad} \\
\quad\quad\quad\quad\quad\; 60)204(3 \\
\quad\quad\quad\quad\quad\;\; 180 \\
\quad\quad\quad\quad\quad\;\; \overline{\quad\quad} \\
\quad\quad\quad\quad\quad\quad\; 24)60(2 \\
\quad\quad\quad\quad\quad\quad\;\; 48 \\
\quad\quad\quad\quad\quad\quad\;\; \overline{\quad\quad} \\
\quad\quad\quad\quad\quad\quad\quad\; 12)24(2 \\
\quad\quad\quad\quad\quad\quad\quad\;\; 24
\end{array}
$$

F

996, the first remainder, becomes the second divisor ; 264, the second remainder, becomes the third divisor, &c.   12, the last divisor, is the required greatest common measure.

101. REASON OF THE RULE.—Before we prove the correct. ness of the rule, it will be necessary for the pupil to be satis. fied that "if any quantity measures another, it will measure any multiple of that other;" thus if 6 go into 30, 5 times, it will evidently go into 9 times 30, 9 times 5 times.

Also, that "if a quantity measure two others, it will measure their sum, and their difference."   First, it will measure their sum, for if 6 go into 24, 4 times, and into 36, 6 times, it will evi. dently go into 24+36, 4+6 times :—that is, if $\frac{24}{6}=4$, and $\frac{36}{6}=6$, $\frac{24}{6}+\frac{36}{6}=4+6$.

Secondly, if 6 goes into 36 oftener than it goes into 24, it is because of the difference between 36 and 24 ; for as the differ. ence between the numbers of times it will go into them is due to this difference, 6 must be contained in it some number of times :—that is, since $\frac{36}{6}=6$, and $\frac{24}{6}=4$, $\frac{36}{6}-\frac{24}{6}\left(\text{or }\frac{36-24}{6}\right)$ $=6-4=2$, a whole number [26]—or, the difference between the quantities is measured by 6, their measure.

This reasoning would be found equally correct with any other similar numbers.

102. Next; to prove the rule from the given example, it is necessary to prove that 12 is a *common* measure ; and that it is the *greatest* common measure.

It is a *common* measure. Beginning at the end of the process, we find that 12 measures 24, its multiple; and 48, because it is a multiple of 24 ; and their sum, 24+48 (because it measures each of them) or 60 ; and 180, because it is a multiple of 60 ; and 180+24 (we have also just seen that it measures each of these) or 204 ; and 204+60 or 264 ; and 792, because a multi- ple of 264 ; and 792+204 or 996 ; and 2988, a multiple of 996 ; and 2988+264 or 3252 (one of the given numbers) and 3252+ 996 or 4248 (the other given number).   Therefore it measures each of the given numbers, and is their *common* measure.

103. It is also their *greatest* common measure.   If not, let some other be greater; then (beginning now at the top of the process) measuring 4248 and 3252 (this is the supposition), it measures their difference, 996 ; and 2988, because a multiple of 996 ; and, because it measures 3252, and 2988, it measures their difference, 264 ; and 792, because a multiple of 264 ; and the difference between 996 and 792 or 204 ; and the difference between 264 and 204 or 60 ; and 180 because a multiple of 60 ; and the difference between 204 and 180 or 24 ; and 48, because a multiple of 24 ; and the difference between 60 and 48 or 12. But measuring 12, it cannot be greater than 12.

In the same way it could be shown, that any other common measure of the given numbers must be less than 12—and consequently that 12 is their *greatest* common measure. As the rule might be proved from any other example equally well, it is true in all cases.

104. We may here remark, that the measure of two or more quantities can sometimes be found by inspection:

Any quantity, the digit of whose lowest denomination is an even number, is divisible by 2 at least.

Any number ending in 5 is divisible by 5 at least.

Any number ending in a cypher is divisible by 10 at least.

Any number which leaves nothing when the threes are cast out of the sum of its digits, is divisible by 3 at least ; or leaves nothing when the nines are cast out of the sum of its digits, is divisible by 9 at least.

EXERCISES.

1. What is the greatest common measure of 464320 and 18945 ? *Ans.* 5.
2. Of 638296 and 33888 ? *Ans.* 8.
3. Of 18996 and 29932 ? *Ans.* 4.
4. Of 260424 and 54423 ? *Ans.* 9.
5. Of 143168 and 2064888 ? *Ans.* 8.
6. Of 1141874 and 19823208 ? *Ans.* 2.

105. To find the greatest common measure of more than two numbers—

RULE—Find the greatest common measure of two of them ; then of this common measure and a third ; next, of this last common measure and a fourth, &c. The last common measure found, will be the greatest common measure of all the given numbers.

EXAMPLE 1.—Find the greatest common measure of 679, 5901, and 6734.

By the last rule we learn that 7 is the greatest common measure of 679 and 5901 ; and by the same rule, that it, the greatest common measure of 7 and 6734 (the remaining number), for $6734 \div 7 = 962$, with no remainder. Therefore 7 is the required number.

EXAMPLE 2.—Find the greatest common measure of 936, 736, and 142.

The greatest common measure of 936 and 736 is 8, and the common measure of 8 and 142 is 2 ; therefore 2 is the greatest common measure of the given numbers.

106. REASON OF THE RULE.—It may be shown to be correct in the same way as the last; except that in proving the number found to be a *common* measure, we are to begin at the end of *all* the processes, and go through all of them in succession ; and in proving that it is the *greatest* common measure, we are to begin at the commencement of the first process, or that used to find the common measure of the two first numbers, and proceed successively through *all*.

EXERCISES.

7. Find the greatest common measure of 29472, 176832, and 1074.   *Ans.* 6.

8. Of 648485, 10810, 3672835, and 473580. *Ans.* 5.

9. Of 16264, 14816, 8600, 75288, and 8472. *Ans.* 8.

THE LEAST COMMON MULTIPLE OF NUMBERS.

107. To find the least common multiple of two quantities—

RULE.—Divide their product by their greatest common measure. Or ; divide one of them by their greatest common measure, and multiply the quotient by the other—the result of either method will be the required least common multiple.

EXAMPLE.—Find the least common multiple of 72 and 84. 12 is their greatest common measure.

$\dfrac{72}{12} = 6$, and $6 \times 84 = 504$, the number sought.

108. REASON OF THE RULE.—It is evident that if we multiply the given numbers together, their product will be a multiple of each by the other [30]. It will be easy to find the smallest part of this product, which will still be their common multiple.—Thus, to learn if, for example, its nineteenth part is such.

From what we have already seen [69], each of the factors of any product divided by any number and multiplied by the product of the other factors, is equal to the product of all the factors divided by the same number. Hence, 72 and 84 being the given numbers—

$\frac{72 \times 84}{19}$ (the nineteenth part of their product)$= \frac{72}{19} \times 84$, or $72 \times \frac{84}{19}$. Now if $\frac{72}{19}$ and $\frac{84}{19}$ be equivalent to integers, $\frac{72}{19} \times 84$ will be a multiple of 84, and $\frac{84}{19} \times 72$, will be a multiple of 72 [29]; and $\frac{72 \times 84}{19}$, $\frac{72}{19} \times 84$, and $72 \times \frac{84}{19}$ will each be the common multiple of 72 and 84 [30]. But unless 19 is a common measure of 72 and 84, $\frac{72}{19}$ and $\frac{84}{19}$ cannot be both equivalent to integers. Therefore the quantity by which we divide the product of the given numbers, or one of them, before we multiply it by the other to obtain a new, and less multiple of them, must be the common measure of both. And the multiple we obtain will, evidently, be the least, when the divisor we select is the greatest quantity we can use for the purpose—that is, the *greatest* common measure of the given numbers.

It follows, that the least common multiple of two numbers, prime to each other, is their product.

EXERCISES.

1. Find the least common multiple of 78 and 93. *Ans.* 2418.

2. Of 19 and 72. *Ans.* 1368.

3. Of 464320 and 18945. *Ans.* 1759308480.

4. Of 638296 and 33888. *Ans.* 2703821856.

5. Of 18996 and 29932. *Ans.* 142147068.

6. Of 260424 and 54423. *Ans.* 1574783928.

109. To find the least common multiple of three or more numbers—

RULE.—Find the least common multiple of two of them; then of this common multiple, and a third; next of this last common multiple and a fourth, &c. The last common multiple found, will be the least common multiple sought.

EXAMPLE.—Find the least common multiple of 9, 3, and 27

3 is the greatest common measure of 9 and 3; therefore $\frac{9}{3} \times 3$, or 9 is the least common multiple of 9 and 3.

9 is the greatest common measure of 9 and 27; therefore $\frac{27}{9} \times 9$, or 27 is the required least common multiple.

110. REASON OF THE RULE.—By the last rule it is evident that 27 is the least common multiple of 9 and 27.   But since 9 is a multiple of 3, 27, which is a multiple of 9, must also be a multiple of 3; 27, therefore, is a multiple of each of the given numbers, or their *common* multiple.

It is likewise their *least* common multiple, because none that is smaller can be common, also, to both 9 and 27, since they were found to have 27 as their *least* common multiple.

EXERCISES.

7. Find the least common multiple of 18, 17, and 43. *Ans.* 13158.

8. Of 19, 78, 84, and 61.   *Ans.* 1265628.

9. Of 51, 176832, 29472, and 5862.   *Ans.* 2937002688.

10. Of 537842, 16819, 4367, and 2473.
Ans. 8881156168989038.

11. Of 21636, 241816, 8669, 97528, and 1847.
Ans. 15288355550537452616.

QUESTIONS.

1. How is the greatest common measure of two quantities found ? [100].

2. What principles are necessary to prove the correctness of the rule ; and how is it proved ? [101, &c.].

3. How is the greatest common measure of three, or more quantities found ? [105].

4. How is the rule proved to be correct ? [106].

5. How do we find the least common multiple of two numbers that are composite ? [107].

6. Prove the rule to be correct [108].

7. How do we find the least common multiple of two prime numbers ? [108].

8. How is the least common multiple of three or more numbers found ? [109].

9. Prove the rule to be correct [110].

In future it will be taken for *granted* that the pupil is to be asked the reasons for each rule, &c.

SECTION III.

REDUCTION AND THE COMPOUND RULES.

The pupil should now be made familiar with most of the tables given at the commencement of this treatise.

REDUCTION.

1. Reduction enables us to change quantities from one denomination to another without altering their value. Taken in its more extended sense, we have often practised it already :—thus we have changed units into tens, and tens into units, &c. ; but, considered as a separate rule, it is restricted to applicate numbers, and is not confined to a change from one denomination to the *next* higher, or lower.

2. Reduction is either descending, or ascending. It is *reduction descending* when the quantities are changed from a higher to a lower denomination ; and *reduction ascending* when from a lower to a higher.

Reduction Descending.

3. RULE.—Multiply the highest given denomination by that quantity which expresses the number of the next lower contained in one of its units ; and add to the product that number of the next lower denomination which is found in the quantity to be reduced.

Proceed in the same way with the result ; and continue the process until the required denomination is obtained.

EXAMPLE.—Reduce £6 16s. 0¼d. to farthings.

$$
\begin{array}{ccc}
£ & s. & d. \\
6 & ,, \ 16 \ ,, & 0¼ \\
20 & &
\end{array}
$$

136 shillings=£6 ,, 16.
12

1632 pence=£6 ,, 16 ,, 0.
4

6529 farthings=£6 ,, 16 ,, 0¼.

We multiply the pounds by 20, and at the *same* time add the shillings. Since multiplying by 2 tens (20) can give no units in the product, there can be no units of shillings in it except those derived from the 6 of the 16*s.* :—we at once, therefore, put down 6 in the shillings' place. Twice (2 tens' times) 6 are 12 (tens of shillings), and one (ten shillings), to be added from the 16*s.*, are 13 (tens of shillings)—which we put down. £6 16*s.* are, consequently, equal to 136*s.*

12 times 6*d.* are 72*d.* :—since there are no pence in the given quantity, there are none to be added to the 72*d.*—we put down 2 and carry 7. 12 times 3 are 36, and 7 are 43. 12 times 1 are 12, and 4 are 16. £6 16*s.* are, therefore, equal to 1632 pence..

4 times 2 are 8, and ¼ (in the quantity to be reduced) to be carried are 9, to be set down. 4 times 3 are 12. 4 times 6 are 24, and 1 are 25. 4 times 1 are 4, and 2 are 6. Hence £6 16*s.* 0¼*d.* are equal to 6529 farthings.

4. REASONS OF THE RULE.—One pound is equal to 20*s.* ; therefore any number of pounds is equal to 20 times as many shillings ; and any number of pounds and shillings is equal to 20 times as many shillings as there are pounds, plus the shillings.

It is easy to multiply by 20, and add the shillings at the *same* time ; and it shortens the process.

Shillings are equal to 12 times as many pence ; pence to 4 times as many farthings ; hundreds to 4 times as many quarters ; quarters to 28 times as many pounds, &c.

EXERCISES.

1. How many farthings in 23328 pence ? *Ans.* 93312.

2. How many shillings in £348 ? *Ans.* 6960.

3. How many pence in £38 10*s.* ? *Ans.* 9240.

4. How many pence in £58 13*s.* ? *Ans.* 14076.

5. How many farthings in £58 13*s.* ? *Ans.* 56304.

6. How many farthings in £59 13*s.* 6¾*d.* ? *Ans.* 57291.

7. How many pence in £63 0*s.* 9*d.* ? *Ans.* 15129.

8. How many pounds in 16 cwt., 2 qrs., 16 ℔. ? *Ans.* 1864.

9. How many pounds in 14 cwt., 3 qrs., 16 ℔. ? *Ans.* 1668.

10. How many grains in 3 ℔., 5 oz., 12 dwt., 16 grains ? *Ans.* 19984.

11. How many grains in 7 ℔., 11 oz., 15 dwt., 14 grains? *Ans.* 45974.

12. How many hours in 20 (common) years? *Ans.* 175200.

13. How many feet in 1 English mile? *Ans.* 5280.

14. How many feet in 1 Irish mile? *Ans.* 6720.

15. How many gallons in 65 tuns? *Ans.* 16380.

16. How many minutes in 46 years, 21 days, 8 hours, 56 minutes (not taking leap years into account)? *Ans.* 24208376.

17. How many square yards in 74 square English perches? *Ans.* 2238·5 (2238 and one half).

18. How many square inches in 97 square Irish perches? *Ans.* 6159888.

19. How many square yards in 46 English acres, 3 roods, 12 perches? *Ans.* 226633.

20. How many square acres in 767 square English miles? *Ans.* 490880.

21. How many cubic inches in 767 cubic feet? *Ans.* 1325376.

22. How many quarts in 767 pecks? *Ans.* 6136.

23. How many pottles in 797 pecks? *Ans.* 3188.

Reduction Ascending.

5. RULE.—Divide the given quantity by that number of its units which is required to make one of the next higher denomination—the remainder, if any, will be of the denomination to be reduced. Proceed in the same manner until the highest required denomination is obtained.

EXAMPLE.—Reduce 856347 farthings to pounds, &c.

$$
\begin{array}{l}
4)856347 \\
\hline
12)214086\tfrac{3}{4} \\
\hline
20)17840 \,,\, 6\tfrac{3}{4} \\
\hline
\quad 892 \,,\, 0 \,,\, 6\tfrac{3}{4} = 856347 \text{ farthings.}
\end{array}
$$

4 divided into 856347 farthings, gives 214086 pence and 3 farthings. 12 divided into 214086 pence, gives 17840 shillings and 6 pence. 20 divided into 17840 shillings, gives £892 and no shillings; there is, therefore, nothing in the shillings' place of the result.

F 2

We divide by 20 if we divide by 10 and 2 [Sec. II. 97]. To divide by 10, we have merely to cut off the units, if any, [Sec. I. 34], which will then be the units of shillings in the result; and the quotient will be tens of shillings:— dividing the latter by 2, gives the pounds as quotient, and the tens of shillings, if there are any in the required quantity, as remainder.

·6. REASONS OF THE RULE.—It is evident that every 4 farthings are equivalent to one penny, and every 12 pence to one shilling, &c.; and that what is left after taking away 4 farthings as often as possible from the farthings, must be farthings, what remains after taking away 12 pence as often as possible from the pence, must be pence, &c.

7. *To prove Reduction.*—Reduction ascending and descending prove each other.

EXAMPLE.—£20 17s. 2¼d. =20025 farthings; and 20025 farthings=£20 17s. 2¼d.

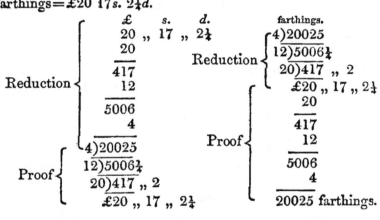

EXERCISES.

24. How many pence in 93312 farthings? *Ans.* 23328.

25. How many pounds in 6960 shillings? *Ans.* £348.

26. How many pounds, &c. in 976 halfpence? *Ans.* £2 0s. 8d.

27. How many pounds, &c. in 7675 halfpence? *Ans.* £15 19s. 9½d.

28. How many ounces, and pounds in 4352 drams? *Ans.* 272 oz., or 17 ℔.

29. How many cwt., qrs., and pounds in 1864 pounds?
Ans. 16 cwt., 2 qrs., 16 ℔.

30. How many hundreds, &c., in 1668 pounds. *Ans.*
14 cwt., 3 qrs., 16 ℔.

31. How many pounds Troy in 115200 grains?
Ans. 20.

32. How many pounds in 107520 oz. avoirdupoise?
Ans. 6720.

33. How many hogsheads in 20658 gallons? *Ans.*
327 hogsheads, 57 gallons.

34. How many days in 8760 hours? *Ans.* 365.

35. How many Irish miles in 1834560 feet? *Ans.*
273.

36. How many English miles in 17297280 inches?
Ans. 273.

37. How many English miles, &c. in 4147 yards?
Ans. 2 miles, 2 furlongs, 34 perches.

38. How many Irish miles, &c. in 4247 yards? *Ans.*
1 mile, 7 furlongs, 6 perches, 5 yards.

39. How many English ells in 576 nails? *Ans.* 28
ells, 4 qrs.

40. How many English acres, &c. in 5097 square
yards? *Ans.* 1 acre, 8 perches, 15 yards.

41. How many Irish acres, &c. in 5097 square yards?
Ans. 2 roods, 24 perches, 1 yard.

42. How many cubic feet, &c., in 1674674 cubic
inches? *Ans.* 969 feet, 242 inches.

43. How many yards in 767 Flemish ells? *Ans.*
575 yards, 1 quarter.

---

44. How many French ells in 576 English? *Ans.* 480.

45. Reduce £46 14s. 6d., the mint value of a pound
of gold, to farthings? *Ans.* 44856 farthings.

46. The force of a man has been estimated as equal
to what, in turning a winch, would raise 256 ℔, in
pumping, 419 ℔, in ringing a bell, 572 ℔, and in row-
ing, 608 ℔, 3281 feet in a day. How many hundreds,
quarters, &c., in the sum of all these quantities? *Ans.*
16 cwt., 2 qrs., 7 ℔.

47. How many lines in the sum of 900 feet, the

length of the temple of the sun at Balbec, 450 feet its
breadth, 22 feet the circumference, and 72 feet the
height of many of its columns ? *Ans.* 207936.

48. How many square feet in 760 English acres, the
inclosure in which the porcelain pagoda, at Nan-King,
in China, 414 feet high, stands ? *Ans.* 33105600.

49. The great bell of Moscow, now lying in a pit,
the beam which supported it having been burned, weighs
360,000 ℔ (some say much more) ; how many tons, &c.,
in this quantity ? *Ans.* 160 tons, 14 cwt., 1 qr., 4 ℔.

QUESTIONS FOR THE PUPIL.

1. What is reduction ? [1].
2. What is the difference between reduction descending and reduction ascending ? [2].
3. What is the rule for reduction descending ? [3].
4. What is the rule for reduction ascending ? [5].
5. How is reduction proved ? [7].

Questions founded on the Table page 3, &c.

6. How are pounds reduced to farthings, and farthings
to pounds, &c. ?
7. How are tons reduced to drams, and drams to
tons, &c. ?
8. How are Troy pounds reduced to grains, and
grains to Troy pounds, &c. ?
9. How are pounds reduced to grains (apothecaries
weight), and grains to pounds, &c. ?
10. How are Flemish, English, or French ells, reduced to inches; or inches to Flemish, English, or French
ells, &c. ?
11. How are yards reduced to ells, or ells to yards,
&c. ?
12. How are Irish or English miles reduced to lines,
or lines to Irish or English miles, &c. ?
13. How are Irish or English square miles reduced
to square inches, or square inches to Irish or English
square miles, &c. ?

14. How are cubic feet reduced to cubic inches, or cubic inches to cubic feet, &c. ?

15. How are tuns reduced to naggins, or naggins to tuns, &c. ?

16. How are butts reduced to gallons, or gallons to butts, &c. ?

17. How are lasts (dry measure) reduced to pints, and pints to lasts, &c. ?

18. How are years reduced to thirds, or thirds to years, &c. ?

19. How are degrees (of the circle) reduced to thirds, or thirds to degrees, &c. ?

THE COMPOUND RULES.

8. The Compound Rules, are those which relate to applicate numbers of more than one denomination.

If the tables of money, weights, and measures, were constructed according to the decimal system, only the rules for Simple Addition, &c., would be required. This would be a considerable advantage, and greatly tend to simplify mercantile transactions.—If 10 farthings were one penny, 10 pence one shilling, and 10 shillings one pound, the addition, for example, of £1 9s. 8$\frac{3}{4}d$. to £6 8s. 6$\frac{1}{2}d$. (a point being used to separate a pound, then the " unit of comparison," from its parts, and 0·005 to express $\frac{1}{2}$ or 5 tenths of a penny), would be as follows—

$$£$$
$$1·983$$
$$6·865$$

Sum,   8·848

The addition might be performed by the ordinary rules, and the sum read off as follows—" eight pounds, eight shillings, four pence, and eight farthings." But even with the present arrangement of money, weights, and measures, the rules already given for addition, subtraction, &c., might easily have been made to include the addition, subtraction, &c., of applicate numbers consisting of more than one denomination; since the

principles of both simple and compound rules are precisely the same—the only thing necessary to bear
carefully in mind, being the number of any one denomination necessary to constitute a unit of the next
higher.

COMPOUND ADDITION.

9. RULE.—I. Set down the addends so that quantities of the same denomination may stand in the same
vertical column—units of pence, for instance, under
units of pence, tens of pence under tens of pence, units
of shillings under units of shillings, &c.

II. Draw a separating line under the addends.

III. Add those quantities which are of the same
denomination together—farthings to farthings, pence to
pence, &c., beginning with the lowest.

IV. If the sum of any column be less than the number of that denomination which makes one of the next
higher, set it down under that column; if not, for each
time it contains that number of its own denomination
which makes one of the next higher, carry one to the
latter and set down the remainder, if any, under the
column which produced it. If in any denomination
there is no remainder, put a cypher under it in the
sum.

10. EXAMPLE.—Add together £52 17s. 3¾d., £47 5s. 6½d.,
and £66 14s. 2¼d.

$$
\begin{array}{ccc}
£ & s. & d. \\
52 & 17 & 3\frac{3}{4} \\
47 & 5 & 6\frac{1}{2} \\
66 & 14 & 2\frac{1}{4} \\
\hline
166 & 17 & 0\frac{1}{2}
\end{array}
$$

addends.

sum.

¼ and ½ make 3 farthings, which, with ¼, make 6 farthings; these are equivalent to one of the next denomination, or that of pence, to be carried, and two of the present,
or one half-penny, to be set down. 1 penny (to be carried)
and 2 are 3, and 6 are 9, and 3 are 12 pence—equal to one

of the next denomination, or that of shillings, to be carried, and no pence to be set down; we therefore put a cypher in the pence' place of the sum. 1 shilling (to be carried) and 14 are 15, and 5 are 20, and 17 are 37 shillings—equal to one of the next denomination, or that of pounds, to be carried, and 17 of the present, or that of shillings, to be set down. 1 pound and 6 are 7, and 7 are 14, and 2 are 16 pounds—equal to 6 units of pounds, to be set down, and 1 ten of pounds to be carried; 1 ten and 6 are 7 and 4 are 11 and 5 are 16 tens of pounds, to be set down.

11. This rule, and the reasons of it, are the same as those already given [Sec. II. 7 and 9]. It is evidently not so necessary to put a cypher where there is no remainder, as in Simple Addition.

12. When the addends are very numerous, we may divide them into parts by horizontal lines, and, adding each part separately, may afterwards find the amount of all the sums.

EXAMPLE:

£	s.	d.		£	s.	d.		£	s.	d.
57	14	2								
32	16	4								
19	17	6	=	151	7	11				
8	14	2								
32	5	9								
							=	404	11	10
17	6	4								
32	17	2								
56	3	9	=	253	3	11				
27	4	2								
52	4	4								
37	8	2								

13. Or, in adding each column, we may put down a dot as often as we come to a quantity which is at least equal to that number of the denomination added which is required to make one of the next—carrying forward what is above this number, if anything, and putting the last remainder, or—when there is nothing left at the end—a cypher under the column :—we carry to the next column one for every dot. Using the same example—

£	s.	d.
57	·14	2
32	16	4
19	·17	·6
8	·14	2
32	5	·9
47	·6	4
32	17	2
56	·3	·9
27	4	2
52	4	4
37	8	2
404	11	10

2 pence and 4 are 6, and 2 are 8, and 9 are 17 pence—equal to 1 shilling and 5 pence; we put down a dot and carry 5. 5 and 2 are 7, and 4 are 11, and 9 are 20 pence—equal to 1 shilling and 8 pence; we put down a dot and carry 8. 8 and 2 are 10 and 6 are 16 pence—equal to 1 shilling and 4 pence; we put down a dot and carry 4. 4 and 4 are 8 and 2 are 10—which, being less than 1 shilling, we set down under the column of pence, to which it belongs, &c. We find, on adding them up, that there are three dots; we therefore carry 3 to the column of shillings. 3 shillings and 8 are 11, and 4 are 15, and 4 are 19, and 3 are 22 shillings—equal to 1 pound and 2 shillings; we put down a dot and carry 1. 1 and 17 are 18, &c.

Care is necessary, lest the dots, not being distinctly marked, may be considered as either too few, or too many. This method, though now but little used, seems a convenient one.

14. Or, lastly, set down the sums of the farthings, shillings, &c., under their respective columns; divide the farthings by 4, put the quotient under the sum of the pence, and the remainder, if any, in a place set apart for it in the sum—under the column of farthings; add together the quotient obtained from the farthings and the sum of the pence, and placing the amount under the pence, divide it by 12; put the quotient under the sum of the shillings, and the remainder, if any, in a place allotted to it in the sum—under the column of pence; add the last quotient and the sum of the shillings, and putting under them their sum, divide the latter by 20, set down the quotient under the sum of

the pounds, and put the remainder, if any, in the sum—under the column of shillings ; add the last quotient and the sum of the pounds, and put the result under the pounds.   Using the following example—

£	s.	d.
47	9	2½
362	4	11¼
51	16	2¾
97	4	6
541	13	2¾
475	6	4
6	11	11¼
72	19	9¾

1651	82	47	13 farthings.
4	4	3	
	86	50	

1655	6	2¼

The sum of the farthings is 13, which, divided by 4, gives 3 as quotient (to be put down under the pence), and one farthing as remainder (to be put in the sum total—under the farthings). 3*d.* (the quotient from the farthings) and 47 (the sum of the pence) are 50 pence, which, being put down and divided by 12, gives 4 shillings (to be set down under the shillings), and 2 pence (to be set down in the sum total—under the pence). 4*s.* (the quotient from the pence) and 82 (the sum of the shillings) are 86 shillings, which, being set down and divided by 20, gives 4 pounds (to be set down under the pounds), and 6 shillings (to be set down in the sum total—under the shillings). £4 (the quotient from the shillings) and 1651 (the sum of the pounds) are 1655 pounds (to be set down in the sum total—under the pounds). The sum of the addends is, therefore, found to be £1655 6*s.* 2¼*d.*

15. In proving the compound rules, we can generally avail ourselves of the methods used with the simple rules [Sec. II. 10, &c.]

EXERCISES FOR THE PUPIL.

Money.

(1)			(2)			(3)			(4)		
£	*s.*	*d.*	£	*s.*	*d.*	£	*s.*	*d.*	£	*s.*	*d.*
76	4	6	58	14	7	75	14	7	84	3	2
57	9	9	69	15	6	67	15	9	96	4	0½
49	10	8	72	14	8	76	19	10	41	0	6
183	4	11									

(5)			(6)			(7)			(8)		
£	*s.*	*d.*	£	*s.*	*d.*	£	*s.*	*d.*	£	*s.*	*d.*
674	14	7	767	15	6	567	14	7	327	8	6
456	17	8	472	14	6	476	16	6	501	2	11¾
676	19	8	567	16	7	547	17	6	864	0	6
527	4	2	423	3	10	527	14	3	121	9	8¼

(9)			(10)			(11)			(12)		
£	*s.*	*d.*	£	*s.*	*d.*	£	*s.*	*d.*	£	*s.*	*d.*
4567	14	6	76	14	7	3767	13	11	5674	17	6½
776	15	7	667	13	6	4678	14	10	4767	16	11½
76	17	9	67	15	7	767	12	9	3466	17	10¾
51	0	10	5	4	2	10	11	5	5984	2	2¼
44	5	6	5	3	4	3	4	11	8762	9	9

(13)			(14)			(15)			(16)		
£	*s.*	*d.*	£	*s.*	*d.*	£	*s.*	*d.*	£	*s.*	*d.*
9767	0	6¼	6767	11	6½	5764	17	6¾	634	7	11¼
7649	11	2½	7676	16	9¼	7457	16	5	65	7	7
4767	16	10¾	5948	17	8½	6743	18	0¼	7	12	10½
164	1	1	5786	7	6	67	6	6½	5678	18	8
92	7	2¼	6325	8	2¼	432	5	9	439	0	0

(17)			(18)			(19)			(20)		
£	*s.*	*d.*	£	*s.*	*d.*	£	*s.*	*d.*	£	*s.*	*d.*
0	14	7¾	5674	16	7½	5674	1	9¼	4767	14	7½
677	1	0	4767	17	6¾	4767	11	10¾	743	13	7¼
5767	2	6	1545	19	7½	78	18	11½	7674	14	6½
3697	14	7¼	3246	17	6	0	19	10¼	7	13	3¼
5634	0	0¾	4766	10	5¾	5044	4	1	750	6	4

(21)			(22)			(23)			(24)		
£	s.	d.	£	s.	d.	£	s.	d.	£	s.	d.
674	11	11½	476	14	7	674	13	3½	674	17	6½
567	14	10¾	576	15	6¾	45	15	7¼	123	12	2
476	4	11	76	17	7¼	476	4	6¾	567	0	7½
347	15	0½	576	11	8	577	16	0¼	579	18	9¾
476	13	9¼	463	14	9¼	578	6	3¾	476	6	6¼

(25)			(26)			(27)			(28)		
£	s.	d.	£	s.	d.	£	s.	d.	£	s.	d.
576	4	7½	549	4	6½	876	0	3	219	0	5
7	7	6	7	19	9¾	0	5	0	32	11	8¾
732	19	0¼	0	16	6¼	56	11	11	0	0	0¼
567	0	9½	734	19	9½	123	5	2¼	127	8	2
754	2	6¼	566	14	4¼	12	0	0	29	6	5½

Avoirdupoise Weight.

(29)			(30)			(31)			(32)		
cwt.	qrs.	lb	cwt.	qrs.	lb	cwt.	qrs.	lb	cwt.	qrs.	lb
76	3	14	44	1	16	14	3	17	56	3	14
37	2	15	56	3	11	37	1	16	57	1	17
14	1	11	47	1	16	47	2	27	58	2	26
128	3	12									

(33)			(34)			(35)			(36)		
cwt.	qrs.	lb	cwt.	qrs.	lb	cwt.	qrs.	lb	cwt.	qrs.	lb
76	1	19	88	2	17	476	3	15	567	2	19
56	3	13	59	2	20	764	1	7	4	1	20
47	2	17	0	3	0	6	3	14	67	3	2
81	2	14	67	1	15	0	1	18	767	1	11

(37)			(38)			(39)			(40)		
cwt.	qrs.	lb	cwt.	qrs.	lb	cwt.	qrs.	lb	cwt.	qrs.	lb
767	1	16	476	1	24½	447	1	7	14	12	12
44	1	17	756	3	21¼	576	1	6	3	4	7
567	3	13	767	1	16	467	1	7½	0	5	15
576	1	0	567	2	15	563	1	6	7	0	3
341	2	11	973	1	12	428	0	0¼	0	0	14

Troy Weight.

(41)					(42)					(43)			
℔	oz.	dwt.	grs.		℔	oz.	dwt.	grs.		℔	oz.	dwt.	grs.
7	0	5	9		5	9	7	0		88	7	9	8
5	6	6	7		0	0	6	7		80	9	8	6
9	5	6	8		8	7	6	4		0	8	7	5
21	11	18	0										

(44)					(45)					(46)			
℔	oz.	dwt.	grs.		℔	oz.	dwt.	grs.		℔	oz.	dwt.	grs.
57	9	12	14		87	3	7	12		57	10	14	11
67	9	11	11		0	11	12	3		0	0	11	10
66	8	10	5		0	0	16	14		46	9	9	8
74	6	5	3		44	12	10	13		22	8	7	5
12	3	5	4		67	8	9	10		11	10	13	14

Cloth Measure.

(47)				(48)				(49)				(50)		
yds.	qrs.	nls.		yds.	qrs.	nls.		yds.	qrs.	nls.		yds.	qrs.	nls.
99	3	1		176	3	3		37	3	2		0	2	1
47	1	3		47	0	2		0	2	3		5	3	2
76	3	2		7	3	3		0	0	2		0	0	3
224	0	2												

(51)				(52)				(53)				(54)		
yds.	qrs.	nls.		yds.	qrs.	nls.		yds.	qrs.	nls.		yds.	qrs.	nls.
567	3	2		147	3	3		157	2	1		156	1	1
476	1	0		173	1	0		143	3	2		176	3	1
72	3	3		148	2	1		0	1	2		54	1	0
5	2	1		92	3	2		54	0	3		573	2	3

Wine Measure.

(55)				(56)				(57)		
ts.	hhds.	gls.		ts.	hhds.	gls.		ts.	hhds.	gls.
99	3	9		89	3	3		76	3	4
80	0	39		7	3	4		67	3	44
98	3	46		76	1	56		0	1	56
87	2	27		44	2	7		5	3	4
41	1	26		54	2	17		602	0	'27
407	3	21								

Time.

(58)				(59)				(60)			
yrs.	ds.	hrs.	ms.	yrs.	ds.	hrs.	ms.	hrs.	ds.	hrs.	ms.
99	359	9	56	60	90	0	50	59	127	7	50
88	0	8	57	6	76	1	57	0	120	9	44
77	120	7	49	0	0	3	58	76	121	11	44
				6	1	2	0	6	47	3	41
265	115	2	42					8	9	11	17

61. What is the sum of the following :—three hundred and ninety-six pounds four shillings and two pence; five hundred and seventy-three pounds and four pence halfpenny; twenty-two pounds and three halfpence; four thousand and five pounds six shillings and three farthings? *Ans.* £4996 10s. 8¾d.

62. A owes to B £567 16s. 7½d.; to C £47 16s.; and to D £56 1d. How much does he owe in all? *Ans.* £671 12s. 8½d.

63. A man has owing to him the following sums :— £3 10s. 7d.; £46 7½d.; and £52 14s. 6d. How much is the entire? *Ans.* £102 5s. 8½d.

64. A merchant sends off the following quantities of butter :—47 cwt., 2 qrs., 7 ℔; 38 cwt., 3 qrs., 8 ℔; and 16 cwt., 2 qrs., 20 ℔. How much did he send off in all? *Ans.* 103 cwt., 7 ℔.

65. A merchant receives the following quantities of tallow, viz., 13 cwt., 1 qr., 6 ℔; 10 cwt., 3 qrs., 10 ℔; and 9 cwt., 1 qr., 15 ℔. How much has he received in all? *Ans.* 33 cwt., 2 qrs., 3 ℔.

66. A silversmith has 7 ℔, 8 oz., 16 dwt.; 9 ℔, 7 oz., 3 dwt.; and 4 ℔, 1 dwt. What quantity has he? *Ans.* 21 ℔, 4 oz.

67. A merchant sells to A 76 yards, 3 quarters, 2 nails; to B, 90 yards, 3 quarters, 3 nails; and to C, 190 yards, 1 nail. How much has he sold in all? *Ans.* 357 yards, 3 quarters, 2 nails.

68. A wine merchant receives from his correspondent 4 tuns, 2 hogsheads; 5 tuns, 3 hogsheads; and 7 tuns, 1 hogshead. How much is the entire? *Ans.* 17 tuns, 2 hogsheads.

69. A man has three farms, the first contains 120 acres, 2 roods, 7 perches; the second, 150 acres, 3 roods, 20 perches; and the third, 200 acres. How much land does he possess in all? *Ans.* 471 acres, 1 rood, 27 perches.

70. A servant has had three masters; with the first he lived 2 years and 9 months; with the second, 7 years and 6 months; and with the third, 4 years and 3 months. What was the servant's age on leaving his last master, supposing he was 20 years old on going to the first, and that he went directly from one to the other? *Ans.* 34 years and 6 months.

71. How many days from the 3rd of March to the 23rd of June? *Ans.* 112 days.

72. Add together 7 tons, the weight which a piece of fir 2 inches in diameter is capable of supporting; 3 tons, what a piece of iron one-third of an inch in diameter will bear; and 1000 ℔, which will be sustained by a hempen rope of the same size. *Ans.* 10 tons, 8 cwt., 3 quarters, 20 ℔.

73. Add together the following :—2*d.*, about the value of the Roman sestertius; 7½*d.*, that of the denarius; 1½*d.*, a Greek obolus; 9*d.*, a drachma; £3 15*s.*, a mina; £225, a talent; 1*s.* 7*d.*, the Jewish shekel; and £342 3*s.* 9*d.*, the Jewish talent. *Ans.* £571 2*s.*

74. Add together 2 dwt. 16 grains, the Greek drachma; 1 ℔, 1 oz., 10 dwt., the mina; 67 ℔, 7 oz., 5 dwt., the talent. *Ans.* 68 ℔, 8 oz., 17 dwt., 16 grains.

QUESTIONS FOR THE PUPIL.

1. What is the difference between the simple and compound rules? [8].

2. Might the simple rules have been constructed so as to answer also for applicate numbers of different denominations? [8].

3. What is the rule for compound addition? [9].

4. How is compound addition proved? [15].

5. How are we to act when the addends are numerous? [12, &c.]

COMPOUND SUBTRACTION.

16. RULE—I. Place the digits of the subtrahend under those of the same denomination in the minuend—farthings under farthings, units of pence under units of pence, tens of pence under tens of pence, &c.

II. Draw a separating line.

III. Subtract each denomination of the subtrahend from that which corresponds to it in the minuend—beginning with the lowest.

IV. If any denomination of the minuend is less than that of the subtrahend, which is to be taken from it, add to it one of the next higher—considered as an equivalent number of the denomination to be increased; and, either suppose unity to be added to the next denomination of the subtrahend, or to be subtracted from the next of the minuend.

V. If there is a remainder after subtracting any denomination of the subtrahend from the corresponding one of the minuend, put it under the column which produced it.

VI. If in any denomination there is no remainder, put a cypher under it—unless nothing is left from any higher denomination.

17. EXAMPLE.—Subtract £56 13s. 4¾d., from £96 7s. 6¼d.

£	s.	d.	
96	7	6¼,	minuend.
56	13	4¾,	subtrahend.
39	14	1½,	difference.

We cannot take ¾ from ¼, but—borrowing one of the pence, or 4 farthings, we add it to the ¼, and then say 3 farthings from 5, and 2 farthings, or one halfpenny, remains : we set down ½ under the farthings. 4 pence from 5 (we have borrowed one of the 6 pence), and one penny remains : we set down 1 under the pence (1½d. is read "three halfpence"). 13 shillings cannot be taken from 7, but (borrowing one from the pounds, or 20 shillings) 13 shillings from 27, and 14 remain : we set down 14 in the shillings' place of the remainder. 6 pounds cannot be taken from 5 (we have borrowed one of the 6 pounds in the minuend) but 6 from 15, and 9 remain: we put 9 under the units of pounds. 5 tens of pounds from 8 tens (we have borrowed one of the 9), and 3 remain: we put 3 in the tens of pounds' place of the remainder.

18. This rule and the reasons of it are substantially the same as those already given for Simple Subtraction [Sec. II. 17, &c.] It is evidently not so necessary to put down cyphers where there is nothing in a denomination of the remainder.

19. Compound may be proved in the same way as simple subtraction [Sec. II. 20].

EXERCISES.

	(1)			(2)			(3)			(4)			(5)		
	£	s.	d.	£	s.	d.	£	s.	d.	£	s.	d.	£	s.	d.
From	1098	12	6	767	14	8	76	15	6	47	16	7	97	14	6
Take	434	15	8	486	13	9	0	14	5	39	17	4	6	15	7
	663	16	10												

	(6)			(7)			(8)			(9)			(10)		
	£	s.	d.	£	s.	d.	£	s.	d.	£	s.	d.	£	s.	d.
From	98	14	2	47	14	6	97	16	6	147	14	4	560	15	6
Take	77	15	3	38	19	9	88	17	7	120	10	8	477	17	7

	(11)			(12)			(13)			(14)		
	£	s.	d.	£	s.	d.	£	s.	d.	£	s.	d.
From	99	13	3	767	14	5$\frac{1}{2}$	891	14	1$\frac{1}{4}$	576	13	7$\frac{3}{4}$
Take	47	16	7	476	6	7$\frac{1}{4}$	677	15	6$\frac{3}{4}$	467	14	9$\frac{1}{2}$

	(15)			(16)			(17)			(18)		
	£	s.	d.	£	s.	d.	£	s.	d.	£	s.	d.
From	567	11	5$\frac{3}{4}$	971	0	0$\frac{1}{2}$	437	15	0	478	10	0
Take	479	10	10$\frac{1}{2}$	0	0	7	0	11	1$\frac{1}{4}$	47	11	0$\frac{1}{2}$

Avoirdupoise Weight.

	(19)			(20)			(21)			(22)		
	cwt.	qrs.	lb	cwt.	qrs.	lb	cwt.	qrs.	lb	cwt.	qrs.	lb
From	200	2	26	275	2	15	9664	2	25	554	0	0
Take	99	3	15	27	2	7	9074	0	27	476	3	5
	100	3	11									

Troy Weight.

	(23)				(24)				(25)		
	℔	oz.	dwt.	gr.	℔	oz.	dwt.	gr.	℔	oz.	dwt.	gr.
From	554	9.	19	4	946	0	10	0	917	0	14	9
Take	97	0	16	15	0	0	17	23	798	0	18	17
	457	9	2	13								

Wine Measure.

	(26)			(27)			(28)			(29)		
	ts.	hhds.	gls.	ts.	hhds.	gls.	ts.	hhds.	gls.	ts.	hhds.	gls.
From	31	3	15	54	0	27	304	0	54	56	0	1
Take	29	2	26	0	3	42	100	3	51	27	2	25
	2	0	52									

Time.

	(30)				(31)				(32)			
	yrs.	ds.	hs.	ms.	yrs.	ds.	hs.	ms.	yrs.	ds.	hs.	ms.
From	767	131	6	30	476	14	14	16	567	126	14	12
Take	476	110	14	14	160	16	13	17	400	0	15	0
	291	20	16	16								

33. A shopkeeper bought a piece of cloth containing 42 yards for £22 10s., of which he sells 27 yards for £15 15s.; how many yards has he left, and what have they cost him? *Ans.* 15 yards; and they cost him £6 15s.

34. A merchant bought 234 tons, 17 cwt., 1 quarter, 23 ℔, and sold 147 tons, 18 cwt., 2 quarters, 24 ℔; how much remained unsold? *Ans.* 86 tons, 18 cwt., 2 qrs., 27 ℔.

35. If from a piece of cloth containing 496 yards, 3 quarters, and 3 nails, I cut 247 yards, 2 quarters, 2 nails, what is the length of the remainder? *Ans.* 249 yards, 1 quarter, 1 nail.

36. A field contains 769 acres, 3 roods, and 20 perches, of which 576 acres, 2 roods, 23 perches are tilled; how much remains untilled? *Ans.* 193 acres, 37 perches.

37. I owed my friend a bill of £76 16s. 9½d., out of which I paid £59 17s. 10¾d.; how much remained due? *Ans.* £16 18s. 10¾d.

G

38. A merchant bought 600 salt ox hides, weighing 561 cwt., 2 ℔ ; of which he sold 250 hides, weighing 239 cwt., 3 qrs., 25 ℔. How many hides had he left, and what did they weigh? *Ans.* 350 hides, weighing 321 cwt., 5 ℔.

39. A merchant has 209 casks of butter, weighing 400 cwt., 2 qrs., 14 ℔ ; and ships off 173 casks, weighing 213 cwt., 2 qrs., 27 ℔. How many casks has he left; and what is their weight? *Ans.* 36 casks, weighing 186 cwt., 3 qrs., 15 ℔.

40. What is the difference between 47 English miles, the length of the Claudia, a Roman aqueduct, and 1000 feet, the length of that across the Dee and Vale of Llangollen? *Ans.* 247160 feet, or 46 miles, 4280 feet.

41. What is the difference between 980 feet, the width of the single arch of a wooden bridge erected at St. Petersburg, and that over the Schuylkill, at Phila-delphia, 113 yards and 1 foot in span? *Ans.* 640 feet.

QUESTIONS FOR THE PUPIL.

1. What is the rule for compound subtraction? [16].
2. How is compound subtraction proved? [19].

COMPOUND MULTIPLICATION.

20. Since we cannot multiply pounds, &c., by pounds, &c., the multiplier must, in compound multiplication, be an abstract number.

21. When the multiplier does not exceed 12—

RULE—I. Place the multiplier to the right hand side of the multiplicand, and beneath it.

II. Put a separating line under both.

III. Multiply each denomination of the multiplicand by the multiplier, beginning at the right hand side.

IV. For every time the number required to make one of the next denomination is contained in any pro-duct of the multiplier and a denomination of the multi-plicand, carry one to the next product, and set down the remainder (if there is any, after subtracting the number equivalent to what is carried) under the denomination to which it belongs; but should there be no remainder, put a cypher in that denomination of the product.

22. EXAMPLE.—Multiply £62 17s. 10d. by 6.

£	s.	d.	
62	17	10,	multiplicand.
		6,	multiplier.

| 377 | 7 | 0, | product. |

Six times 10 pence are 60 pence; these are equal to 5 shillings (5 times 12 pence) to be carried, and no pence to be set down in the product—we therefore write a cypher in the pence place of the product. 6 times 7 are 42 shillings, and the 5 to be carried are 47 shillings—we put down 7 in the units' place of shillings, and carry 4 tens of shillings. 6 times 1 (ten shillings) are 6 (tens of shillings), and 4 (tens of shillings) to be carried, are 10 (tens of shillings), or 5 pounds (5 times 2 tens of shillings) to be carried, and nothing (no ten of shillings) to be set down. 6 times 2 pounds are 12, and 5 to be carried are 17 pounds—or 1 (ten pounds) to be carried, and 7 (units of pounds) to be set down. 6 times 6 (tens of pounds) are 36, and 1 to be carried are 37 (tens of pounds).

23. The reasons of the rule will be very easily understood from what we have already said [Sec. II. 41]. But since, in compound multiplication, the value of the multiplier has no connexion with its position in reference to the multiplicand, where we set it down is a mere matter of convenience; neither is it so necessary to put cyphers in the product in those denominations in which there are no significant figures, as it is in simple multiplication.

24. Compound multiplication may be proved by reducing the product to its lowest denomination, dividing by the multiplier, and then reducing the quotient.

EXAMPLE.—Multiply £4 3s. 8d. by 7.

£	s.	d.		PROOF:		
4	3	8		29	5	8
		7		20		
29	5	8, product.		585		
				12		

7)7028, product reduced.

12)1004

20)83   8

quotient reduced   4   3   8=multiplicand.

£29 5s. 8d. are 7 times the multiplicand; if, therefore, the process has been rightly performed, the seventh part of this should be equal to the multiplicand.

The quantities are to be "reduced," before the division by 7, since the learner is not supposed to be able as yet to divide £29 5s. 8d.

EXERCISES.

	£	s.	d.		£	s.	d.
1.	76	14	$7\frac{1}{2}\times$ 2=	153	9	3.
2.	97	13	$6\frac{1}{2}\times$ 3=	293	0	$7\frac{1}{2}$.
3.	77	10	$7\frac{1}{4}\times$ 4=	310	2	5.
4.	96	11	$7\frac{1}{4}\times$ 5=	482	18	$1\frac{1}{4}$.
5.	77	14	$6\frac{1}{4}\times$ 6=	466	7	$1\frac{1}{2}$.
6.	147	13	$3\frac{1}{4}\times$ 7=	1033	13	$0\frac{1}{4}$.
7.	428	12	$7\frac{1}{2}\times$ 8=	3429	1	0.
8.	572	16	6 $\times$ 9=	5155	8	6.
9.	428	17	3 $\times10=$	4288	12	6.
10.	672	14	4 $\times11=$	7399	17	8.
11.	776	15	5 $\times12=$	9321	5	0.

12. 7 ℔ at 5s. $2\frac{1}{4}d.$ ℔, will cost £1 16s. $3\frac{3}{4}d.$
13. 9 yards at 10s. $11\frac{1}{4}d.$ ℔, will cost £4 18s. $5\frac{1}{4}d.$
14. 11 gallons at 13s. 9d. ℔, will cost £7 11s. 3d.
15. 12 ℔ at £1 3s. 4d. ℔, will cost £14.

25. When the multiplier exceeds 12, and is a composite number—

RULE.—Multiply successively by its factors.

EXAMPLE 1.—Multiply £47 13s. 4d. by 56.

$$\begin{array}{ccc} \pounds & s. & d. \\ 47 & 13 & 4 \end{array}$$

56=7×8

$$\begin{array}{ccc} & & 7 \\ \hline 333 & 13 & 4 \end{array} = 47 \ 13 \ 4\times7.$$

$$\begin{array}{ccc} & & 8 \\ \hline 2669 & 6 & 8 \end{array} = 47 \ 13 \ 4\times7\times8, \text{ or } 56.$$

EXAMPLE 2.—Multiply 14s. 2d. by 100.

$$\begin{array}{cc} s. & d. \\ 14 & 2 \end{array}$$

100=10×10

$$\begin{array}{cc} & 10 \\ \hline \pounds7 \ 1 & 8 \end{array} = 14 \ 2\times10.$$

$$\begin{array}{cc} & 10 \\ \hline \pounds70 \ 16 & 8 \end{array} = 14 \ 2\times10\times10, \text{ or } 100.$$

EXAMPLE 3.—Multiply £8 2s. 4d. by 700.

```
    £   s.   d.
    8   2    4
              10
    ─────────────        £   s.   d.
    81  3    4   =8      2    4×10.
              10
    ─────────────
    811 13   4   =8      2    4×10×10, or 100.
              7
    ─────────────
    5681 13  4   =8      2    4×10×10×7, or 700.
```

The reason of this rule has been already given [Sec. II. 60].

26. When the multiplier is the sum of composite numbers—

RULE.—Multiply by each, and add the results.

EXAMPLE.—Multiply £3 14s. 6d. by 430.

```
    £    s.   d.
    3    14   6
               10
    ─────────────         £    s.   d.   £  s.   d.
    37   5    0  ×3=111   15   0, or 3  14   6×30.
               10
    ─────────────
    372  10   0×4=1490     0   0, or 3  14   6×400.
    ─────────────────────────────────────────────────
                1601  15   0, or 3  14   6×430.
```

The reason of the rule is the same as that already given [Sec. II. 52]. The sum of the products of the multiplicand by the parts of the multiplier, being equal to the product of the multiplicand by the whole multiplier.

EXERCISES.

```
    £   s.   d.           £    s.   d.
16. 3   7    6  ×   18=  60   15   0.
17. 4   16   7  ×   20=  96   11   8.
18. 5   14   6½×    22=125   19   11.
19. 2   17   6  ×   36=103   10   0.
20. 3   16   7  ×   56=214    8   8.
21. 2   3    6  ×   64=139    4   0.
22. 3   4    7  ×   81=261   11   3.
23. 0   9    4  ×  100=  46   13   4.
24. 0   16   4  ×1000=816   13   4.
```

25. 100 yards at 9s. 4½d. ℔, will cost £46 17 6
26. 700 gallons at 13s. 4d. ℔, will cost 466 13 4.
27. 240 gallons at 6s. 8d ℔, will cost 80 0 0.
28. 360 yards at 13s. 4d. ℔, will cost 240 0 0.

27. If the multiplier is not a composite number—

RULE.—Multiply successively by the factors of the nearest composite, and add to or subtract from the product so many times the multiplicand as the assumed composite number is less, or greater than the given multiplier.

EXAMPLE 1.—Multiply £62 12s. 6d. by 76.

```
              £    s.   d.
              62   12   6
                         8
76=8×9+4    ─────────────
              501   0 . 0
                         9
                        ─────    £    s.    d.
              4509   0   0 = 62  12   6×8×9, or 72.
               250  10   0 = 62  12   6×4.
              ─────────────
              4759  10   0 = 62  12   6×8×9+4 or 76
```

EXAMPLE 2.—Multiply £42 3s. 4d. by 27.

```
              £    s.   d.
              42   3    4
                        4
27=4×7−1    ─────────────
              168  13   4
                        7
                       ─────    £    s.   d.
              1180 13   4 = 42   3   4×4×7, or 28.
                42  3   4 = 42   3   4×1.
              ─────────────
              1138 10   0 = 42   3   4×4×7−1, or 27.
```

The reason of the rule is the same as that already given Sec. II. 61].

EXERCISES.

```
       £    s.   d.              £     s.   d.
29.    12   2    4  × 83 =     1005   13   8.
30.    15   0    0¼ × 146 =    2190    3   0¼.
31.   122   5    0  × 102 =   12469   10   0.
32.   963   0    0¾ × 999 =  962040    2   5¼.
```

28. When the multiplier is large, we may often conveniently proceed as follows—

RULE.—Write once, ten times, &c., the multiplicand, and, multiplying these respectively by the units, tens, &c. of the multiplier, add the results.

· Example.—Multiply £47 16*s.* 2*d.* by 5783.

$$5783 = 5 \times 1000 + 7 \times 100 + 8 \times 10 + 3 \times 1$$

	£	s.	d.		£	s.	d.
Units of the multiplicand,	47	16	2×3=		143	8	6.
			10				
Tens of the multiplicand,	478	1	8×8=		3824	13	4.
			10				
Hundreds of the multiplicand,	4780	16	8×7=		33465	16	8.
			10				
Thousands of the multiplicand,	47808	6	8×5=		239041	13	4.

Product of multiplicand and multiplier = 276475 11 10.

EXERCISES.

	£	s.	d.		£	s.	d.
33.	76	14	4	× 92=	7057	18	8.
34.	974	14	2	× 76=74077	16	8.	
35.	780	17	4	× 92=71839	14	8.	
36.	73	17	7¼× 122=	9013	10	3.	
37.	42	7	7¼× 162=	6865	11	10¼.	
38.	76 gallons at £0 13 4 ℔, will cost £50 13 4.						
39.	92 gallons at 0 14 2 ℔, will cost 65 3 4.						

40. What is the difference between the price of 743 ounces of gold at £3 17*s.* 10½*d.* per oz. Troy, and that of the same weight of silver at 62*d.* per oz. ? *Ans.* £2701 2*s.* 3½*d.*

41. In the time of King John (money being then more valuable than at present) the price, per day, of a cart with three horses was fixed at 1*s.* 2*d.*; what would be the hire of such a cart for 272 days? *Ans.* £15 17*s.* 4*d.*

42. Veils have been made of the silk of caterpillars, a square yard of which would weigh about 4 grains; what would be the weight of so many square yards of this texture as would cover a square English mile? *Ans.* 2151 ℔, 1 oz., 6 dwt., 16 grs., Troy.

QUESTIONS TO BE ANSWERED BY THE PUPIL.

1. Can the multiplier be an applicate number ? [20].

2. What is the rule for compound multiplication when the multiplier does not exceed 12 ? [21].

· 3. What is the rule when it exceeds 12, and is a composite number ? [25].

4. When it is the sum of composite numbers? [26].

5. When it exceeds 12, and not a composite number? [27].

6. How is compound multiplication proved? [24].

COMPOUND DIVISION.

29. Compound Division enables us, if we divide an applicate number into any number of equal parts, to ascertain what each of them will be; or to find out how many times one applicate number is contained in another.

If the divisor be an applicate, the quotient will be an abstract number—for the quotient, when multiplied by the divisor, must give the dividend [Sec. II. 79]; but two applicate numbers cannot be multiplied together [20]. If the divisor be abstract, the quotient will be applicate—for, multiplied by the quotient, it must give the dividend—an applicate number. Therefore, either divisor or quotient must be abstract.

30. When the divisor is abstract, and does not exceed 12—

RULE—I. Set down the dividend, divisor, and separating line—as directed in simple division [Sec. II. 72].

II. Divide the divisor, successively, into all the denominations of the dividend, beginning with the highest.

III. Put the number expressing how often the divisor is contained in each denomination of the dividend under that denomination—and in the quotient.

IV. If the divisor is not contained in a denomination of the dividend, multiply that denomination by the number which expresses how many of the next lower denomination is contained in one of its units, and add the product to that next lower in the dividend.

V. " Reduce" each succeeding remainder in the same way, and add the product to the next lower denomination in the dividend.

VI. If any thing is left after the quotient from the lowest denomination of the dividend is obtained, put it down, with the divisor under it, and a separating line between :—or omit it, and if it is not less than half the divisor, add unity to the lowest denomination of the quotient.

31. EXAMPLE 1.—Divide £72 6s. 9¼d. by 5.

$$
\begin{array}{ccc}
£ & s. & d. \\
5)\overline{72} & 6 & 9\frac{1}{2} \\
\hline
14 & 9 & 4\frac{1}{4}
\end{array}
$$

5 will go into 7 (tens of pounds) once (ten times), and leave 2 tens. 5 will go into 22 (units of pounds) 4 times, and leave two pounds or 40s. 40s. and 6s. are 46s., into which 5 will go 9 times, and leave one shilling or 12d. 12d. and 9d. are 21.l., into which 5 will go 4 times, and leave 1d., or 4 farthings. 4 farthings and 2 farthings are 6 farthings, into which 5 will go once, and leave 1 farthing—still to be divided; this would give ⅕, or the fifth part of a farthing as quotient, which, being less than half the divisor, may be neglected.

A knowledge of fractions will hereafter enable us to understand better the nature of these remainders.

EXAMPLE 2.—Divide £52 4s. 1¾d. by 7.

$$
\begin{array}{ccc}
£ & s. & d. \\
7)\overline{52} & 4 & 1\frac{3}{4} \\
\hline
7 & 9 & 2
\end{array}
$$

One shilling or 12d. are left after dividing the shillings, which, with the 1d. already in the dividend, make 13d. 7 goes into 13 once, and leaves 6d., or 24 farthings, which, with ¾, make 27 farthings. 7 goes into 27 3 times and 6 over ; but as 6 is more than the half of 7, it may be considered, with but little inaccuracy, as 7—which will add one farthing to the quotient, making it 4 farthings, or one to be added to the pence.

32. This rule, and the reasons of it, are substantially the same as those already given [Sec. II. 72 and 77]. The remainder, after dividing the farthings, may, from its insignificance, be neglected, if it *is not* greater than half the divisor. If it *is* greater, it is evidently more accurate to consider it as giving one farthing to the quotient, than 0, and therefore it is proper to add a farthing to the quotient. If it is exactly half the divisor, we may consider it as equal either to the divisor, or 0.

33. Compound division may be proved by multiplication—since the product of the quotient and divisor, plus the remainder, ought to be equal to the dividend [Sec. II. 79].

EXERCISES.

	£	s.	d.	÷	£	s.	d.
1.	96	7	6	÷ 2 =	48	3	9.
2.	76	14	7	÷ 3 =	25	11	$6\frac{1}{4}$.
3.	47	17	6	÷ 4 =	11	19	$4\frac{1}{2}$.
4.	96	19	4	÷ 5 =	19	7	$10\frac{1}{2}$.
5.	77	16	7	÷ 6 =	12	19	$5\frac{1}{4}$.
6.	32	12	2	÷ 7 =	4	13	2.
7.	44	16	7	÷ 8 =	5	12	1.
8.	97	14	3	÷ 9 =	10	17	$1\frac{3}{4}$.
9.	147	14	6	÷ 10 =	14	15	$5\frac{1}{2}$.
10.	157	16	7	÷ 11 =	14	6	$11\frac{1}{2}$.
11.	176	14	6	÷ 12 =	14	14	$6\frac{1}{2}$.

The above quotients are true to the nearest farthing.

34. When the divisor exceeds 12, and is a composite number—

RULE.—Divide successively by the factors.

EXAMPLE.—Divide £12 17*s.* 9*d.* by 36.

```
                3)12 17  9
                 ---------
36=3×12         12)4  5 11
                 ---------
                      7  2
```

This rule will be understood from See II 97.

EXERCISES.

	£	s.	d.	÷	£	s.	d.
12.	24	17	6	÷ 24 =	1	0	$8\frac{1}{4}$.
13.	576	13	3	÷ 36 =	16	0	$4\frac{1}{2}$.
14.	447	12	2	÷ 48 =	9	6	6.
15.	547	12	4	÷ 56 =	9	15	7.
16.	9740	14	6	÷ 120 =	81	2	$5\frac{1}{4}$.
17.	740	13	4	÷ 49 =	15	2	$3\frac{3}{4}$.

35. When the divisor exceeds 12, and is not a composite number—

RULE.—Proceed by the method of long division; but in performing the multiplication of the remainders by the numbers which make them respectively a denomination lower, and adding to the products of that next lower denomination whatever is already in the dividend, set down the multipliers, &c. obtained. Place the quotient as directed in long division [Sec. II. 89].

EXAMPLE.—Divide £87 16s. 4d. by 62.

$$
\begin{array}{ccccccc}
 & & £ & s. & d. & £ & s. & d. \\
62)87 & 16 & 4 & (1 & 8 & 4 \\
62 & & & & & \\
\hline
25 & & & & & \\
20 & \text{multiplier.} & & & & \\
\hline
\end{array}
$$

shillings 516(=25×20+16)
       496
      ————
       20
       12 multiplier.
      ————

pence   244(=20×12+4)
      186
      ————
       58
       4 multiplier.
      ————

farthings 232 (=58×4)
      186
      ————
      46

62 goes into £87 once (that is, it gives £1 in the quotient), and leaves £25. £25 are equal to 500s. (25×20), which, with 16s. in the dividend, make 516s. 62 goes into 516s. 8 times (that is, it gives 8s. in the quotient), and leaves 20s., or 240d. (20×12) as remainder. 62 goes into 240, &c.

Were we to put ¾ in the quotient, the remainder would be 46, which is more than half the divisor; we consider the quotient, therefore, as 4 farthings, that is, we add one penny to (3) the pence supposed to be already in the quotient. £1 8s. 4d. is nearer to the true quotient than £1 8s. 3¾d. [32].

This is the same in principle as the rule given above [30]—but since the numbers are large, it is more convenient actually to set down the sums of the different denominations of the dividend and the preceding remainders (reduced), the products of the divisor and quotients, and the numbers by which we multiply for the necessary reductions: this prevents the memory from being too much burdened [Sec. II. 93].

36. When the divisor and dividend are both applicate numbers of one and the same denomination, and no reduction is required—

RULE.—Proceed as already-directed [Sec. II. 70, 72, or 89]. .

EXAMPLE.—Divide £45 by £5.

```
     £
  £5)45
     ——
      9
```

That is £5 is the ninth part of £45.

37. When the divisor and dividend are applicate, but not of the same denomination; or more than one denomination is found in either, or both—

RULE.—Reduce both divisor and dividend to the lowest denomination contained in either [3], and then proceed with the division.

EXAMPLE.—Divide £37 5*s.* $9\frac{1}{4}$*d.* by 3*s.* $6\frac{1}{2}$*d.*

```
 s.  d.                 £    s.  d.
 3   6½                37    5   9¼
12                     20
——                    ———
42                    745
 4                     12
———                  ————
170 farthings.       8949
                        4
                    —————
               170)35797(211
                   340
                   ———
                    179
                    170
                    ———
                     97
```

Therefore 3*s.* $6\frac{1}{2}$*d.* is the 211th part of £37 5*s.* $9\frac{1}{4}$*d.*

97 not being less than the half of 170 [32], we consider it as equal to the divisor, and therefore add 1 to the 0 obtained as the last quotient.

EXERCISES.

	£	s.	d.			£	s.	d.
18.	176	12	2	÷	191 =	0	18	6.
19.	134	17	8	÷	183 =	0	14	9.
20.	4736	14	7	÷	443 =	10	13	$10\frac{1}{4}$
21.	73	16	7	÷	271 =	0	5	$5\frac{1}{2}$
22.	147	14	6	÷	973 =	0	3	$0\frac{1}{2}$
23.	157	16	7	÷	487 =	0	6	$5\frac{3}{4}$
24.	58	15	2	÷	751 =	0	1	$6\frac{3}{4}$.
25.	62	10	$6\frac{1}{4}$	÷	419 =	0	2	$11\frac{3}{4}$.
26.	8764	4	$0\frac{1}{4}$	÷	468 =	18	14	$6\frac{1}{2}$.
27.	4728	16	2	÷	317 =	14	18	$4\frac{1}{4}$.
28.	8234	0	$5\frac{1}{4}$	÷	261 =	31	10	$11\frac{1}{2}$.
29.	5236	2	$7\frac{3}{4}$	÷	875 =	5	19	$8\frac{1}{4}$.
30.	4598	4	2	÷	9842 =	0	9	$4\frac{1}{4}$.

31. A cubic foot of distilled water weighs 1000 ounces, what will be the weight of one cubic inch? *Ans.* 253·1829 grains, nearly.

32. How many Sabbath days' journeys (each 1155 yards) in the Jewish days' journey, which was equal to 33 miles and 2 furlongs English? *Ans.* 50·66, &c.

33. How many pounds of butter at $11\frac{3}{4}d.$ per lb. would purchase a cow, the price of which is £14 15*s.*? *Ans.* 301·2766.

QUESTIONS FOR THE PUPIL.

1. What is the use of compound division? [29].

2. What kind is the quotient when the divisor is an abstract, and what kind is it when the divisor is an applicate number? [29].

3. What are the rules when the divisor is abstract, and does not exceed 12? [30];

4. When it exceeds 12, and is composite? [34];

5. When it exceeds 12, and is not composite? [35];

6. And when the divisor is an applicate number? [36 and 37].

SECTION IV.

FRACTIONS.

1. If one or more units are divided into equal parts, and one or more of these parts are taken, we have what is called a *fraction*.

Any example in division—before the process has been performed—may be considered as affording a fraction:— thus $\frac{5}{6}$ (which means 5 *to be* divided by 6 [Sec. II. 68]) is a fraction of 5—its sixth part; that is, 5 being divided into six equal parts, $\frac{5}{6}$ will express one of them; or (as we shall see presently), if unity is divided into six equal parts, five of them will be represented by $\frac{5}{6}$.

2. When the dividend and divisor constitute a fraction, they change their names—the former being then termed the *numerator*, and the latter the *denominator*; for while the denominator tells the *denomination* or kind of parts into which the unit is supposed to be divided, the numerator *numerates* them, or indicates the number of them which is taken. Thus $\frac{3}{7}$ (read three-sevenths) means that the parts are "sevenths," and that "three" of them are represented. The numerator and denominator are called the *terms* of the fractions.

3. The greater the numerator, the greater the value of the fraction—because the quotient obtained when we divide the numerator by the denominator is its real value; and the greater the dividend the larger the quotient. On the contrary, the greater the denominator the less the fraction—since the larger the divisor the smaller the quotient [Sec. II. 78]:—hence $\frac{6}{7}$ is greater than $\frac{5}{7}$—which is expressed thus, $\frac{6}{7} > \frac{5}{7}$; but $\frac{5}{8}$ is less than $\frac{5}{7}$—which is expressed by $\frac{5}{8} < \frac{5}{7}$.

4. Since the fraction is equal to the quotient of its numerator divided by its denominator, as long as this quotient is unchanged, the value of the fraction is the same, though its form may be altered. Hence we can multiply or divide both terms of a fraction by the same number without affecting its value; since this is equally

to increase or diminish both the dividend and divisor—which does not affect the quotient.

5. The following will represent unity, seven-sevenths, and five-sevenths.

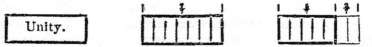

The very faint lines indicate what $\frac{5}{7}$ wants to make it *equal* to unity, and *identical with* $\frac{7}{7}$. In the diagrams which are to follow, we shall, in this manner, generally subjoin the difference between the fraction and unity.

The teacher should impress on the mind of the pupil that he might have chosen any *other* unity to exemplify the nature of a fraction.

6. The following will show that $\frac{5}{7}$ may be considered as either the $\frac{5}{7}$ of 1, or the $\frac{1}{7}$ of 5, both—though not identical—being perfectly equal.

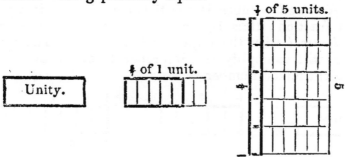

In the one case we may suppose that the five parts belong to but one unit; in the other, that each of the five beiongs to different units of the same kind

Lastly, $\frac{5}{7}$ may be considered as the $\frac{1}{7}$ of one unit five times as large as the former; thus—

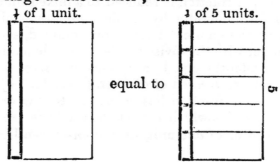

equal to

7. If its numerator is equal to, or greater than its denominator, the fraction is said to be *improper ;* because, although it has the fractional form, it is equal to, or greater than an integer. Thus $\frac{7}{5}$ is an improper fraction, and means that each of its seven parts is equal to one of those obtained from a unit divided into five equal parts. When the numerator of a proper-fraction is divided by its denominator, the quotient will be expressed by decimals; but when the numerator of an improper fraction is divided by its denominator, part, at least, of the quotient will be an integer.

It is not inaccurate to consider $\frac{7}{5}$ as a fraction, since it consists of " parts" of an integer. It would not, however, be true to call it *part* of an integer ; but this is not required by the definition of a fraction—which, as we have said, consists of " part," or " parts" of a unit [1].

8. A *mixed* number is one that contains an integer and a fraction ; thus $1\frac{2}{5}$—which is *equivalent* to, but not *identical* with the improper fraction $\frac{7}{5}$. The following will exemplify the improper fraction, and its equivalent mixed number—

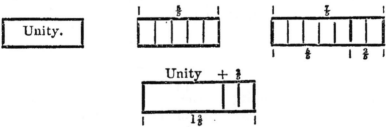

9. To reduce an improper fraction to a mixed number.

An improper fraction is reduced to a mixed number if we divide the numerator by the denominator, and, after the units in the quotient have been obtained, set down the remainder with the divisor under it, for denominator ; thus $\frac{7}{5}$ is evidently equal to $1\frac{2}{5}$—as we have already noticed when we treated of division [Sec. II. 71].

10. A *simple* fraction has reference to one or more integers ; thus $\frac{5}{7}$—which means, as we have seen [6], the *five*-sevenths of *one* unit, or the *one*-seventh of *five* units

11. A *compound* fraction supposes one fraction to refer to another; thus $\frac{4}{9}$ of $\frac{3}{4}$—represented also by $\frac{3}{4} \times \frac{4}{9}$ (three-fourths multiplied by four-ninths), means not the four-ninths of unity, but the four-ninths of the three-fourths of unity :—that is, unity being divided into four parts, three of these are to be divided into nine parts, and then four of these nine are to be taken; thus—

12. A *complex* fraction has a fraction, or a mixed number in its numerator, denominator, or both; thus $\frac{\frac{2}{3}}{4}$, which means that we are to take the fourth part, not of unity, but of the $\frac{2}{3}$ of unity. This will be exemplified by—

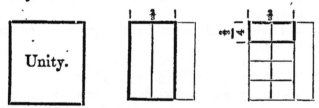

$\frac{8}{4}$, $\frac{\frac{2}{3}}{\frac{5}{6}}$, $\frac{1\frac{1}{2}}{4}$, $\frac{1\frac{4}{5}}{5\frac{4}{6}}$, are complex fractions, and will be better understood when we treat of the division of fractions.

13. Fractions are also distinguished by the nature of their denominators. When the denominator is *unity*, followed by one or more cyphers, it is a *decimal fraction*—thus, $\frac{5}{10}$, $\frac{6}{1000}$, &c.; all other fractions are *vulgar*—thus, $\frac{4}{9}$, $\frac{5}{6}$, $\frac{3}{200}$, &c.

Arithmetical processes may often be performed with fractions, without *actually* dividing the numerators by the denominators. Since a fraction, like an integer, may be increased or diminished, it is capable of addition, subtraction, &c.

14. To reduce an integer to a fraction of any denomination.

An integer may be considered as a fraction if we make unity its denominator :—thus $\frac{5}{1}$ may be taken for 5; since $\frac{5}{1}=5$.

We may give an integer any denominator we please if we previously multiply it by that denominator ;

thus, $5 = \frac{25}{5}$, or $\frac{30}{6}$, or $\frac{35}{7}$, &c., for $\frac{25}{5}=\frac{5 \times 5}{1 \times 5}=\frac{5}{1}=5$;

and $\frac{30}{6}=\frac{5 \times 6}{1 \times 6}=\frac{5}{1}=5$, &c.

EXERCISES.

1. Reduce 7 to a fraction, having 4 as denominator. *Ans.* $\frac{28}{4}$.

2. Reduce 13 to a fraction, having 16 as denominator. *Ans.* $\frac{208}{16}$.

3. $4=\frac{28}{7}$. | 4. $19=\frac{57}{3}$. | 5. $42=\frac{504}{12}$. | 6. $71=\frac{6674}{94}$.

15. To reduce fractions to lower terms.

Before the addition, &c., of fractions, it will be often convenient to reduce their terms as much as possible. For this purpose—

RULE.—Divide each term by the greatest common measure of both.

EXAMPLE.— $\frac{40}{72}=\frac{5}{9}$. For $\frac{40}{72}=\frac{40 \div 8}{72 \div 8}=\frac{5}{9}$.

We have already seen that we do not alter the quotient— which is the real value of the fraction [4]—if we multiply or divide the numerator and denominator by the same number.

What has been said, Sec. II. 104, will be usefully remembered here.

EXERCISES.

Reduce the following to their lowest terms.

7. $\frac{5744}{10080}=\frac{287}{503}$. | 13. $\frac{63}{72}=\frac{7}{8}$. | 19. $\frac{133133}{133133}=\frac{1331}{1331}$.

8. $\frac{418}{848}=\frac{51}{81}$. | 14. $\frac{144}{156}=\frac{12}{13}$. | 20. $\frac{7188}{7188}=\frac{1}{1}$.

9. $\frac{914}{914}=\frac{914}{914}$. | 15. $\frac{33}{99}=\frac{11}{33}$. | 21. $\frac{8698}{8698}=\frac{1179}{1179}$.

10. $\frac{5448}{71123}=\frac{1483}{2481}$. | 16. $\frac{48}{60}=\frac{4}{5}$. | 22. $\frac{425}{755}=\frac{5}{1557}$.

11. $\frac{316}{322}=\frac{139}{139}$. | 17. $\frac{42}{62}=\frac{3}{5}$. | 23. $\frac{412}{162}=\frac{286}{233}$.

12. $\frac{126}{126}=\frac{1}{1}$. | 18. $\frac{24}{112}=\frac{1}{5}$. | 24. $\frac{511}{117}=\frac{354}{354}$.

In the answers to questions given as exercises, we shall, in future, generally reduce fractions to their lowest denominations.

16. To find the value of a fraction in terms of a lower denomination—

RULE.—Reduce the numerator by the rule already given [Sec. III. 3], and place the denominator under it.

EXAMPLE.—What is the value, in shillings, of $\frac{3}{4}$ of a pound? £3 reduced to shillings $= 60s.$; therefore $£\frac{3}{4}$ reduced to shillings $= \frac{60}{4}s.$

The reason of the rule is the same as that already given [Sec. III. 4]. The $\frac{3}{4}$ of a pound becomes 20 times as much if the "unit of comparison" is changed from a pound to a shilling.

We may, if we please, obtain the value of the resulting fraction by actually performing the division [9]; thus $\frac{60}{4}s. = 15s.$:—hence $£\frac{3}{4} = 15s.$

EXERCISES.

25. $£\frac{23}{40} = 14s.\ 6d.$ 28. $£\frac{3}{4} = 15s.$
26. $£\frac{13}{15} = 17s.\ 4d.$ 29. $£\frac{3}{12} = 5s.$
27. $£\frac{19}{20} = 19s.$ 30. $£\frac{1}{240} = 1d.$

17. To express one quantity as the fraction of another—

RULE.—Reduce both quantities to the lowest denomination contained in either—if they are not already of the same denomination; and then put that which is to be the fraction of the other as numerator, and the remaining quantity as denominator.

EXAMPLE.—What fraction of a pound is $2\frac{1}{4}d.$? £1 = 960 farthings, and $2\frac{1}{4}d. = 9$ farthings; therefore $\frac{9}{960}$ is the required fraction, that is, $2\frac{1}{4}d. = £\frac{9}{960}.$

REASON OF THE RULE.—One pound, for example, contains 960 farthings, therefore one farthing is $£\frac{1}{960}$ (the 960th part of a pound), and 9 times this, or $2\frac{1}{4}$, is $£9 \times \frac{1}{960} = \frac{9}{960}.$

EXERCISES.

31. What fraction of a pound is $14s.\ 6d.$? $Ans.\ \frac{29}{40}.$
32. What fraction of £100 is $17s.\ 4d.$? $Ans.\ \frac{13}{1500}.$
33. What fraction of £100 is £32 10s.? $Ans.\ \frac{13}{40}.$
34. What fraction of 9 yards, 2 quarters is 7 yards, 3 quarters? $Ans.\ \frac{31}{38}.$
35. What part of an Irish is an English mile? $Ans.\ \frac{11}{14}.$
36. What fraction of $6s.\ 8d.$ is $2s.\ 1d.$? $Ans.\ \frac{5}{16}.$
37. What part of a pound avoirdupoise is a pound Troy? $Ans.\ \frac{144}{175}.$

QUESTIONS.

1. What is a fraction? [1].

2. When the divisor and dividend are made to constitute a fraction, what do their names become? [2].

3. What are the effects of increasing or diminishing the numerator, or denominator? [3].

4. Why may the numerator and denominator be multiplied or divided by the same number without altering the value of the fraction? [4].

5. What is an improper fraction? [7].

6. What is a mixed number? [8].

7. Show that a mixed number is not *identical* with the equivalent improper fraction? [8].

8. How is an improper fraction reduced to a mixed number? [9].

9. What is the difference between a simple, a compound, and a complex fraction? [10, 11, and 12];

10. Between a vulgar and decimal fraction? [13].

11. How is an integer reduced to a fraction of any denomination? [14].

12. How is a fraction reduced to a lower term? [15].

13. How is the value of a fraction found in terms of a lower denomination? [16].

14. How do we express one quantity as the fraction of another? [17].

VULGAR FRACTIONS.

ADDITION.

18. If the fractions to be added have a common denominator—

RULE.—Add all the numerators, and place the common denominator under their sum.

EXAMPLE.—$\frac{5}{7} + \frac{6}{7} = \frac{11}{7}$.

REASON OF THE RULE.—If we add together 5 and 6 of any kind of individuals, their sum must be 11 of the same kind of individuals—since the process of addition has not changed their nature. But the units to be added were, in the present instance, sevenths; therefore their sum consists of sevenths. Addition may be illustrated as follows :—

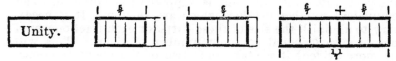

EXERCISES.

1. $\frac{3}{7}+\frac{4}{7}+\frac{5}{7}=\frac{12}{7}=1\frac{5}{7}$.
2. $\frac{1}{9}+\frac{3}{9}+\frac{4}{9}=\frac{8}{9}$.
3. $\frac{4}{13}+\frac{10}{13}+\frac{9}{13}=\frac{30}{13}=2\frac{4}{13}$.
4. $\frac{4}{14}+\frac{9}{14}+\frac{2}{14}=\frac{25}{14}=1\frac{11}{14}$.
5. $\frac{4}{12}+\frac{5}{12}+\frac{6}{12}=\frac{13}{12}=1\frac{1}{12}$.
6. $\frac{2}{8}+\frac{3}{8}+\frac{6}{8}=\frac{9}{8}=1\frac{1}{8}$.
7. $\frac{8}{20}+\frac{11}{20}+\frac{13}{20}=\frac{43}{20}=2\frac{3}{20}$.
8. $\frac{7}{17}+\frac{11}{17}+\frac{13}{17}=\frac{47}{17}=2\frac{13}{17}$.

9. $\frac{3}{13}+\frac{4}{13}+\frac{11}{13}=\frac{18}{13}=2$.
10. $\frac{1}{21}+\frac{9}{21}+\frac{8}{21}=\frac{57}{21}=2\frac{10}{21}$.
11. $\frac{4}{23}+\frac{11}{23}+\frac{13}{23}=\frac{31}{23}=1\frac{14}{23}$.
12. $\frac{3}{15}+\frac{5}{15}+\frac{11}{15}=\frac{18}{15}=1\frac{1}{15}$.
13. $\frac{11}{31}+\frac{12}{31}+\frac{27}{31}=\frac{51}{31}=1\frac{7}{31}$.
14. $\frac{7}{8}+\frac{4}{8}+\frac{12}{8}=\frac{2}{8}$.
15. $\frac{6}{23}+\frac{7}{23}+\frac{5}{23}=\frac{4}{23}$.
16. $\frac{7}{71}+\frac{11}{71}+\frac{77}{71}=\frac{32}{71}$.

19. If the fractions to be added have not a common denominator, and all the denominators are prime to each other—

RULE.—Multiply the numerator and denominator of each fraction by the product of the denominators of all the others, and then add the resulting fractions—by the last rule.

EXAMPLE.—What is the sum of $\frac{2}{3}+\frac{3}{4}+\frac{4}{7}$?

$$\frac{2}{3}+\frac{3}{4}+\frac{4}{7}=\frac{2\times4\times7}{3\times4\times7}+\frac{3\times3\times7}{4\times3\times7}+\frac{4\times3\times4}{7\times3\times4}=\frac{56}{84}+\frac{63}{84}+\frac{48}{84}=\frac{167}{84}$$

Having found the denominator of one fraction, we may at once put it as the common denominator; since the same factors (the given denominators) must necessarily produce the same product.

20. REASON OF THE RULE.—To bring the fractions to a common denominator we have merely multiplied the numerator and denominator of each by the same number, which [4] does not alter the fraction. It is necessary to find a common denominator; for if we add the fractions without so doing, we cannot put the denominator of any one of them as the denominator of their sum;—thus $\frac{2+3+4}{3}$ for instance, would not be correct—since it would suppose all the quantities to be thirds, while some of them are fourths and sevenths, which are *less* than thirds; neither would $\frac{2+3+4}{7}$ be correct—since it would suppose all of them to be

sevenths, although some of them are thirds and fourths, which are *greater* than sevenths.

21. In altering the denominators, we have only changed the parts into which the unit is supposed to be divided, to an equivalent number of others which are smaller. It is necessary to diminish the size of these parts, or each fraction would not be *exactly* equal to some number of them. This will be more evident if we take only two of the above fractions. Thus, to add $\frac{2}{3}$ and $\frac{3}{4}$,

$$\frac{2}{3}+\frac{3}{4}=\frac{2\times4}{3\times4}+\frac{3\times3}{4\times3}=\frac{8}{12}+\frac{9}{12}=\frac{17}{12}$$

These fractions, before and after they receive a common denominator, will be represented as follows :—

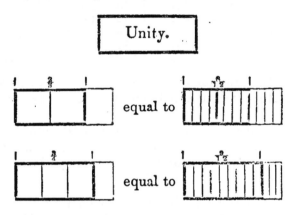

We have increased the number of the parts just as much as we have diminished their size; if we had taken parts larger than twelfths, we could not have found any numbers of them exactly equivalent, respectively, to *both* $\frac{2}{3}$ and $\frac{3}{4}$.

EXERCISES.

17. $\frac{1}{2}+\frac{2}{3}+\frac{4}{5}=\frac{63}{30}=1\frac{33}{30}$.

18. $\frac{1}{3}+\frac{1}{4}+\frac{1}{6}=\frac{9}{12}$.

19. $\frac{2}{3}+\frac{5}{8}+\frac{7}{9}=1\frac{48}{72}=1\frac{7}{10}\frac{7}{5}$.

20. $\frac{3}{4}+\frac{5}{8}+\frac{9}{7}=\frac{131}{14}=1\frac{121}{140}$.

21. $\frac{5}{8}+\frac{4}{7}+\frac{3}{5}=\frac{213}{110}=1\frac{113}{115}$.

22. $\frac{13}{15}+\frac{29}{21}+\frac{4}{27}=2\frac{111}{110}$.

23. $\frac{33}{35}+\frac{43}{47}+\frac{65}{57}=1\frac{131}{138}\frac{81}{57}$.

24. $\frac{83}{84}+\frac{107}{107}+\frac{103}{103}=2\frac{211}{312}\frac{83}{83}$.

22. If the fractions to be added have not a common denominator, and all the denominators are *not* prime to each other—

Proceed as directed by the last rule ; or—

RULE.—Find the least common multiple of all the denominators [Sec. II. 107, &c.], this will be the common denominator ; multiply the numerator of each fraction into the quotient obtained on dividing the common multiple by its denominator—this will give the new numerators; then add the numerators as already directed [18].

EXAMPLE.—Add $\frac{5}{32}+\frac{4}{48}+\frac{3}{72}$. 288 is the least common multiple of 32, 48, and 72; therefore $\frac{5}{32}+\frac{4}{48}+\frac{3}{72}=\frac{288\div32\times5}{288}$ $+\frac{288\div48\times4}{288}+\frac{288\div72\times3}{288}=\frac{45}{288}+\frac{24}{288}+\frac{12}{288}=\frac{81}{288}$.

23. REASON OF THE RULE.—We have multiplied each numerator and denominator by the same number (the least common multiple of the denominators [4])—since $\frac{5\times288\div32}{288}$ (for

instance)$=\frac{5\times288}{32\times288}$. For we obtain the same quotient, whether we multiply the divisor or divide the dividend by the same number—as in both cases we to the very same amount, diminish the number of times the one can be subtracted from the other.

When the denominators are *not* prime to each other the fractions we obtain have lower terms if we make the least common *multiple* of the denominators, rather than the *product* of the denominators, the common denominator. In the present instance, had we proceeded according to the *last* rule [19], we would have found $\frac{5}{32}+\frac{8}{48}+\frac{3}{72}=\frac{17280}{110592}+\frac{18432}{110592}+\frac{4608}{110592}=$ $\frac{40320}{110592}$: but $\frac{40320}{110592}$ is evidently a fraction containing larger terms than $\frac{81}{288}$.

EXERCISES.

25. $\frac{3}{4}+\frac{5}{6}+\frac{4}{5}=\frac{143}{60}=2\frac{23}{60}$.

26. $\frac{3}{4}+\frac{3}{5}+\frac{5}{8}=\frac{43}{20}=2\frac{3}{20}$.

27. $\frac{7}{8}+\frac{5}{6}+\frac{4}{9}=\frac{133}{60}=2\frac{47}{60}$.

28. $\frac{5}{6}+\frac{7}{9}+\frac{5}{8}=\frac{365}{168}=2\frac{29}{168}$.

29. $\frac{1}{4}+\frac{5}{6}+\frac{4}{7}=\frac{17}{12}=1\frac{5}{12}$.

30. $\frac{3}{4}+\frac{3}{5}+\frac{4}{7}=\frac{51}{40}=1\frac{9}{40}$.

31 $1\frac{5}{8}+1\frac{7}{12}+\frac{4}{9}=\frac{3611}{1008}=2\frac{601}{1008}$.

32. $\frac{4}{5}+\frac{11}{20}+\frac{5}{6}=\frac{188}{80}=2\frac{23}{80}$.

33. $\frac{1}{4}+\frac{23}{33}+\frac{5}{8}=\frac{41}{44}=1\frac{7}{8}$.

34. $\frac{7}{24}+\frac{4}{7}+\frac{1}{8}=\frac{46}{48}=1\frac{2}{24}$.

35. $\frac{3}{8}+\frac{5}{6}+\frac{4}{5}=\frac{91}{88}=1\frac{11}{88}$.

36. $\frac{1}{4}+\frac{1}{12}+\frac{1}{48}+\frac{1}{24}=\frac{33}{12}=\frac{2}{3}$.

37. $\frac{4}{5}+\frac{5}{8}+\frac{6}{9}+\frac{3}{8}+\frac{7}{11}=\frac{33109}{9240}$

$=3\frac{5419}{9240}$.

24. To reduce a mixed number to an improper fraction—

RULE.—Change the integral part into a fraction, having the same denominator as the fractional part [14], and add it to the fractional part.

EXAMPLE.—What fraction is equal to $4\frac{5}{9}$? $4\frac{5}{9}=\frac{4}{1}+\frac{5}{9}=$ $\frac{36}{9}+\frac{5}{9}=\frac{41}{9}$.

25. REASON OF THE RULE.—We have already seen that an integer may be expressed as a fraction having any denominator we please:—the reduction of a mixed number, therefore, is really the addition of fractions, previously reduced to a common denominator.

EXERCISES.

- 38. $16\frac{1}{7}=\frac{113}{7}$.
- 39. $18\frac{5}{8}=\frac{149}{8}$.
- 40. $79\frac{1}{8}=\frac{633}{8}$.
- 41. $47\frac{1}{4}=\frac{189}{4}$.
- 42. $74\frac{5}{9}=\frac{667}{9}$.
- 43. $95\frac{1}{5}=\frac{476}{5}$.

- 44. $99\frac{1}{11}=\frac{1090}{11}$.
- 45. $12\frac{1}{12}=\frac{145}{12}$.
- 46. $15\frac{1}{6}=\frac{91}{6}$.
- 47. $46\frac{5}{8}=\frac{373}{8}$.
- 48. $13\frac{7}{9}=\frac{120}{9}$.
- 49. $27\frac{1}{18}=\frac{487}{18}$.

26. To add mixed numbers—

RULE.—Add together the fractional parts; then, if the sum is an improper fraction, reduce it to a mixed number [9], and to its integral part add the integers in the given addends; if it is not an improper fraction, set it down along with the sum of the given integers.

EXAMPLE 1.—What is the sum of $4\frac{5}{8}+18\frac{7}{8}$?

$$\frac{7}{8}+\frac{5}{8}=\frac{12}{8}=1\frac{4}{8} \qquad \begin{array}{r} 4\frac{5}{8} \\ 18\frac{7}{8} \\ \hline \text{sum} \quad 23\frac{4}{8} \end{array}$$

5 eighths and 7 eighths are 12 eighths; but, as 8 eighths make one unit, 12 eighths are equal to one unit and 4 eighths—that is, one to be carried, and $\frac{4}{8}$ to be set down. 1 and 18 are 19, and 4 are 23.

EXAMPLE 2.—Add $12\frac{2}{6}$ and $29\frac{11}{15}$.

$$\frac{11}{15}+\frac{5}{6}=\frac{47}{30}=1\frac{17}{30} \qquad \begin{array}{r} 12\frac{5}{6}=12\frac{25}{30} \\ 29\frac{11}{15}=29\frac{22}{30} \\ \hline \text{sum} \quad 42\frac{17}{30} \end{array}$$

In this case it is necessary, before performing the addition [19 and 22], to reduce the fractional parts to a common denominator.

27. REASON OF THE RULE.—The addition of mixed numbers is performed on the same principle as simple addition but, in the first example, for instance, *eight* of one denomination is equal to *one* of the next—while in simple addition [Sec. II. 3], *ten* of one denomination is equal to *one* of the next.

EXERCISES.

- 50. $4\frac{2}{7}+3\frac{3}{7}=8\frac{5}{7}$.
- 51. $8\frac{11}{12}+2\frac{2}{3}=11\frac{7}{12}$.
- 52. $19\frac{5}{11}+7\frac{2}{4}=26\frac{31}{44}$.
- 53. $10\frac{5}{8}+11\frac{7}{12}=22\frac{5}{24}$.
- 54. $11\frac{3}{4}+8\frac{1}{4}=19\frac{2}{4}$.

- 55. $3\frac{5}{7}+11\frac{1}{4}+14\frac{17}{22}=29\frac{131}{54}$.
- 56. $40\frac{3}{4}+38\frac{1}{2}+40\frac{3}{8}=119\frac{5}{8}$.
- 57. $81\frac{3}{4}+6\frac{3}{8}+11=99\frac{1}{12}$.
- 58. $92\frac{5}{11}+37\frac{9}{13}+7\frac{1}{4}=137\frac{355}{143}$.
- 59. $173\frac{5}{12}+8\frac{5}{7}+91\frac{11}{13}=273\frac{231}{84}$.

1. What is the rule for adding fractions which have a common denominator? [18].

2. How are fractions brought to a common denominator? [19 and 22].

3. What is the rule for addition when the fractions have different denominators, all prime to each other? [19].

4. What is the rule when the denominators are not the same, but are not all prime to each other? [22].

5. How is a mixed number reduced to an improper fraction? [24].

6. How are mixed numbers added? [26].

SUBTRACTION.

28. To subtract fractions, when they have a common denominator—

RULE.—Subtract the numerator of the subtrahend from that of the minuend, and place the common denominator under the difference.

EXAMPLE.—Subtract $\frac{4}{9}$ from $\frac{7}{9}$.

$$\frac{7}{9} - \frac{4}{9} = \frac{7-4}{9} = \frac{3}{9}.$$

29. REASON OF THE RULE —If we take 4 individuals of any kind, from 7 of the same kind, three of them will remain. In the example, we take 4 (ninths) from 7 (ninths), and 3 are left—which must be ninths, since the process of subtraction cannot have changed their nature. The following will exemplify the subtraction of fractions :—

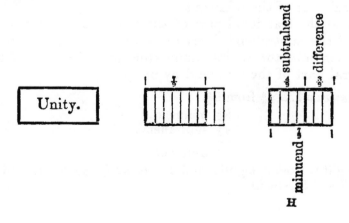

H

EXERCISES.

1. $1\frac{1}{2} - \frac{5}{12} = \frac{1}{2}$.

2. $1\frac{5}{8} - \frac{7}{16} = \frac{1}{2}$.

3. $1\frac{9}{20} - \frac{17}{20} = \frac{7}{10}$.

4. $1\frac{5}{8} - \frac{5}{16} = \frac{3}{8}$.

5. $2\frac{3}{22} - \frac{7}{22} = \frac{1}{11}$.

6. $\frac{5}{8} - \frac{7}{23} = 1\frac{1}{1}$.

7. $2\frac{19}{21} - \frac{8}{21} = \frac{1}{2}\frac{1}{1}$.

8. $\frac{7}{8} - \frac{1}{8} = \frac{3}{4}$.

9. $\frac{7}{11} - \frac{4}{11} = \frac{3}{11}$.

10. $\frac{14}{27} - \frac{8}{27} = \frac{2}{9}$.

30. If the subtrahend and minuend have not a common denominator—

Rule.—Reduce them to a common denominator [19 and 22]; then proceed as directed by the last rule.

Example.—Subtract $\frac{5}{9}$ from $\frac{7}{8}$.

$$\frac{7}{8} - \frac{5}{9} = \frac{63}{72} - \frac{40}{72} = \frac{23}{72}.$$

31. Reason of the Rule.—It is similar to that already given [20] for reducing fractions to a common denominator, previously to *adding* them.

EXERCISES.

11. $\frac{3}{4} - \frac{5}{9} = \frac{7}{36}$.

12. $1\frac{1}{2} - \frac{5}{16} = \frac{23}{48}$.

13. $\frac{7}{8} - \frac{3}{4} = \frac{1}{8}$.

14. $1\frac{4}{5} - 1\frac{3}{4} = \frac{3}{155}$.

15. $1\frac{14}{15} - 1\frac{11}{12} = \frac{63}{6094}$.

16. $\frac{47}{85} - \frac{5}{18} = \frac{3201}{1028}$.

17. $\frac{36}{45} - \frac{48}{81} = \frac{6}{25}$.

18. $1\frac{16}{84} - \frac{310}{846} = \frac{73}{72}$.

32. To subtract mixed numbers, or fractions from mixed numbers.

If the fractional parts have a common denominator—

Rule—I. Subtract the fractional part of the subtrahend from that of the minuend, and set down the difference with the common denominator under it: then subtract the integral part of the subtrahend from the integral part of the minuend.

II. If the fractional part of the minuend is less than that of the subtrahend, increase it by adding the common denominator to its numerator, and decrease the integral part of the minuend by unity.

Example 1.—$4\frac{3}{8}$ from $9\frac{5}{8}$.

$$9\frac{5}{8} \text{ minuend.}$$
$$4\frac{3}{8} \text{ subtrahend.}$$
$$\overline{5\frac{1}{4}} \text{ difference.}$$

3 eighths from 5 eighths and 2 eighths ($=\frac{1}{4}$) remain. 4 from 9 and 5 remain.

EXAMPLE 2.—Subtract $12\frac{3}{4}$ from $18\frac{1}{4}$.

$18\frac{1}{4}$ minuend.

$12\frac{3}{4}$ subtrahend.

$5\frac{1}{2}$ difference.

3 fourths cannot be taken from 1 fourth; but (borrowing one from the next denomination, considering it as 4 fourths, and adding it to the 1 fourth) 3 fourths from 5 fourths and 2 fourths $(=\frac{1}{2})$ remain. 12 from 17, and 5 remain.

If the minuend is an integer, it may be considered as a mixed number, and brought under the rule.

EXAMPLE 3.—Subtract $3\frac{4}{5}$ from 17.

17 may be supposed equal to $17\frac{5}{5}$; therefore $17-3\frac{4}{5}=17\frac{5}{5}-3\frac{4}{5}$. But, by the rule, $17\frac{5}{5}-3\frac{4}{5}=16\frac{5}{5}-3\frac{4}{5}=13\frac{1}{5}$.

33. REASON OF THE RULE.—The principle of this rule is the same as that already given for simple subtraction [Sec. II. 19]:—but in example 3, for instance, *five* of one denomination make *one* of the next, while in simple subtraction *ten* of one, make *one* of the next denomination.

34. If the fractional parts have *not* a common denominator—

RULE.—Bring them to a common denominator, and then proceed as directed in the last rule.

EXAMPLE 1.—Subtract $42\frac{1}{4}$ from $56\frac{1}{3}$.

$56\frac{1}{3}=56\frac{4}{12}$, minuend.

$42\frac{1}{4}=42\frac{3}{12}$, subtrahend.

$14\frac{1}{12}$, difference.

35. REASON OF THE RULE.—We are to subtract the different denominations of the subtrahend from those which correspond in the minuend [Sec. II. 19]—but we cannot subtract fractions unless they have a common denominator [30].

EXERCISES.

19. $27\frac{4}{5}-3\frac{1}{2}=24\frac{3}{10}$.

20. $15\frac{2}{3}-7\frac{4}{5}=7\frac{13}{15}$.

21. $12\frac{1}{8}-12\frac{1}{6}=\frac{1}{2}$.

22. $84\frac{11}{12}-\frac{11}{12}=84$.

23. $147\frac{15}{16}-\frac{7}{8}=147\frac{1}{24}$.

24. $82\frac{11}{13}-7\frac{12}{13}=74\frac{11}{13}$.

25. $76\frac{7}{8}-72\frac{9}{10}=3\frac{39}{40}$.

26. $67\frac{1}{4}-34\frac{3}{16}=32\frac{13}{16}$.

27. $97\frac{1}{2}-32\frac{13}{15}=64\frac{9}{16}$.

28. $60\frac{4}{5}-41\frac{13}{16}=19\frac{1}{4}$.

29. $92\frac{1}{6}-90\frac{7}{12}=2\frac{7}{12}$.

30. $100\frac{1}{2}-9\frac{5}{8}=90\frac{7}{8}$.

31. $60-\frac{3}{17}=59\frac{14}{17}$.

32. $12\frac{1}{4}-10\frac{7}{8}=1\frac{3}{8}$.

1. What is the rule for the subtraction of fractions, when they have a common denominator ? [28].

2. What is the rule, when they have not a common denominator ? [30].

3. How are mixed numbers, or fractions, subtracted from mixed numbers, or integers? [32 and 34].

MULTIPLICATION.

36. To multiply a fraction by a whole number; or the contrary—

RULE.—Multiply the numerator by the whole number, and put the denominator of the fraction under the product.

EXAMPLE.—Multiply $\frac{4}{7}$ by 5.

$$\frac{4}{7} \times 5 = \frac{20}{7}.$$

37. REASON OF THE RULE.—To multiply by any number, we are to add the multiplicand [Sec. II. 33] so many times as are indicated by the multiplier; but to add fractions having a common denominator we must add the numerators [18], and put the common denominator under the product. Hence—

$$\frac{4}{7} \times 5 = \frac{4}{7} + \frac{4}{7} + \frac{4}{7} + \frac{4}{7} + \frac{4}{7} = \frac{4+4+4+4+4}{7} = \frac{4 \times 5}{7} = \frac{20}{7}.$$

We increase the *number* of those "parts" of the integer which constitute the fraction, to an amount expressed by the multiplier—their *size* being unchanged. It would evidently be the same thing to increase their *size* to an equal extent without altering their *number*—this would be effected by dividing the denominator by the given multiplier; thus $\frac{4}{15} \times 5 = \frac{4}{3}$. This will become still more evident if we reduce the fractions resulting from both methods to others having a common denominator—for $\frac{20}{15} \left(= \frac{4 \times 5}{15} \right)$, and $\frac{4}{3} \left(= \frac{4}{15 \div 5} \right)$ will then be found equal.

As, very frequently, the multiplier is not contained in the denominator *any* number of times expressed by an integer, the method given in the *rule* is more generally applicable.

The rule will evidently apply if an integer is to be multiplied by a fraction—since the same product is obtained in whatever order the factors are taken [Sec. II. 35].

38. The integral quantity which is to form one of the factors may consist of more than one denomination.

EXAMPLE.—What is the $\frac{2}{3}$ of £5 2s. 9d.?

$$\begin{array}{ccc} £ & s. & d. \\ 5 & 2 & 9 \end{array} \times \frac{2}{3} = \frac{\begin{array}{ccc} £ & s. & d. \\ 5 & 2 & 9 \end{array} \times 2}{3} = \begin{array}{ccc} £ & s. & d. \\ 3 & 8 & 6. \end{array}$$

EXERCISES.

1. $\frac{4}{5} \times 2 = 1\frac{3}{5}$.
2. $\frac{5}{6} \times 8 = 6\frac{2}{3}$.
3. $\frac{7}{10} \times 12 = 10\frac{4}{5}$.
4. $\frac{3}{4} \times 12 = 9\frac{1}{4}$.
5. $\frac{7}{15} \times 30 = 14$.

6. $27 \times \frac{4}{9} = 12$.
7. $\frac{3}{17} \times 18 = 3\frac{3}{7}$.
8. $\frac{15}{16} \times 8 = 7\frac{1}{2}$.
9. $21 \times \frac{3}{7} = 9$.
10. $15 \times \frac{1}{5} = 3$.

11. $\frac{17}{18} \times 36 = 34$.
12. $\frac{19}{20} \times 20 = 19$.
13. $22 \times \frac{2}{9} = 4\frac{8}{9}$.
14. $\frac{1}{15} \times 17 = 1\frac{2}{15}$.
15. $143 \times \frac{3}{7} = 61\frac{2}{7}$.

16. How much is $\frac{83}{106}$ of 26 acres 2 roods? *Ans.* 20 acres 3 roods.

17. How much is $\frac{14}{49}$ of 24 hours 30 minutes? *Ans.* 7 hours.

18. How much is $\frac{870}{2219}$ of 19 cwt., 3 qrs., 7 ℔? *Ans.* 7 cwt., 3 qrs., 2 ℔.

19. How much is $\frac{13}{42}$ of £29? *Ans.* £$\frac{377}{42}$ = £8 19s. $6\frac{1}{4}d.$

39. To multiply one fraction by another—

RULE.—Multiply the numerators together, and under their product place the product of the denominators.

EXAMPLE.—Multiply $\frac{4}{9}$ by $\frac{5}{6}$.

$$\frac{4}{9} \times \frac{5}{6} = \frac{4 \times 5}{9 \times 6} = \frac{20}{54}.$$

40. REASON OF THE RULE.—If, in the example given, we were to multiply $\frac{4}{9}$ by 5, the product ($\frac{20}{9}$) would be 6 times too great—since it was by the *sixth* part of 5 ($\frac{5}{6}$), we should have multiplied.—But the product will become what it ought to be (that is, 6 times smaller), if we multiply its denominator by 6, and thus cause the *size* of the parts to become 6 times less.

We have already illustrated this subject when explaining the nature of a compound fraction [11].

EXERCISES.

20. $\frac{7}{12} \times \frac{5}{6} = \frac{35}{72}$.
21. $\frac{14}{15} \times \frac{5}{6} = \frac{7}{12}$.
22. $\frac{3}{7} \times \frac{4}{1} \times \frac{3}{2} = 2\frac{4}{7}$.
23. $\frac{1}{4} \times \frac{2}{3} = \frac{1}{6}$.

24. $\frac{14}{17} \times \frac{14}{17} = \frac{49}{33}$.
25. $\frac{7}{8} \times \frac{8}{9} \times \frac{7}{16} = \frac{343}{44}$.
26. $\frac{6}{8} \times \frac{4}{6} = \frac{1}{2}$.
27. $\frac{314}{453} \times \frac{117}{312} = \frac{9363}{23656}$.

28. $\frac{13}{18} \times \frac{27}{9} = \frac{2}{3}$.
29. $\frac{1}{4} \times \frac{1}{2} = \frac{1}{8}$.
30. $\frac{5}{15} \times \frac{5}{24} = \frac{1}{18}$.
31. $\frac{2}{12} \times \frac{5}{8} = \frac{2}{19}$.

32. How much is the $\frac{2}{3}$ of $\frac{3}{4}$? *Ans.* $\frac{1}{2}$.

33. How much is the $\frac{2}{3}$ of $\frac{7}{8}$? *Ans.* $\frac{7}{12}$.

41. When we multiply one proper fraction by another, we obtain a product smaller than either of the factors.— Nevertheless such multiplication is a species of addition ; for when we add a fraction *once*, (that is, when we take the *whole* of it,) we get the fraction itself as result ; but when we add it *less than once*, (that is, take *so much of it* as is indicated by the fractional multiplier,) we must necessarily get a result which is less than when we took the *whole* of it. Besides, the multiplication of a fraction by a fraction supposes multiplication by one number—the numerator of the multiplier, and (which will be seen presently) division by another—the denominator of the multiplier. Hence, when the division exceeds the multiplication—which is the case when the multiplier is a proper fraction—the result is, in reality, that of division ; and the number *said* to be multiplied must be made less than before.

42. To multiply a fraction, or a mixed number by a mixed number.

RULE.—Reduce mixed numbers to improper fractions [24], and then proceed according to the last rule.

EXAMPLE 1.—Multiply $\frac{3}{4}$ by $4\frac{5}{9}$.

$4\frac{5}{9} = \frac{41}{9}$; therefore $\frac{3}{4} \times 4\frac{5}{9} = \frac{3}{4} \times \frac{41}{9} = \frac{123}{36}$.

EXAMPLE 2.—Multiply $5\frac{7}{8}$ by $6\frac{2}{5}$.

$5\frac{7}{8} = \frac{47}{8}$, and $6\frac{2}{5} = \frac{32}{5}$; therefore $5\frac{7}{8} \times 6\frac{2}{5} = \frac{47}{8} \times \frac{32}{5} = \frac{1504}{40}$.

43. REASON OF THE RULE.—We merely put the mixed numbers into a more convenient form, without altering their value.

To obtain the required product, we might multiply each part of the multiplicand by each part of the multiplier.—Thus, taking the first example.

$\frac{3}{4} \times 4\frac{5}{9} = \frac{3}{4} \times 4 + \frac{3}{4} \times \frac{5}{9} = \frac{12}{4} + \frac{15}{36} = \frac{108}{36} + \frac{15}{36} = \frac{123}{36}$.

EXERCISES.

34. $8\frac{3}{4} \times \frac{4}{5} = 7\frac{1}{2}$.

35. $5\frac{6}{15} \times \frac{2}{7} = 2\frac{11}{21}$.

36. $4\frac{1}{4} \times 7\frac{1}{2} \times 3 = 101\frac{1}{4}$.

37. $\frac{7}{10} \times 8\frac{3}{4} \times \frac{9}{11} \times 1\frac{1}{2} = 5\frac{3}{2}$.

38. $5\frac{4}{5} \times 16 \times 10\frac{1}{2} = 880\frac{8}{5}\frac{4}{5}$.

39. $3\frac{3}{17} \times 19\frac{1}{4} \times \frac{5}{8} = 50\frac{11}{8}$.

40. $6\frac{3}{4} \times \frac{1}{7} \times \frac{3}{8} \times \frac{4}{5} = 2\frac{7}{10}$.

41. $12\frac{1}{4} \times 13\frac{1}{4} \times 6\frac{3}{8} = 1097\frac{11}{4}$.

42. $3\frac{3}{4} \times 14\frac{1}{4} \times 15 = 818\frac{1}{4}$.

43. $14 \times 15\frac{1}{17} \times 3\frac{5}{8} = 749\frac{11}{17}\frac{3}{5}$.

44. What is the product of 6, and the $\frac{2}{3}$ of 5 ? *Ans.* 20.

45. What is the product of $\frac{2}{9}$ of $\frac{3}{5}$, and $\frac{5}{8}$ of $3\frac{2}{7}$? *Ans.* $\frac{23}{84}$.

44. If we perceive the numerator of one fraction to be the same as the denominator of the other, we may, to perform the multiplication, omit the number which is common. Thus $\frac{5}{6} \times \frac{6}{9} = \frac{5}{9}$.

This is the same as dividing both the numerator and denominator of the product by the same number—and therefore does not alter its value; since

$$\frac{5}{6} \times \frac{6}{9} = \frac{5 \times 6}{6 \times 9} = \frac{5 \times 6 \div 6}{6 \times 9 \div 6} = \frac{5}{9}$$

45. Sometimes, before performing the multiplication, we can reduce the numerator of one fraction and the denominator of another to lower terms. by dividing both by the same number :—thus, to multiply $\frac{3}{8}$ by $\frac{4}{7}$.

Dividing both 8 and 4, by 4, we get in their places, 2 and 1 ; and the fractions then are $\frac{3}{2}$ and $\frac{1}{7}$, which, multiplied together, become $\frac{3}{2} \times \frac{1}{7} = \frac{3}{14}$.

This is the same as dividing the numerator and denominator of the product by the same number; for

$$\frac{3}{8} \times \frac{4}{7} = \frac{3 \times 4 \div 4}{8 \times 7 \div 4} = \frac{3 \times 1}{2 \times 7} \left(= \frac{3}{2} \times \frac{1}{7} \right) = \frac{3}{14}.$$

QUESTIONS.

1. How is a fraction multiplied by a whole number; or the contrary? [36].

2. Is it necessary that the integer which constitutes one of the factors should consist of a single denomination? [38].

3. What is the rule for multiplying one fraction by another? [39].

4. Explain how it is that the product of two *proper* fractions is less than either? [41].

5. What is the rule for multiplying a fraction or a mixed number by a mixed number? [42].

6. How may fractions sometimes be reduced, before they are multiplied? [44 and 45].

DIVISION.

46. To divide a vulgar fraction by a whole number—
RULE.—Multiply the denominator of the fraction by the whole number, and put the product under its numerator.

EXAMPLE.—$\frac{2}{3} \div 4 = \dfrac{2}{3 \times 4} = \frac{2}{12}$.

47. REASON OF THE RULE.—To divide a quantity by 3, for instance, is to make it 3 times smaller than before. But it is evident that if, while we leave the *number* of the parts the same, we make their *size* 3 times less, we make the fraction itself 3 times less—hence to multiply the denominator by 3, is to divide the fraction by the same number.

A similar effect will be produced if we divide the numerator by 3; since the fraction is made 3 times smaller, if, while we leave the *size* of the parts the same, we make their *number* 3 times less; thus $\dfrac{8}{9} \div 4 = \dfrac{8 \div 4}{9} = \dfrac{2}{9}$. But since the numerator is not *always* exactly divisible by the divisor, the method given in the *rule* is more generally applicable.

The division of a fraction by a whole number has been already illustrated, when we explained the nature of a complex fraction [12].

EXERCISES.

1. $\frac{8}{9} \div 2 = \frac{4}{9}$.
2. $\frac{14}{15} \div 8 = \frac{7}{60}$.
3. $\frac{18}{19} \div 19 = \frac{1}{20}$.
4. $\frac{1}{7} \div 9 = \frac{1}{63}$.

5. $\frac{11}{12} \div 3 = \frac{11}{36}$.
6. $\frac{7}{8} \div 8 = \frac{7}{64}$.
7. $\frac{7}{13} \div 14 = \frac{1}{38}$.
8. $\frac{9}{17} \div 3 = \frac{3}{17}$.

9. $\frac{11}{12} \div 5 = \frac{2}{15}$.
10. $\frac{5}{8} \div 11 = \frac{5}{93}$.
11. $\frac{1}{14} \div 42 = \frac{1}{837}$.
12. $\frac{7}{15} \div 14 = \frac{1}{30}$.

48. It follows from what we have said of the multiplication and division of a fraction by an integer, that, when we multiply or divide its numerator and denominator by the same number, we do not alter its value—since we then, at the same time, equally increase and decrease it.

49. To divide a fraction by a fraction—
RULE.—Invert the divisor (or suppose it to be inverted), and then proceed as if the fractions were to be multiplied.

EXAMPLE.—Divide $\frac{5}{7}$ by $\frac{3}{4}$.

$$\frac{5}{7} \div \frac{3}{4} = \frac{5}{7} \times \frac{4}{3} = \frac{5 \times 4}{7 \times 3} = \frac{20}{21}.$$

REASON OF THE RULE.—If, for instance, in the example just given, we divide $\frac{5}{7}$ by 3 (the numerator of the divisor), we use a quantity 4 times too great, since it is not by 3, but the fourth part of 3 ($\frac{3}{4}$) we are to divide, and the quotient ($\frac{5}{21}$) is 4 times too small.—It is, however, made what it ought to be, if we multiply its numerator by 4—when it becomes $\frac{20}{21}$, which was the result obtained by the rule.

50. The division of one fraction by another may be illustrated as follows—

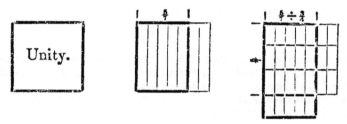

The quotient of $\frac{5}{7} \div \frac{3}{4}$ must be some quantity, which, taken three-fourth times (that is, multiplied by $\frac{3}{4}$), will be equal to $\frac{5}{7}$ of unity. For since the quotient multiplied by the divisor ought to be equal to the dividend [Sec. II. 79], $\frac{5}{7}$ is $\frac{3}{4}$ of the quotient. Hence, if we divide the five-sevenths of unity into three equal parts, each of these will be *one*-fourth of the quotient—that is, precisely what the dividend wants to make it four-fourths of the quotient, or the quotient itself.

51. When we divide one proper fraction by another, the quotient is greater than the dividend. Nevertheless such division is a species of subtraction. For the quotient expresses *how often* the divisor can be taken from the dividend; but were the fraction to be divided by unity, the dividend itself would express how often the divisor could be taken from it; when, therefore, the divisor is less than unity, the number of times it can be taken from the dividend must be expressed by a quantity *greater* than the dividend [Sec. II. 78]. Besides, dividing one fraction by another supposes the multiplication of the dividend by one number and the division of it by another—but when the multiplication is by a greater

H 2

number than the division, the result is, in reality, that of multiplication, and the quantity *said* to be divided must be increased.

EXERCISES.

13. $\frac{7}{8} \div \frac{4}{5} = 1\frac{3}{32}$.
14. $\frac{1}{5} \div \frac{2}{5} = \frac{1}{2}$.
15. $\frac{5}{6} \div \frac{5}{9} = 1\frac{2}{3}$.

16. $\frac{4}{5} \div \frac{5}{8} = \frac{32}{35}$.
17. $\frac{3}{4} \div \frac{3}{8} = 2$.
18. $\frac{11}{16} \div \frac{8}{16} = 1\frac{1}{2}$.

19. $\frac{11}{12} \div \frac{9}{11} = 1\frac{25}{114}$.
20. $\frac{5}{9} \div \frac{8}{9} = \frac{5}{8}$.
21. $\frac{4}{9} \div \frac{5}{6} = \frac{8}{15}$.

52. To divide a whole number by a fraction—

RULE.—Multiply the whole number by the denominator of the fraction, and make its numerator the denominator of the product.

EXAMPLE.—Divide 5 by $\frac{3}{7}$.

$$5 \div \frac{3}{7} = \frac{5 \times 7}{3} = \frac{35}{3}.$$

This rule is a consequence of the last; for every whole number may be considered as a fraction having unity for denominator [14]; hence $5 \div \frac{3}{7} = \frac{5}{1} \div \frac{3}{7} = \frac{5}{1} \times \frac{7}{3} = \frac{35}{3}$.

It is not necessary that the whole number should consist of but one denomination [38].

EXAMPLE.—Divide 17s. 3¼d. by $\frac{3}{5}$.

$$17s.\ 3\frac{1}{4}d. \div \frac{3}{5} = 17s.\ 3\frac{1}{4}d. \times \frac{5}{3} = £1\ 8s.\ 9\frac{1}{4}d.$$

EXERCISES.

22. $3 \div \frac{4}{9} = 6\frac{3}{4}$.
23. $11 \div \frac{5}{9} = 19\frac{4}{5}$.
24. $42 \div \frac{1}{12} = 864$.

25. $5 \div \frac{15}{16} = 5\frac{1}{3}$.
26. $19 \div \frac{19}{20} = 20$.
27. $9 \div \frac{1}{7} = 63$.

28. $8 \div \frac{14}{15} = 8\frac{4}{7}$.
29. $14 \div \frac{7}{19} = 38$.
30. $16 \div \frac{1}{2} = 32$.

31. Divide £7 16s. 2d. by $\frac{4}{9}$.   *Ans.* £17 11s. 4½d.
32. Divide £8 13s. 4d. by $\frac{5}{6}$.   *Ans.* £10 8s.
33. Divide £5 0s. 1d. by $\frac{11}{12}$.   *Ans.* £5 9s. 2¼d.

53. To divide a mixed number by a whole number, or a fraction—

RULE.—Divide each part of the mixed number according to the rules already given [46 and 49], and add the quotients. Or reduce the mixed number to an improper fraction [24], and then divide, as already directed [46 and 49].

EXAMPLE 1.—Divide $9\frac{3}{7}$ by 3.

$$9\frac{3}{7} \div 3 = 9 \div 3 + \frac{3}{7} \div 3 = 3 + \frac{1}{7} = 3\frac{1}{7}.$$

EXAMPLE 2.—Divide $14\frac{3}{11}$ by 7.

$14\frac{3}{11} = \frac{157}{11}$; therefore $14\frac{3}{11} \div \frac{7}{8} = \frac{157}{11} \div \frac{7}{8} = \frac{157}{11} \times \frac{8}{7} = \frac{1256}{77} = 16\frac{24}{77}$.

54. REASON OF THE RULE.—In the first example we have divided each part of the dividend by the divisor and added the results—which [Sec. II. 77] is the same as dividing the whole dividend by the divisor.

In the second example we have put the mixed number into a more convenient form, without altering its value.

EXERCISES.

34. $8\frac{3}{4} \div 17 = \frac{35}{68}$.

35. $51\frac{4}{9} \div 3 = 17\frac{4}{27}$.

36. $187\frac{9}{16} \div \frac{8}{17} = 398\frac{13}{24}$.

37. $19\frac{21}{22} \div 41 = \frac{431}{902}$.

38. $16\frac{100}{121} \div 4\frac{8}{9} = 17\frac{7}{1112}$.

39. $4\frac{325}{327} \div 4\frac{1}{8} = 5\frac{645}{13107}$.

40. $84\frac{19}{11} \div 22 = 3\frac{161}{484}$.

41. $18\frac{6}{316} \div 2\frac{5}{7} = 19\frac{173}{5373}$.

42. $106\frac{9}{231} \div \frac{8}{15} = 198\frac{251}{308}$.

43. $18\frac{4}{5} \div 11 = 1\frac{67}{55}$.

55. To divide an integer by a mixed number—

RULE.—Reduce the mixed number to an improper fraction [24]; and then proceed as already directed [52].

EXAMPLE.—Divide 8 by $4\frac{3}{5}$.

$4\frac{3}{5} = \frac{23}{5}$, therefore $8 \div 4\frac{3}{5} = 8 \div \frac{23}{5} = 8 \times \frac{5}{23} = 1\frac{17}{23}$.

REASON OF THE RULE.—It is evident that the improper fraction which is equal to the divisor, is contained in the dividend the same number of times as the divisor itself.

EXERCISES.

44. $5 \div 3\frac{4}{7} = 1\frac{2}{5}$.

45. $16 \div 11\frac{2}{23} = 1\frac{133}{331}$.

46. $14 \div 1\frac{8}{9} = 7\frac{7}{17}$.

47. $21 \div 14\frac{4}{11} = 1\frac{73}{158}$.

48. Divide £7 16s. 7d. by $3\frac{1}{3}$. *Ans.* £2 6s. $11\frac{3}{4}d$.

49. Divide £3 3s. 3d. by $4\frac{1}{2}$. *Ans.* 14s. $0\frac{3}{4}d$.

56. To divide a fraction, or a mixed number, by a mixed number—

RULE.—Reduce mixed numbers to improper fractions [24]; and then proceed as already directed [49].

EXAMPLE 1.—Divide $\frac{3}{4}$ by $5\frac{7}{9}$.

$5\frac{7}{9} = \frac{52}{9}$, therefore $\frac{3}{4} \div 5\frac{7}{9} = \frac{3}{4} \div \frac{52}{9} = \frac{3}{4} \times \frac{9}{52} = \frac{27}{208}$.

EXAMPLE 2.—Divide $8\frac{9}{11}$ by $7\frac{5}{6}$.

$8\frac{9}{11} = \frac{97}{11}$, and $7\frac{5}{6} = \frac{47}{6}$, therefore $8\frac{9}{11} \div 7\frac{5}{6} = \frac{97}{11} \div \frac{47}{6} = \frac{97}{11} \times \frac{6}{47} = \frac{562}{517}$.

57. REASON OF THE RULE.—We (as in the last rule) merely change the mixed numbers into others more conveniently divided—without, however, altering their value.

EXERCISES.

50. $\frac{6}{11} \div 5\frac{5}{7} = \frac{28}{209}$.   55. $82\frac{1}{17} \div 26\frac{5}{11} = 3\frac{858}{969}$.

51. $3\frac{1}{3} \div 4\frac{1}{2} = \frac{13}{18}$.   56. $\frac{103}{8} \div 81\frac{7}{12} = \frac{4366}{7825}$.

52. $\frac{5}{23} \div 3\frac{5}{11} = \frac{213}{5456}$.   57. $8\frac{5}{8} \div 8\frac{4}{7} = \frac{483}{480}$.

53. $\frac{15}{12} \div 1\frac{5}{8} = \frac{25}{6}$.   58. $1\frac{3}{7} \div 2\frac{1}{2} + 5\frac{1}{2} \div 3\frac{1}{2} = \frac{7}{100}$.

54. $6\frac{1}{3} \div 5\frac{1}{3} = 1\frac{3}{52}$.   59. $2\frac{1}{4} \div \frac{3}{4} + \frac{5}{8} = 1\frac{9}{11}$.

58. When the divisor, dividend, or both, are compound, or complex fractions—

RULE.—Reduce compound and complex to simple fractions—by performing the multiplication, in those which are compound, and the division, in those which are complex; then proceed as already directed [49, &c.]

EXAMPLE 1.—Divide $\frac{4}{7}$ of $\frac{6}{8}$ by $\frac{3}{4}$.

$\frac{4}{7}$ of $\frac{6}{8} = \frac{30}{56}$ [39], therefore $\frac{4}{7} \times \frac{6}{8} \div \frac{3}{4} = \frac{30}{56} \div \frac{3}{4} = \frac{30}{56} \times \frac{4}{3} = \frac{120}{168}$.

EXAMPLE 2. — Divide $\frac{\frac{4}{7}}{6}$ by $\frac{4}{5}$.

$\frac{\frac{4}{7}}{6} = \frac{4}{42}$ [46], therefore $\frac{\frac{4}{7}}{6} \div \frac{4}{5} = \frac{4}{42} \div \frac{4}{5} = \frac{4}{42} \times \frac{5}{4} = \frac{32}{210}$.

EXERCISES.

60. $\frac{4}{7} \times \frac{3}{8} \div \frac{5}{9} = \frac{9}{28}$.

61. $4\frac{11}{12} \div \frac{5}{13} \times \frac{5}{11} = 50\frac{45}{99}$.

62. $\frac{5}{13} \div \frac{\frac{3}{4}}{6} = 2\frac{2}{9}$.

63. $\frac{\frac{21}{22}}{97} \div \frac{2}{3} \times \frac{7}{13} = \frac{117}{4268}$.

64. $\frac{\frac{3}{4}}{\frac{3}{8}} \div \frac{\frac{3}{4}}{\frac{5}{5}} = 25$.

65. $\frac{\frac{27}{5}}{\frac{1}{9}} \div \frac{21}{13} \times \frac{6}{23} = 243\frac{7}{8}$.

66. $\frac{\frac{4}{5}}{\frac{3}{7}} \div \frac{3}{4} \times \frac{5}{8} = 3\frac{121}{225}$.

QUESTIONS.

1. How is a fraction divided by an integer? [46].

2. How is a fraction divided by a fraction? [49].

3. Explain how it occurs that the quotient of two fractions is sometimes greater than the dividend? [51].

4. How is a whole number divided by a fraction? [52].

5. What is the rule for dividing a mixed number by an integer, or a fraction? [53].

6. What are the rules for dividing an integer, a fraction, or mixed number, by a mixed number? [55 and 56.]

7. What is the rule when the divisor, dividend, or both are compound, or complex fractions? [58].

MISCELLANEOUS EXERCISES IN VULGAR FRACTIONS.

1. How much is $\frac{1}{9}$ of 186 acres, 3 roods? *Ans.* 20 acres, 3 roods.

2. How much is $\frac{4}{9}$ of 15 hours, 45 minutes? *Ans.* 7 hours.

3. How much is $\frac{870}{2219}$ of 19 cwt., 3 qrs., 7 ℔? *Ans.* 7 cwt., 3 qrs., 2 ℔.

4. How much is $\frac{729}{2000}$ of £100? *Ans.* £36 9s.

5. If one farm contains 20 acres, 3 roods, and another 26 acres, 2 roods, what fraction of the former is the latter? *Ans.* $\frac{83}{106}$.

6. What is the simplest form of a fraction expressing the comparative magnitude of two vessels—the one containing 4 tuns, 3 hhds., and the other 5 tuns, 2 hhds.? *Ans.* $\frac{19}{22}$.

7. What is the sum of $\frac{2}{3}$ of a pound, and $\frac{5}{9}$ of a shilling? *Ans.* 13s. $10\frac{2}{3}d$.

8. What is the sum of $\frac{2}{3}s$. and $\frac{4}{15}d$.? *Ans.* $7\frac{7}{15}d$.

9. What is the sum of £$\frac{1}{7}$, $\frac{2}{9}s$., and $\frac{5}{12}d$.? *Ans.* 3s. $1\frac{31}{84}d$.

10. Suppose I have $\frac{3}{8}$ of a ship, and that I buy $\frac{5}{16}$ more; what is my entire share? *Ans.* $\frac{11}{16}$.

11. A boy divided his marbles in the following manner: he gave to A $\frac{1}{3}$ of them, to B $\frac{1}{10}$, to C $\frac{1}{8}$, and to D $\frac{1}{6}$, keeping the rest to himself; how much did he give away, and how much did he keep? *Ans.* He gave away $\frac{87}{120}$ of them, and kept $\frac{33}{120}$.

12. What is the sum of $\frac{1}{3}$ of a yard, $\frac{1}{7}$ of a foot, and $\frac{1}{7}$ of an inch? *Ans.* 7 inches.

13. What is the difference between the $\frac{2}{3}$ of a pound, and $5\frac{1}{4}d$.? *Ans.* 11s. $6\frac{3}{4}d$.

14. If an acre of potatoes yield about 82 barrels of 20 stone each, and an acre of wheat 4 quarters of 460 ℔—but the wheat gives three times as much nourishment as the potatoes; what will express the subsistence given by each, in terms of the other? *Ans.* The potatoes will give $4\frac{11}{69}$ times as much as the wheat; and the wheat the $\frac{69}{287}$ part of what is given by the potatoes.

15. In Fahrenheit's thermometer there are 180 degrees between the boiling and freezing points; in that

of Reaumar only 80 ; what fraction of a degree in the latter expresses a degree of the former ? *Ans.* $\frac{4}{9}$.

16. The average fall of rain in the United Kingdom is about 34 inches in depth during the year in the plains ; but in the hilly countries about 50 inches ; what fraction of the latter expresses the former? *Ans.* $\frac{17}{25}$.

17. Taking Chimborazo as 21,000 feet high, and Purgeool, in the Himalayas, as 22,480; what fraction of the height of Purgeool expresses that of ·Chimborazo? *Ans.* $\frac{525}{562}$.

18. Taking 4200 feet as the depth of a fissure or crevice at Cutaco, in the Andes, and 5000 feet as the depth of that at Chota, in the same range of mountains ; how will the depth of the former be expressed as a fraction of the latter ? *Ans.* $\frac{21}{25}$.

DECIMAL FRACTIONS.

59. A decimal fraction, as already remarked [13], has unity with one, or more cyphers to the right hand, for its denominator ; thus, $\frac{5}{1000}$ is a decimal fraction. Since the division of the numerator of a decimal fraction by its denominator—from the very nature of notation [Sec. I. 34]—is performed by moving the decimal point, the quotient of a decimal fraction—the equivalent *decimal*—is obtained with the greatest facility. Thus $\frac{5}{1000} = ·005$; for to divide any quantity by a thousand, we have only to move the decimal point three places to the right.

60. It is as inaccurate to confound a decimal fraction with the corresponding decimal, as to confound a vulgar fraction with its quotient.—For if 75 is the *quotient* of $\frac{300}{4}$, or of $\frac{7500}{100}$, and is distinct from either ; so also is ·75 the quotient of $\frac{3}{4}$ or of $\frac{75}{100}$, and equally distinct from either.

61. A decimal is changed into its corresponding decimal fraction by putting unity with as many cyphers as it contains decimal places, under it, for denominator— having first taken away its decimal point. Thus ·5646 = $\frac{5646}{10000}$; ·008 = $\frac{8}{1000}$, &c.

62. Decimal fractions follow exactly the same rules as vulgar fractions.—It is, however, generally more convenient to obtain their quotients [59], and then perform on them the required processes of addition, &c., by the methods already described [Sec. II. 11, &c.]

63. To reduce a vulgar fraction to a *decimal*, or to a *decimal fraction*—

RULE.—Divide the numerator by the denominator—this will give the required *decimal;* the latter may be changed into its corresponding decimal fraction—as already described [61].

EXAMPLE 1.—Reduce $\frac{3}{4}$ to a decimal fraction.

$$4)3$$
$$0\cdot75= \tfrac{75}{100}.$$

EXAMPLE 2.—What decimal of a pound is $7\frac{3}{4}d.$?

$7\frac{3}{4}d.=$ [17] $£\frac{31}{960}$; but $£\frac{31}{960}=£\cdot0032$, &c.

This rule requires no explanation.

EXERCISES.

1. $\frac{7}{8}=\frac{875}{1000}$.	5. $\frac{5}{8}=\cdot625$.	9. $\frac{9.5}{10.5}=\cdot90476$, &c.
2. $\frac{3}{8}=\cdot375$.	6. $\frac{71}{73}=\cdot973$&c.	10. $\frac{4}{5}=\cdot8$.
3. $\frac{9}{25}=\cdot36$.	7. $\frac{1}{2}=\cdot5$.	11. $\frac{9}{16}=\cdot5625$.
4. $\frac{1}{4}=\frac{25}{100}$.	8. $\frac{5}{16}=\cdot3125$.	12. $\frac{43}{80}=\cdot5375$.

13. Reduce 12*s.* 6*d.* to the decimal of a pound. *Ans.* ·625.

14. Reduce 15*s.* to the decimal of a pound. *Ans.* ·75.

15. Reduce 3 quarters, 2 nails, to the decimal of a yard. *Ans.* ·875·

16. Reduce 3 cwt., 1 qr., 7 ℔s, to the decimal of a ton. *Ans.* ·165625·

64. To reduce a decimal to a lower denomination—

RULE.—Reduce it by the rule already given [Sec. III. 3] for the reduction of integers.

EXAMPLE 1.—Express £·6237 in terms of a shilling.

$$\cdot6237$$
$$20$$

Answer, $\overline{12\cdot4740}$ shillings$=£\cdot6237$.

EXAMPLE 2.—Reduce £·9734 to shillings, &c.

$$·9734$$
$$20$$

19·4680 shillings = £·9734.

$$12$$

5·6160 pence = ·468s.

$$4$$

2·4640 farthings = ·616d.

Answer, £·9734 = 19s. 5½d.

65. This rule is founded on the same reasons as were given for the mode of reducing integers [Sec. III. 4].

Multiplying the decimal of a pound by 20, reduces it to shillings and the decimal of a shilling. Multiplying the decimal of a shilling by 12, reduces it to pence and the decimal of a penny. Multiplying the decimal of a penny by 4, reduces it to farthings and the decimal of a farthing.

EXERCISES.

23. What is the value of £·86875 ?   *Ans.* 17s. 4½d.

24. What is the value of £·5375 ?   *Ans.* 10s. 9d.

25. How much is ·875 of a yard? *Ans.* 3 qrs., 2 nails.

26. How much is ·165625 of a ton ?   *Ans.* 3 cwt., 1 qr., 7 ℔.

27. What is the value of £·05 ?   *Ans.* 1s.

28. How much is ·9375 of a cwt. ?   *Ans.* 3 qrs., 21 ℔.

29. What is the value of £·95 ?   *Ans.* 19s.

30. How much is ·95 of an oz. Troy? *Ans.* 19 dwt.

31. How much is ·875 of a gallon ?   *Ans.* 7 pints.

32. How much is ·3945 of a day ?   *Ans.* 9 hours, 28′, 4″, 48‴.

33. How much is ·09375 of an acre ?   *Ans.* 15 perches.

66. The following will be found useful, and—being intimately connected with the doctrine of fractions—may be advantageously introduced here :

To find at once what decimal of a pound is equivalent to any number of shillings, pence, &c.

When there is an even number of shillings—

RULE.—Consider them to be half as many tenths of a pound.

EXAMPLE.—16s.=£·8.

Every two shillings are equal to one-tenth of a pound; therefore 8 times 2s. are equal to 8 tenths.

67. When the number of shillings is odd—

RULE.—Consider half the next lower even number, as so many tenths of a pound, and with these set down 5 hundredths.

EXAMPLE.—15s.=£·75.

For, 15s.=14s.+1s.; but by the last rule 14s.=£·7; and since 2s.=1 tenth—or, as is evident, 10 hundredths of a pound—1s.=5 hundredths.

68. When there are pence and farthings—

RULE.—If, when reduced to farthings, they exceed 24, add 1 to the number, and put the sum in the second and third decimal places. After taking 25 from the number of farthings, divide the remainder by 3, and put the nearest quantity to the true quotient, in the fourth decimal place.

If, when reduced to farthings, they are less than 25, set down the number in the third, or in the second and third decimal places; and put what is nearest to one-third of them in the fourth.

EXAMPLE 1.—What decimal of a pound is equal to 8¾d.?

8¾=35 farthings. Since 35 contains 25, we add one to the number of farthings, which makes it 36—we put 36 in the second and third decimal places. The number nearest to the third of 10 (35−25 farthings) is 3—we put 3 in the fourth decimal place. Therefore, 8¾=£·0363.

EXAMPLE 2.—What decimal of a pound is equal to 1¾d.?

1¾=7 farthings; and the nearest number to the third of 7 is 2. Therefore 1¾d.=£·0072.

EXAMPLE 3.—What decimal of a pound is equal to 5¼d.?

5¼d.=21 farthings; and the third of 21 is 7. Therefore 5¼d.=£·0217.

69. REASON OF THE RULE.—We consider 10 farthings as the one hundredth, and one farthing as the one thousandth of a pound—because a pound consists of nearly one thousand farthings. This, however, in 1000 farthings (taken as so many thousandths of a pound) leads to a mistake of about 40—since £1=(not 1000, but) 1000−40 farthings. Hence, to a thousand farthings (considered as thousandths of a pound),

forty, or one in 25, must be added; that is, about the one-*thirtieth* of the number of farthings. It is evident that, as those *above* 25 have not been allowed for when we added *one* to the farthings, one-*thirtieth* of *their* number, also, must be added—or, which is the same thing, one-*third* of their number, in the *fourth* or next lower decimal place.

If the farthings are less than 25, it is evident that the correction should still be about the *thirtieth* of their number, or one-*third* of it, in the *fourth* decimal place.

EXERCISES.

17. 19s. 11$\frac{1}{2}$d. = £·9977.	20. 14s. 3$\frac{3}{4}$d. = £·7155.
18. 7$\frac{3}{4}$d. = £·0322·	21. 19s. 11$\frac{3}{4}$d. = £·9987·
19. £27 5s. 10d. = £27·2915.	22. £42 11s. 6$\frac{1}{2}$d. = £42·577.

70. To find *at once* the number of shillings, pence, &c., in any decimal of a pound—

RULE.—Double the number of tenths for shillings— to which, if the hundredths are not less than 5, add one. Consider the digit in the second place (after subtracting 5, if it is not less than 5), as tens, and that in the third as units of farthings; and subtract unity from the result if it exceeds 25.

EXAMPLE.—£·6874 = 13s. 9d.

6 tenths are equal to *twelve* shillings; as the hundredths are not less than 5, there is an *additional* shilling—which makes 13s. Subtracting 5 from the hundredths and adding the remainder (reduced to thousandths) to the thousandths, we have 37 thousandths from which—since they exceed 25, we subtract unity; this leaves 36 as the number of farthings. £·6874, therefore, is equal to 13s. and 36 farthings—or 13s. 9d.

This rule follows from the last three—being the reverse of them.

CIRCULATING DECIMALS.

71. We cannot, as already noticed [Sec. II. 72], always obtain an exact quotient, when we divide one number by another :—in such a case, what is called an *in-terminate* or (because the same digit, or digits, constantly recur, or circulate) a *recurring*, or *circulating*

decimal is produced.—The decimal is said to be *termi-nate* if there is an exact quotient—or one which leaves no remainder.

72. An interminate decimal, in which only a single figure is repeated, is called a *repetend;* if two or more digits constantly recur, they form a *periodical* decimal. Thus ·77, &c., is a repetend ; but ·597597, &c. is a periodical. For the sake of brevity, the repeated digit, or period is set down but once, and may be marked as follows, ·5′ (=·555, &c.) or ·493′ (=·493493493, &c.)

The ordinary method of marking the period is some-what different—what is here given, however, seems preferable, and can scarcely be mistaken, even by those in the habit of using the other.

When the decimal contains only an *infinite* part—that is, only the repeated digit, or period—it is a *pure* repetend, or a *pure* periodical. But when there is *both* a finite and an infinite part, it is a *mixed* repetend *or* *mixed* circulate. Thus

·3′ (=·333, &c.) is a pure repetend.
·578′ (=·57888, &c.) is a mixed repetend.
·397′ (=·397397397, &c.) is a pure circulate.
865′64271′ (=·865642716427164271,&c) is a mixed cir·ulate.

73. The number of digits in a period must always be less than the divisor. For, different digits in the period suppose different remainders during the division; but the number of remainders can never exceed—nor even be equal to the divisor. Thus, let the latter be *seven* : the only remainders possible are 1, 2, 3, 4, 5, and 6 ; any other than one of these would contain the divisor at least once—which would indicate [Sec. II. 71] that the quotient figure is not sufficiently large.

74. It is sometimes useful to change a decimal into its equivalent vulgar fraction—as, for instance, when in adding, &c., those which circulate, we desire to obtain an exact result. For this purpose—

RULE—I. If the decimal is a *pure repetend,* put the repeated digit for numerator, and 9 for denominator.

II. If it is a *pure periodical,* put the period for numerator, and so many nines as there are digits in the period, for denominator.

EXAMPLE 1.—What vulgar fraction is equivalent to ·3′ ? *Ans.* ⅓.

EXAMPLE 2. — What vulgar fraction is equivalent to ·7854′ ? *Ans.* $\frac{7854}{9999}$.

75. REASON OF I.—⅑ will be found equal to ·111, &c.—or ·1′; therefore ⅓ (=3×⅑)=·333, &c.=(3×111, &c.) For if we multiply two equal quantities by the same, or by equal quantities, the products will still be equal.

In the same way it could be shown that *any other* digit divided by 9 would give that other digit as a repetend.—And, consequently, a repetend of any digit will be equal to a vulgar fraction having the same digit for numerator, and 9 for denominator.

REASON OF II.—$\frac{1}{99}$ will give ·0101, &c.—or ·01′ as quotient. For before unity can be divided by 99, it must be considered as 100 *hundredths;* and the quotient [Sec. II. 77] will be one *hundredth,* or ·01. One hundredth, the remainder, must be made 100 *ten thousandths* before it will contain 99; and the quotient will be one *ten thousandth,* or ·0001. One ten thousandth, the remainder, must, in the same way, be considered as ten *million-eths;* and the next quotient will be one *millioneth,* or ·000001— and so ou with the other quotients, which, taken together, will be ·01+·0001+·000001+&c., or ·010101, &c.—represented by ·01′.

$\frac{37}{99}$ (=37×$\frac{1}{99}$=37×·01′) will give ·373737, &c.—or ·37′ as quotient. Thus

$$
\begin{array}{r}
010101, \&c. \\
37 \\
\hline
70707 \\
30303 \\
\hline
373737, \&c.=37×·01′.
\end{array}
$$

In the same way it could be shown that *any other* two digits divided by 99 would give those other digits as the period of a circulate.—And, consequently, a circulate having any two digits as a period, will be equal to a vulgar fraction having the same digits for numerator, and 2 nines for denominator.

For similar reasons $\frac{1}{999}$ will give ·001001, &c., or ·001′ as quotient. But ·001001, &c., × (for instance) 563=·563563, &c. Thus

$$
\begin{array}{r}
001001001, \&c. \\
563 \\
\hline
3003003003 \\
6006006006 \\
5005005005 \\
\hline
563563563563, \&c.=563×·001′.
\end{array}
$$

In the same way it could be shown that *any other* three digits divided by 999 would give a circulating decimal having these digits as a period.—And, consequently, a circulating decimal having any three digits as period will be equal to a vulgar fraction having the same digits for numerator, and 3 nines for denominator.

We might, in a similar way, show that *any* number of digits divided by an equal number of nines must give a circulate, each period of which would consist of those digits.—And, consequently, a circulate whose periods would consist of any digits must be equal to a vulgar fraction having one of its periods for numerator, and a number of nines equal to the number of digits in the period, for denominator.

76. If the decimal is a mixed repetend or a *mixed circulate*—

RULE.—Subtract the finite part from the whole, and set down the difference for numerator; put for denominator so many cyphers as there are digits in the *finite* part, and to the left of the cyphers so many nines as there are digits in the *infinite* part.

EXAMPLE.—What is the vulgar fraction equivalent to ·97'8734'?

There are 2 digits in 97, the finite part, and 4 in 8734, the infinite part. Therefore

$$\frac{978734 - 97}{999900} = \frac{978637}{999900},$$ is the required vulgar fraction.

77. REASON OF THE RULE.—If, for example, we multiply ·97'8734' by 100, the product is 97·8734 = 97 + 8734. This (by the last rule) is equal to $97 + \frac{8734}{9999}$, which (as we multiplied by 100) is one hundred times greater than the original quantity—but if we divide it by 100 we obtain $\frac{97}{100} + \frac{8734}{999900}$, which is *equal* the original quantity. To perform the addition of $\frac{97}{100}$ and $\frac{8734}{999900}$, we must [19 and 22] reduce them to a common denominator—when they become

$$\frac{97 \times 999900}{99990000} + \frac{873400}{99990000} = \frac{97 \times 9999}{999900} + \frac{8734}{999900} = \text{(since } 9999 =$$

$$10000 - 1)\ \frac{97 \times \overline{10000 - 1}}{999900} + \frac{8734}{999900} = \frac{97 \times 10000 - 97}{999900} + \frac{8734}{999900} =$$

$$\frac{970000 - 97}{999900} + \frac{8734}{999900} = \frac{978734 - 97}{999900} = \frac{978637}{999900},$$ which is exactly the result obtained by the rule. The same reasoning would hold with any other example.

EXERCISES.

1. ·5' = $\frac{5}{9}$.

2. ·8' = $\frac{8}{9}$.

3. ·73' = $\frac{73}{99}$.

4. ·145' = $\frac{145}{999}$.

5. ·057' = $\frac{57}{999}$.

6. ·45632' = $\frac{45632}{99999}$.

7. ·574' = $\frac{574}{999}$.

8. ·83'25' = $\frac{8242}{9900}$.

9. ·147'658' = $\frac{147511}{999000}$.

10. ·432'0075' = $\frac{4320643}{9999000}$.

11. 875·49'65' = $875\frac{4916}{9900}$.

12. 301·82'756' = $301\frac{82674}{99900}$.

78. Except where great accuracy is required, it is not necessary to reduce circulating decimals to their equivalent vulgar fractions, and we may add, and subtract them, &c., like other decimals—merely taking care to put down so many of them as will secure sufficient accuracy.

79. It may be here remarked, that no vulgar fraction will give a *finite* decimal if, when reduced to its lowest terms, the denominator contains any *prime* factors (factors that are prime numbers—and all the factors, can be reduced to such) except *twos* or *fives*. For neither 10, 100, 1000, &c., nor any multiples of these—as 30, 400, 5000, &c., nor the sum of any of their multiples—as 6420 (5000+400+20), &c., will exactly contain any prime numbers, but 2 or 5. Thus $\frac{3}{5}$ (considered as $\dfrac{30 \text{ tenths}}{5}$) *will* give an exact quotient; so also will $\frac{7}{2}$ (considered as $\dfrac{70 \text{ tenths}}{2}$). But $\frac{1}{7}$ *will not* give one; for $\frac{1}{7}$ (considered as $\dfrac{10 \text{ tenths}}{7}$, or $\dfrac{100 \text{ hundredths,}}{7}$ &c.) does not contain 7 exactly.

For a similar reason $\frac{4}{7}$ *will not* give an exact quotient; since $\frac{4}{7}$ (considered as $\dfrac{40 \text{ tenths,}}{7}$ or $\dfrac{400 \text{ hundredths,}}{7}$ &c.) does not exactly contain 7.

80. A finite decimal must have so many decimal places as will be equal to the greatest number of twos, or fives, contained as factors in the denominator of the original vulgar fraction, reduced to its lowest terms.

Thus $\frac{1}{2}$ will give one decimal place; for 2 (found *once* in its denominator) is contained in 10 (5×2); and therefore $\dfrac{10 \text{ tenths}}{2}$ ($=\frac{1}{2}$) will give some digit (in the tenths' place [Sec. II. 77]), that is, *one* decimal as quotient.

$\frac{3}{4}$ $\left(=\dfrac{3}{2\times2}\right)$ will give two decimal places; because 2 being found twice as a factor in its denominator, it will not be enough to consider the numerator as so many tenths; for $\dfrac{30 \text{ tenths}}{4}$ $(=\frac{3}{4})$ cannot give an exact quotient—30 being equal to $3 \times 2 \times 5$, which contains 2, but not 2×2. It will, however, be sufficient to reduce the numerator to hundredths; because $\dfrac{300 \text{ hundredths}}{4}$ *will* give an exact quotient—for 300 is equal to $3 \times 2 \times 2 \times 5 \times 5$, and consequently contains 2×2. But 300 *hundredths* divided by an integer will give *hundredths*— or *two* decimals as quotient. Hence, when there are *two* twos found as factors in the denominator of the vulgar fraction, there are also *two* decimal places in the quotient.

$\frac{6}{40}$ $\left(=\dfrac{6}{2 \times 2 \times 2 \times 5}\right)$ contains 2 repeated *three* times as a factor, in its denominator, and will give *three* decimal places. For though 10 tenths—and therefore 6×10 tenths—contains 5, one of the factors of 40, it does not contain $2 \times 2 \times 2$, the others; consequently it will not give an exact quotient.—Nor, for the same reason, will 6×100 hundredths. 6×1000 thousandths *will* give one—that is, $\dfrac{6 \times 1000 \text{ thousandths}}{40}$ $(=\frac{6}{40})$ will leave no remainder; for 6×1000 $(=6 \times 2 \times 2 \times 2 \times 5 \times 5 \times 5)$ contains $2 \times 2 \times 2 \times 5$. But 6×1000 *thousandths* divided by an integer will give *thousandths*—or *three* decimals as quotient. Hence, when there are *three* twos found as factors in the denominator of the vulgar fraction, there are also *three* decimal places in the quotient.

81. Were the *fives* to constitute the larger number of factors—as, for instance, in $\frac{4}{50}$ $\frac{6}{500}$, &c., the same reasoning would show that the number of decimal places would be equal to the number of *fives*.

It might also be proved, in the same way, that were the greatest number of twos or fives, in the denominator of the vulgar fraction, any *other* than one of those numbers given above, there would be an equal number of decimal places in the quotient.

82. A pure circulate will have so many digits in its period as will be equal to the least number of nines, which would represent a quantity measured by the denomina-

tor of the original vulgar fraction, reduced to its lowest terms. For we have seen [74] that such a circulate will be equal to a fraction having some period for its numerator, and some number of nines for its denominator—that is, it will be equal to some fraction, the numerator of which (the *period* of the circulate) will be *as many times* the numerator of the given vulgar fraction, as the quantity represented by the nines is of its denominator. For if a fraction having a given denominator is equal to another which has a larger, it is because the numerator of the latter is to the same amount larger than that of the former—in which case the increased size of the numerator counteracts the effect of the increased size of the denominator.   Thus $\frac{5}{6} = \frac{25}{30}$; because, if the numerator of $\frac{25}{30}$ is 5 times greater than that of $\frac{5}{6}$, the denominator of $\frac{25}{30}$, also, is five times greater than that of $\frac{5}{6}$.

Let the given fraction be $\frac{5}{13}$.   Since $\frac{5}{13} = \cdot 384615'$; and $\cdot 384615' = \frac{384615}{999999}$; $\frac{5}{13}$, also, is equal to $\frac{384615}{999999}$;—and, therefore, whatever multiple 384615 is of 5, 999999 is the same of 13.—But 999999 is the least multiple of 13, consisting of nines.   If not, let some other be less. Then take for numerator such a multiple of 5, as that lesser number of nines is of 13—and put that lesser number of nines for its denominator.   The numerator of this new fraction will [75] form the period of a circulate equal to the original fraction.   But as this new period is different from 384615 (the former one), the circulate of which it is an element, is also different from the former circulate; there are, therefore, two different circulates equal to $\frac{5}{13}$—that is two different values, or quotients for the same fraction—which is impossible.   Hence it is absurd to suppose that any *less* number of nines is a multiple of 13.

83. The periodical obtained does not contain a finite part, when neither 2 nor 5 is found in the denominator of the vulgar fraction—reduced to its lowest terms.

For [76] a finite part would add cyphers to the right hand of the nines in the denominator of the vulgar fraction, obtained from the circulate. But cyphers would suppose the denominator of the original fraction to contain twos, or fives—since no other prime factors could give cyphers in their multiple—the denominator of the vulgar fraction obtained from the circulate.

84. If there is a finite part in the decimal, it will contain as many digits as there are units in the greatest number of twos or fives found in the denominator of the original vulgar fraction, reduced to its lowest terms.

Let the original fraction be $\frac{5}{56}$. Since $56 = 2 \times 2 \times 2 \times 7$, the equivalent fraction must have as many nines as will just contain the 7 (cyphers would not *cause* a number of nines to be a multiple of 7), multiplied by as many tens as form a product which will just contain the twos as factors. But we have seen [80] that one ten (which adds one cypher to the nines) contains one *two,* or *five ;* that the product of two tens (which add two cyphers to the nines), contains the product of two *twos* or *fives ;* that the product of three tens (which add three cyphers to the nines), contains the product of three *twos* or *fives,* &c. That is, there will be so many cyphers in the denominator as will be equal to the greatest number of twos or fives, found among the factors in the denominator of the original vulgar fraction.

But as the digits of the finite part of the decimal add an equal number of cyphers to the denominator of the new vulgar fraction [76], the cyphers in the denominator, on the other hand, evidently suppose an equal number of places in the finite part of a circulate :—there will therefore be in the finite part of a circulate so many digits as will be equal to the greatest number of *twos* or *fives* found among the factors in the denominator of a vulgar fraction containing, also, *other* factors than 2 or 5.

85. It follows from what has been said, that there is no number which is not *exactly* contained in some quantity expressed by one or more nines, or by one or more nines followed by cyphers, or by unity followed by cyphers.

CONTRACTIONS IN MULTIPLICATION AND DIVISION
(derived from the properties of fractions.)

86. To multiply any number by 5—
RULE.—Remove it one place to the left hand, and divide the result by 2.

EXAMPLE.—$736 \times 5 = {}^{7}\frac{3\,6\,0}{2} = 3680$.

REASON.—$5 = \frac{1\,0}{2}$; therefore $736 \times 5 = 736 \times \frac{1\,0}{2} = {}^{7}\frac{3\,6\,0}{2} = 3680$.

87. To multiple by 25—

RULE.—Remove the quantity two places to the left, and divide by 4.

EXAMPLE.—$6732 \times 25 = {}^{6\,7}\frac{3\,2\,0\,0}{4} = 168300$.

REASON.—$25 = \frac{1\,0\,0}{4}$; therefore $6732 \times 25 = 6732 \times \frac{1\,0\,0}{4}$.

88. To multiply by 125—

RULE.—Remove the quantity three places to the left, and divide the result by 8.

EXAMPLE.—$7865 \times 125 = {}^{7\,8\,6}\frac{5\,0\,0\,0}{8} = 983125$.

REASON.—$125 = \frac{1\,0\,0\,0}{8}$; therefore $7865 \times 125 = 7865 \times \frac{1\,0\,0\,0}{8}$.

89. To multiply by 75—

RULE.—Remove the quantity two places to the left, then multiply the result by 3, and divide the product by 4.

EXAMPLE.—$685 \times 75 = \frac{68500 \times 3}{4} = \frac{205500}{4} = 51375$.

REASON.—$75 = \frac{3\,0\,0}{4} = 100 \times \frac{3}{4}$; therefore $685 \times 75 = 685 \times 100 \times \frac{3}{4}$.

90. To multiply by 35—

RULE.—To the multiplicand removed two places to the left and divided by 4, add the multiplicand removed one place to the left.

EXAMPLE 1.—$67896 \times 35 = {}^{6\,7\,8}\frac{9\,6\,0\,0}{4} + 678960 = 1697400 + 678960 = 2376360$.

REASON.—$35 = \frac{1\,0\,0}{4} + 10$; therefore $67896 \times 35 = 67896 \times \frac{1\,0\,0}{4} + 10$.

Many similar abbreviations will easily suggest themselves to both pupil and teacher.

91. To divide by any one of the above multipliers—

RULE.—Multiply by the equivalent fraction, inverted.

EXAMPLE.—Divide 847 by 5. $847 \div 5 = 847 \div \frac{1\,0}{2} = 847 \times \frac{2}{1\,0} = 169\cdot4$.

REASON.—We divide by any number when we divide by the fraction equivalent to it; but we divide by a fraction when we invert it, and then consider it as a multiplier [49].

92. Sometimes what is convenient as a multiplier will not be equally so as a divisor: thus 35. For it is not so easy to divide, as to multiply by $\frac{1\,0\,0}{4} + 10$, its equivalent mixed number.

QUESTIONS FOR THE PUPIL.

1. Show that a decimal fraction, and the corresponding decimal are not identical [59].

2. How is a decimal changed into a decimal fraction? [61].

3. Are the methods of adding, &c., vulgar and decimal fractions different? [62].

4. How is a vulgar reduced to a decimal fraction? [63].

5. How is a decimal reduced to a lower denomination? [64].

6. How are pounds, shillings, and pence changed, *at once*, into the corresponding decimal of a pound? [66, 67, and 68].

7. How is the decimal of a pound changed, *at once*, into shillings, pence, &c.? [70].

8. What are *terminate* and *circulating* decimals? [71].

9. What are a repetend and a periodical, a pure and a mixed circulate? [72].

10. Why cannot the number of digits in a period be equal to the number of units contained in the divisor? [73].

11. How is a pure circulate or pure repetend changed into an equivalent vulgar fraction? [74].

12. How is a mixed repetend or mixed circulate reduced to an equivalent vulgar fraction? [76].

13. What kind of vulgar fraction can produce no equivalent *finite* decimal? [79].

14. What number of decimal places must necessarily be found in a finite decimal? [80].

15. How many digits must be found in the periods of a *pure* circulate? [82].

16. When is no finite part found in a repetend, or circulate? [83].

17. How many digits must be found in the finite part of a *mixed* circulate? [84].

---

18. On what principal can we use the properties of fractions as a means of abbreviating the processes of multiplication and division? [86, &c.]

SECTION V.

PROPORTION.

1. The rule of Proportion is called also the *golden rule*, from its extensive utility ; in some cases it is termed the *rule of three*—because, by means of it, when three numbers are given, a fourth, which is unknown, may be found.

2. The rule of proportion is divided into the *simple*, and the *compound*. Sometimes also it is divided into the *direct*, and *inverse*—which is not accurate, as was shown by Hatton, in his arithmetic published nearly one hundred years ago.

3. The pupil, to have accurate ideas of the rule of proportion, must be acquainted with a few simple but important principles, connected with the nature of *ratios*, and the *doctrine* of proportion.

The following truths are self-evident :—

If the same, or equal quantities are *added* to equal quantities, the sums are equal. Thus, if we add the *same* quantity, 4 for instance, to 5×6 and 3×10, which are equal, we shall have $5 \times 6 + 4 = 3 \times 10 + 4$.

Or if we add *equal* quantities to those which are equal, the sums will be equal. Thus, since

$$5 \times 6 = 3 \times 10, \text{ and } 2 + 2 = 4$$
$$5 \times 6 + 2 \times 2 = 3 \times 10 + 4.$$

4. If the same, or equal quantities are *subtracted* from others which are equal, the remainders will be equal. Thus, if we subtract 3 from each of the equal quantities 7, and $5 + 2$, we shall have

$$7 - 3 = 5 + 2 - 3.$$

And since $8 = 6 + 2$, and $4 = 3 + 1$.

$$8 - 4 = 6 + 2 - \overline{3 + 1}.$$

5. If equal quantities are *multiplied* by the same, or by equal quantities, the products will be equal. Thus,

if we multiply the equals $5+6$, and $10+1$ by 3, we shall have

$$\overline{5+6}\times 3=\overline{10+1}\times 3.$$

And since $4+9=13$, and $3\times 6=18$.

$$\overline{4+9}\times 3\times 6=13\times 18.$$

6. If equal quantities are *divided* by the same, or by equal quantities, the quotients will be equal. Thus if we divide the equals 8 and $4+4$ by 2, we shall have

$$\frac{8}{2}=\frac{4+4}{2}$$

And since $20=17+3$, and $10=2\times 5$.

$$\frac{20}{10}=\frac{17+3}{2\times 5}.$$

7. *Ratio* is the relation which exists between two quantities, and is expressed by two dots (:) placed between them—thus $5:7$ (read, 5 is to 7); which means that 5 has a certain relation to 7. The former quantity is called the *antecedent*, and the latter the *consequent*.

8. If we invert the terms of a ratio, we shall have their *inverse ratio ;* thus $7:5$ is the inverse of $5:7$.

9. The relation between two quantities may consist in one being *greater* or *less* than the other— then the ratio is termed *arithmetical ;* or in one being some *multiple* or *part* of the other—and then it is *geometrical.*

If two quantities are equal, the ratio between them is said to be that of equality; if they are unequal it is a ratio of *greater inequality* when the antecedent is greater than the consequent, and of *lesser inequality* when it is less.

10. As the *arithmetical* ratio between two quantities is measured by their *difference,* so long as this difference is not altered, the ratio is unchanged. Thus the ratio of $7:5$ is equal to that $15:13$—for 2 is, in each case, the difference between the antecedent and consequent.

Hence we may *add* the same quantity to both the antecedent and consequent of an arithmetical ratio, or may *subtract* it from them, without changing the ratio. Thus $7:5$, $7+3:5+3$, and $7-2:5-2$, are *equal* arithmetical ratios.

But we cannot *multiply* or *divide* the terms of an arith

metical ratio by the same number. Thus $12 \times 2 : 10 \times 2$, $12 \div 2 : 10 \div 2$, and $12 : 10$ are not equal arithmetical ratios; for $12 \times 2 - 10 \times 2 = 4$, $12 \div 2 - 10 \div 2 = 1$, and $12 - 10 = 2$.

11. A geometrical ratio is measured by the quotient obtained if we divide its antecedent by its consequent;— therefore, so long as this quotient is unaltered the ratio is not changed. Hence ratios expressed by equal fractions are equal; thus $10 : 5 = 12 : 6$, for $\frac{10}{5} = \frac{12}{6}$.—Hence, also, we may *multiply* or *divide* both terms of a geometrical ratio by the same number without altering the ratio; thus $7 \times 2 : 14 \times 2 = 7 : 14$—because $\dfrac{7 \times 2}{14 \times 2} = \dfrac{7}{14}$.

But we cannot *add* the same quantity to both terms of a geometrical ratio, nor *subtract* it from them, without altering the ratio.

12. When the pupil [Sec. IV. 17] was taught how to express one quantity as the fraction of another, he in reality learned how to discover the geometrical ratio between the two quantities. Thus, to repeat the question formerly given, " What fraction of a pound is $2\frac{1}{4}d.$?"—which in reality means, " What *relation* is there between $2\frac{1}{4}d.$ and a pound;" or " What must we consider $2\frac{1}{4}d.$, if we consider a pound as unity ;" " or," in fine, " What is the value of $2\frac{1}{4} : 1$ "—

We have seen [Sec. I. 40] that the relation between quantities cannot be ascertained, unless they are made to have the same " unit of comparison:" but a farthing is the only unit of comparison which can be applied to *both* $2\frac{1}{4}d.$ and £1 ; we must therefore reduce them to farthings—when the ratio of one to the other will become that of $9 : 960$. But we have also seen that a geometrical ratio is not altered, if we divide both its terms by the same number; therefore $9 : 960$ is the same ratio as $\frac{9}{960} : \frac{960}{960}$, or $\frac{9}{960} : 1$.—That is, the ratio between $2\frac{1}{4}d.$ and £1 may be expressed by $2\frac{1}{4}d. : $ £1, or $9 : 960$, or $\frac{9}{960} : 1$; or, the pound being considered as unity, the farthing will be represented by $\frac{9}{960}$.

13. The geometrical ratio between two numbers is the same as that which exists between the quotient of the fraction which represents their ratio, and unity. Thus,

in the last example $9 : 960$ and $\frac{9}{960} : 1$ are equal ratios.
It is not necessary that we should be able to express by
integers, nor even by a finite decimal, what part or multiple one of the terms is of the other ; for a geometrical
ratio may be considered to exist between *any* two quantities. Thus, if the ratio is $10 : 2$, 5 $(\frac{10}{2})$ is the quantity
by which we must multiply one term to make it equal
to the other ; if $1 : 2$, it is $0\cdot5$ $(\frac{1}{2})$, a *finite* decimal ; but
if $3 : 7$, it is '428571' $(\frac{3}{7})$, an *infinite* decimal—in which
case we obtain only an approximation to the value of
the ratio. But though the measure of the ratio is expressed by an *infinite* decimal, when there is no quantity
which will *exactly* serve as the multiplier, or divisor of
one quantity so as to make it equal to the other—since
we may obtain as near an approximation as we please—
there is no inconvenience in supposing that *any* one
number is some part or multiple of any other ; that is,
that any number may be expressed in terms of another—
or may form one term of a geometrical ratio, unity
being the other.

14. *Proportion*, or *analogy*, consists in the equality
of ratios, and is indicated by putting $\doteq$, or $::$, between
the equal ratios; thus $5 : 7 \doteq 9 : 11$, or $5 : 7 :: 9 : 11$ (read,
5 is to 7 as 9 : 11), means that the two ratios $5 : 7$ and
$9 : 11$ are equal; or that 5 bears the same relation to 7
that 9 does to 11. Sometimes we express the equality
of more than two ratios; thus $4 : 8 :: 6 : 12 :: 18 : 36$,
(read, 4 is to 8, as 6 is to 12, as 18 is to 36), means
there is the same relation between 4 and 8, as between
6 and 12; and between 18 and 36, as between either 4
and 8, or 6 and 12—it follows that $4 : 8 :: 18 : 36$—for
two ratios which are equal to the same, are equal to
each other. When the equal ratios are arithmetical, they
constitute an *arithmetical* proportion ; when geometrical, a *geometrical* proportion.

15. The quantities which form the proportion are
called *proportionals ;* and a quantity that, along with
three others, constitutes a proportion, is called a *fourth
proportional* to those others. In a proportion, the two
outside terms are called the *extremes*, and the two middle
terms the *means*; thus in $5 : 6 :: 7 : 8$, 5 and 8 are the

extremes, 6 and 7 the means. When the same quantity is found in *both* means, it is called *the* mean of the extremes; thus, since $5 : 6 :: 6 : 7$, 6 is *the mean* of 5 and 7. When the proportion is arithmetical, *the mean* of two quantities is called their *arithmetical* mean; when the proportion is geometrical, it is termed their *geometrical* mean. Thus 7 is the arithmetical mean of 4 and 10; for, since $7-4=10-7, 4: 7 :: 7 : 10$. And 8 is the geometrical mean of 2 and 32; for, since $\frac{2}{8}=\frac{8}{32}$, $2 : 8 :: 8 : 32$.

16. In an *arithmetical* proportion, "the *sum* of the means is equal to the *sum* of the extremes." Thus, since $11 : 9 :: 17 : 15$ is an arithmetical proportion, $11-9= 17-15$; but, adding 9 to both the equal quantities, we have $11-9+9=17-15+9$ [3]; and, adding 15 to these, we have $11-9+9+15=17-15+9+15$; but $11-9+9+15$ is equal to $11+15$—since 9 to be subtracted and 9 to be added $=0$; and $17-15+9+15= 17+9$—since 15 to be subtracted and 15 to be added $=0$: therefore $11+15$ (the sum of the extremes) $=17+9$ (the sum of the means).—The same thing might be proved from *any other* arithmetical proportion; and, therefore, it is true in every case.

17. This *equation* (as it is called), or the equality which exists between the sum of the means and the sum of the extremes, is the *test* of an arithmetical proportion:—that is, it shows us whether, or not, four given quantities constitute an arithmetical proportion. It also enables us to find a fourth arithmetical proportional to three given numbers—since any mean is evidently the difference between the sum of the extremes and the other mean; and any extreme, the difference between the sum of the means and the other extreme—

For if $4 : 7 :: 8 : 11$ be the arithmetical proportion, $4+11=7+8$ [16]; and, subtracting 4 from the equals, we have 11 (one of the extremes) $=7+8-4$ (the sum of the means, minus the other extreme); and, subtracting 7, we have $4+11-7$ (the sum of the extremes minus one of the means) $=8$ (the other mean). We might in the same way find the remaining extreme, or the remaining mean. *Any other* arithmetical proportion would have answered just as well—hence what we have said is true in all cases.

18. EXAMPLE.—Find a fourth proportional to 7, 8, 5.

Making the required number one of the extremes, and putting the note of interrogation in the place of it, we have $7 : 8 :: 5 : ?$; then $7 : 8 :: 5 : 8+5-7$ (the sum of the means minus the given extreme, $=6$); and the proportion completed will be

$$7 : 8 :: 5 : 6.$$

Making the required number one of the means, we shall have $7 : 8 :: ? : 5$, then $7 : 8 :: 7+5-8$ (the sum of the extremes minus the given mean, $=4$) $: 5$; and the proportion completed will be

$$7 : 8 :: 4 : 5.$$

As the sum of the means will be found equal to the sum of the extremes, we have, in each case, completed the proportion.

19. The *arithmetical mean* of two quantities is half the sum of the extremes. For the sum of the means is equal to the sum of the extremes; or—since the means are equal—twice *one* of the means is equal to the sum of the extremes; consequently, half the sum of the means—or *one* of them, will be equal to half the sum of the extremes. Thus the arithmetical mean of 19 and 27 is $\dfrac{19+27}{2}$ $(=23)$; and the proportion completed is

$$19 : 23 :: 23 : 27, \text{ for } 19+27=23+23.$$

20. If with any four quantities the sum of the means is equal to the sum of the extremes, these quantities are in arithmetical proportion. Let the quantities be

$$8 \quad 6 \quad 7 \quad 5.$$

As the sum of the means is equal to the sum of the extremes

$$8+5=6+7.$$

Subtracting 6 from each of the equal quantities, we have $8+5-6=6+7-6$; and subtracting 5 from each of these, we have $8+5-6-5=6+7-6-5$. But $8+5-6-5$ is equal to $8-6$, since 5 to be added and 5 to be subtracted are $=0$; and $+6+7-6-5=7-5$, since 6 to be added and 6 to be subtracted $=0$;

therefore $8+5-6-5=6+7-6-5$ is the same as $8-6=7-5$; but if $8-6=7-5$, $8:6$ and $7:5$, are two equal arithmetical ratios; and if they are two *equal* arithmetical ratios, they constitute an arithmetical proportion. It might in the same way be proved that any *other* four quantities are in arithmetical proportion, if the sum of the means is equal to the sum of the extremes.

21. In a *geometrical* proportion, "the *product* of the means is equal to the *product* of the extremes." Thus, since $14:7::16:8$ is a geometrical proportion, $\frac{14}{7}=\frac{16}{8}$ [11]; but, multiplying each of the equal quantities by 7, we have 14 ($\frac{14}{7}\times7$)$=\frac{16}{8}\times7$; and multiplying each of these by 8, we have $14\times8=16\times7$ ($\frac{16}{8}\times7\times8$):— but 14×8 is the product of the extremes; and 16×7 is the product of the means. The same reasoning would hold with *any other* geometrical proportion, and therefore it is true in all cases.

22. This *equation* (as it is called), or the equality of the product of the means and the product of the extremes, is the test of a *geometrical* proportion: that is, it shows us whether or not four given quantities constitute a geometrical proportion. It also enables us to find a fourth geometrical proportional to three given quantities—which is the object of the *rule of three;* since any mean is, evidently, the quotient of the product of the extremes divided by the other mean; and any extreme, is the quotient of the product of the means divided by the other extreme.

For if $7:14::11:22$ be the geometrical proportion, $7\times22=14\times11$; and, dividing the equals by 7, we have 22 (one of the extremes) $=\dfrac{14\times11}{11}$ (the product of the means divided by the other extreme); and, dividing these by 11, we have $\dfrac{7\times22}{11}$ (the product of the extremes divided by one mean) $=14$ (the other mean). We might in the same way find the remaining mean or the remaining extreme. *Any other* proportion would have answered just as well—and therefore what we have said is true in every case.

23. EXAMPLE.—Find a fourth proportional to 8, 10, and 14.

Making the required quantity one of the extremes, we shall have $8 : 10 :: 14 : ?$; and $8 : 10 :: 14 : \dfrac{10 \times 14}{8}$ (the product of the means divided by the given extreme, $= 17 \cdot 5$).

And the proportion completed will be

$$8 : 10 :: 14 : 17 \cdot 5.$$

Making the required number one of the means, we shall have $8 : 10 :: ? : 14$; and $8 : 10 :: \dfrac{8 \times 14}{10}$ (the product of the extremes divided by the *given* mean, $= 11 \cdot 2$) $: 14$.

And the proportion completed will be

$$8 : 10 :: 11 \cdot 2 : 14.$$

EXERCISES.

Find fourth proportionals

1.	To	3,	6,	and	12	.	*Ans.*	24.
2.	,,	6,	8	,,	3	.	.	4.
3.	,,	3,	6	,,	8	:	.	16.
4.	,,	6,	12	,,	4	.	.	8.
5.	,,	10,	150	,,	68	.	.	1020.
6.	,,	1020,	68	,,	150	.	.	10.
7.	,,	150,	10	,,	1020	.	.	68.
8.	,,	68,	1020	,,	10	.	.	150.

24. If with any four quantities the product of the means is equal to the product of the extremes, these quantities are in geometrical proportion. Let the quantities be

$$5 \quad 20 \quad 6 \quad 24,$$

As the product of the means is equal to the product of the extremes,

$$5 \times 24 = 20 \times 6.$$

Dividing the equals by 24, we have $\dfrac{5 \times \cancel{24}}{\cancel{24}} = \dfrac{20 \times 6}{24}$;

and, dividing these by 20, we have $\dfrac{5 \times \cancel{24}}{\cancel{20} \times \cancel{24}} = \dfrac{20 \times 6}{20 \times 24}$.

But $\dfrac{5 \times 24}{20 \times 24} = \dfrac{5}{20}$; and $\dfrac{20 \times 6}{20 \times 24} = \dfrac{6}{24}$; therefore $\dfrac{5}{20} = \dfrac{6}{24}$; consequently the geometrical relation between 5 and 20 is the same as that between 6 and 24; hence there are two equal geometrical ratios—or a geometrical proportion. It might, in the same way, be proved that *any other* four quantities are in geometrical proportion, if the product of the means is equal to the product of the extremes.

25. When the first term is unity, to find a fourth proportional—

RULE.—Find the product of the second and third.

EXAMPLE.—What is the fourth proportional to 1, 12, and 27 ?

$$1 : 12 :: 27 : 12 \times 27 = 324$$

We are to divide the product of the means by the given extreme; but we may neglect the divisor when it is unity—since dividing a number by unity does not alter it.

EXERCISES.

Find fourth proportionals

9.	To	1,	17, and	8	.	*Ans.*	136.
10.	,,	1,	23 ,,	20	.	.	460.
11.	,,	1,	100 ,,	73	.	.	7300.
12.	,,	1,	53 ,,	110	.	.	5830.
13.	,,	1,	15 ,,	1234	.	.	18510.

26. When either the second, or third term is unity—

RULE.—Divide that one of them which is not unity, by the first.

EXAMPLE.—Find a fourth proportional to 8, 1, and 5.

$$8 : 1 :: 5 : \tfrac{5}{8}.$$

We are to divide the product of the means by the given extreme; but one of the means may be considered as the product of both, when the other is unity. For, since multiplication by unity produces no effect, it may be omitted.

EXERCISES.

Find fourth proportionals

14.	To	5,	20, and	1	.	*Ans.*	4.
15.	,,	5,	1 ,,	20	.	.	4.
16.	,,	7,	21 ,,	1	.	.	3.
17.	,,	8,	24 ,,	1	.	.	3.
18.	,,	6,	1 ,,	50	.	.	$8\tfrac{1}{3}$.
19.	,,	17,	1 ,,	68	.	.	4.
20.	,,	200,	1000 ,,	1	.	.	5.
21.	,,	200,	1 ,,	1000	.	.	5.

27. When the means are equal, each is said to be *the* geometrical mean of the extremes; and the product

of the extremes is equal to *the mean* multiplied by itself. Hence, to discover the *geometrical mean* of two quantities, we have only to find some number which, multiplied by itself, will be equal to their product—that is, to find, what we shall term hereafter, the *square root* of their product. Thus 6 is the geometrical mean of 3 and 12; for $6 \times 6 = 3 \times 12$. And $3 : 6 :: 6 : 12$.

28. It will be useful to make the pupil acquainted with the following properties of a geometrical proportion—

We may consider the same quantity either as a mean, or an extreme. Thus, if $5 : 10 :: 15 : 30$ be a geometrical proportion, so also will $10 : 5 :: 30 : 15$; for we obtain the same equal products in both cases—in the former, $5 \times 30 = 10 \times 15$; and in the latter, $10 \times 15 = 5 \times 30$—which are the same thing. This change in the proportion is called *inversion.*

29. The product of the means will continue equal to the product of the extremes—or, in other words, the proportion will remain unchanged—

If we *alternate* the terms; that is, if we say, "the first is to the third, as the second is to the fourth"—

If we " *multiply,* or *divide* the first and second, or the first and third terms, by the same quantity"—

If we " read the proportion *backwards*"—

If we say " the first term plus the second is to the second, as the third plus the fourth is to the fourth"—

If we say " the first term plus the second is to the first, as the third plus the fourth is to the third"—&c.

RULE OF SIMPLE PROPORTION.

30. This rule, as we have said, enables us, when three quantities are given, to find a fourth proportional.

The only difficulty consists in *stating* the question; when this is done, the required term is easily found.

In the rule of *simple* proportion, *two* ratios are given, the one perfect, and the other imperfect.

31. RULE—I. Put that given quantity which belongs to the *imperfect* ratio in the third place.

II. If it appears from the nature of the question that the *required* quantity must be greater than the other,

or given term of the same ratio, put the larger term
of the *perfect* ratio in the second, and the smaller in
the first place. But if it appears that the required
quantity must be less, put the larger term of the *perfect*
ratio in the first, and the smaller in the second place.

III. Multiply the second and third terms together,
and divide the product by the first.—The answer will
be of the same kind as the third term.

32. EXAMPLE 1.—If 5 men build 10 yards of a wall in one
day, how many yards would 21 men build in the same time?
. It will facilitate the stating, if the pupil puts down the
question briefly, as follows—using a note of interrogation to
represent the required quantity—

> 5 men.
> 10 yards.
> 21 men.
> ? yards.

10 yards is the *given* term of the *imperfect* ratio—it must,
therefore, be put in the third place.

5 men, and 21 men are the quantities which form the
perfect ratio; and, as 21 will build a greater number of yards
than 5 men, the required number of yards will be greater
than the *given* number—hence, in this case, we put the larger
term of the perfect ratio in the second, and the smaller in
the first place—

$$5 : 21 :: 10 : ?$$

And, completing the proportion,

$$5 : 21 :: 10 : \frac{21 \times 10}{5} = 42, \text{ the required number.}$$

Therefore, if 5 men build 10 yards in a day, 21 men will
build 42 yards in the same time.

33. EXAMLPE 2.—If a certain quantity of bread is sufficient
to last 3 men for 2 days; for how long a time ought it to
last 5 men? This is set down briefly as follows:

> 3 men.
> 2 days.
> 5 men.
> ? days.

2 days is the *given* term of the *imperfect* ratio—it must,
therefore, be put in the third place.

The larger the number of men, the shorter the time a given
quantity of bread will last them; but this *shorter* time is the

required quantity—hence, in this case, the greater term of the perfect ratio is to be put in the first, and the smaller in the second place—

$$5 : 3 :: 2 : ?$$

And, completing the proportion,

$$5 : 3 :: 2 : \frac{3 \times 2}{5} = 1\tfrac{1}{5}, \text{ the required term.}$$

34. EXAMPLE 3.—If 25 tons of coal cost £21, what will be the price of 1 ton?

$$25 : 1 :: 21 : \frac{1 \times 21}{25} \text{ pounds } £\frac{21}{25} = 16s. \ 9\tfrac{1}{2}d.$$

It is necessary in this case to reduce the pounds to lower denominations, in order to divide them by 25; this causes the answer, also, to be of *different* denominations.

35. REASON OR I.—It is convenient to make the required quantity the fourth term of the proportion—that is, one of the extremes. It could, however, be found equally well, if considered as a mean [23].

REASON OR II.—It is also convenient to make quantities of the same kind the terms of the same ratio; because, for instance, we can compare men with men, and days with days—but we cannot compare *men* with *days*. Still there is nothing inaccurate in comparing the *number* of one, with the *number* of the other; nor in comparing the *number of men* with the *quantity of work* they perform, or with the *number of loaves* they eat; for these things are proportioned to each other. Hence we shall obtain the same result whether we state example 2, thus

$$5 : 3 :: 2 : ?$$
$$\text{or thus } 5 : 2 :: 3 : ?$$

When diminishing the *kind* of quantity which is in the perfect ratio increases *that kind* which is in the imperfect—or the reverse—the question is sometimes said to belong to the *inverse* rule of three; and different methods are given for the solution of the two species of questions. But Hatton, in his Arithmetic, (third edition, London, 1753,) suggests the above general mode of solution. It is not accurate to say "the *inverse* rule of three" or "*inverse* rule of proportion;" since, although there is an inverse *ratio*, there is no inverse *proportion*.

REASON OR III.—We multiply the second and third terms, and divide their product by the first, for reasons already given [22].

The answer is of the same kind as the third term, since neither the multiplication, nor the division of this term has changed its nature;—20s. the payment of 5 days divided by 5 gives $\dfrac{20s.}{5}$ as the payment of one day; and $\dfrac{20s.}{5}$, the payment of one day multiplied by 9 gives $\dfrac{20s.}{5} \times 9$ as the payment of 9 days.

If the fourth term were not of the same kind as the third, it would not complete the *imperfect* ratio, and therefore it would not be the required *fourth proportional*.

36. It will often be convenient to divide the first and second, or first and third terms, by their greatest common measure, when these terms are composite to each other [29].

EXAMPLE.—If 36 cwt. cost £24, what will 27 cwt. cost?

$$36 : 27 :: 24 : ?$$

Dividing the first and second by 9 we have

$$4 : 3 :: 24 : ?$$

And, dividing the first and third by 4,

$$1 : 3 :: 6 : 3 \times 6 = £18.$$

EXERCISES FOR THE PUPIL.

Find a fourth proportional to

1. 5 pieces of cloth : 50 pieces :: £27. *Ans.* £270.
2. 1 cwt. : 215 cwt. :: 50s. *Ans.* 10750s.
3. 10 ℔ : 150℔ :: 5s. *Ans.* 75s.
4. 6 yards : 1 yard :: 27s. *Ans.* 4s. 6d.
5. 9 yards : 36 yards :: 18s. *Ans.* 72s.
6. 5 ℔ : 1 ℔ :: 15s. *Ans.* 3s.
7. 4 yards : 18 yards :: 1s. *Ans.* 4s. 6d.

---

8. What will 17 tons of tallow come to at £25 per ton? *Ans.* £425.

9. If one piece of cloth cost £27, how much will 50 pieces cost? *Ans.* £1350.

10. If a certain quantity of provisions would last 40 men for 10 months, how long would they suffice for 32? *Ans.* 12½ months.

11. What will 215 cwt. of madder cost at 50s. per cwt.? *Ans.* 10750s.

12. I desire to have 30 yards of cloth 2 yards wide, with baize 3 yards in breadth to line it, how much of the latter shall I require? *Ans.* 20 yards.

13. At 10*s*. per barrel, what will be the price of 130 barrels of barley ? *Ans.* £65.

14. At 5*s*. per ℔, what will be the price of 150 ℔ of tea ? *Ans.* 750*s*.

15. A merchant agreed with a carrier to bring 12 cwt. of goods 70 miles for 13 crowns, but his waggon being heavily laden, he was obliged to unload 2 cwt. ; how far should he carry the remainder for the same money ? *Ans.* 84 miles.

16. What will 150 cwt. of butter cost at £3 per cwt. ? *Ans.* £450.

17. If I lend a person £400 for 7 months, how much ought he to lend me for 12 ? *Ans.* £233 6*s*. 8*d*.

18. How much will a person walk in 70 days at the rate of 30 miles per day ? *Ans.* 2100.

19. If I spend £4 in one week, how much will I spend in 52 ? *Ans.* £208.

20. There are provisions in a town sufficient to support 4000 soldiers for 3 months, how many must be sent away to make them last 8 months ? *Ans.* 2500.

21. What is the rent of 167 acres at £2 per acre ? *Ans.* £334.

22. If a person travelling 13 hours per day would finish a journey in 8 days, in what time will he accomplish it at the .rate of 15 hours per day ? *Ans.* 6$\frac{14}{15}$ days.

23. What is the cost of 256 gallons of brandy at 12*s*. per gallon ? *Ans.* 3072*s*.

24. What will 156 yards of cloth come to, at £2 per yard ? *Ans.* £312.

25. If one pound of sugar cost 8*d*., what will 112 pounds come to ? *Ans.* 896*d*.

26. If 136 masons can build a fort in 28 days, how many men would be required to finish it in 8 days ? *Ans.* 476.

27. If one yard of calico cost 6*d*., what will 56 yards come to ? *Ans.* 336*d*.

28. What will be the price of 256 yards of tape at 2*d*. per yard ? *Ans.* 512*d*.

29. If £100 produces me £6 interest in 365 days, what would bring the same amount in 30 days ? *Ans.* £1216 13*s*. 4*d*.

30. What shall I receive for 157 pair of gloves, at 10*d*. per pair ? *Ans*. 1570*d*.

31. What would 29 pair of shoes come to, at 9*s*. per pair ? *Ans*. 261*s*.

32. If a farmer lend his neighbour a cart horse which draws 15 cwt. for 30 days, how long should he have a horse in return which draws 20 cwt. ? *Ans*. 22½ days.

33. What sum put to interest at £6 per cent. would give £6 in one month ? *Ans*. £1200.

34. If I lend £400 for 12 months, how long ought £150 be lent to me, to return the kindness ? *Ans*. 32 months.

35. Provisions in a garrison are found sufficient to last 10,000 soldiers for 6 months, but it is resolved to add as many men as would cause them to be consumed in 2 months ; what number of men must be sent in ? *Ans*. 20,000.

36. If 8 horses subsist on a certain quantity of hay for 2 months, how long will it last 12 horses ? *Ans*. 1⅓ months.

37. A shopkeeper is so dishonest as to use a weight of 14 for one of 16 oz. ; how many pounds of just will be equal to 120 of unjust weight ? *Ans*. 105 ℔.

38. A meadow was to be mowed by 40 men in 10 days ; in how many would it be finished by 30 men ? *Ans*. 13⅓ days.

37. When the first and second terms of the proportion are not of the same denomination ; or one, or both of them contain different denominations—

RULE.—Reduce both to the lowest denomination contained in either, and then divide the product of the second and third by the first term.

EXAMPLE 1.—If three ounces of tea cost 15*d*. what will 87 pounds cost ?
The lowest denomination contained in either is ounces.

$$
\begin{array}{ccc}
\text{oz.} & \text{℔} & d. \\
3 : & 87 & :: \quad 15 : \quad \dfrac{1392 \times 15}{3} \overset{d.}{=} 6960 = £29. \\
& 16 &
\end{array}
$$

—————
1392 ounces.

There is evidently the same ratio between 3 oz. and 87 ℔, as between 3 oz. and 1392 oz. (the equal of 87 ℔).

EXAMPLE 2.—If 3 yards of any thing cost 4s. 9¾d., what can be bought for £2?

The lowest denomination in either is farthings.

$$
\begin{array}{cc}
s. \quad d. & £ \\
4 \quad 9\tfrac{3}{4} \;\; : & 2 \;::\; 3 \;:\; \dfrac{1920 \times 3}{231} \overset{\text{yds.}}{=} 24 \overset{\text{q. nls.}}{3\;\;3.} \\
12 & 20 \\
\hline
57 \text{ pence.} & 40 \text{ shillings.} \\
4 & 12 \\
\hline
231 \text{ farthings.} & 480 \text{ pence.} \\
& 4 \\
& \hline \\
& 1920 \text{ farthings.}
\end{array}
$$

There is evidently the same ratio between 4s. 9¾d. and £2, as between the numbers of farthings they contain, respectively. For there is the same ratio between any two quantities, as between two others which are equal to them.

EXAMPLE 3.—If 4 cwt., 3 qrs., 17 ℔, cost £19, how much will 7 cwt. 2 qrs. cost?

The lowest denomination in either is pounds.

$$
\begin{array}{ccc}
\text{cwt. qr. } ℔ & \text{cwt. qr.} & £ \\
4 \quad 3 \quad 17 \;:\; & 7 \quad 2 \;::\; & 19 \;:\; \dfrac{840 \times 19}{549} = £29 \; 1s. \; 5d. \\
4 & 4 & \\
\hline
19 \text{ qrs.} & 30 \text{ qrs.} & \\
28 & 28 & \\
\hline
549 \text{ ℔bs.} & 840 \text{ ℔bs.} &
\end{array}
$$

EXERCISES.

Find fourth proportionals to

39. 1 cwt. : 17 tons :: £5. Ans. £1700.
40. 5s. : £20 :: 1 yard. Ans. 80 yards.
41. 80 yards : 1 qr. :: 400s. Ans. 1s. 3d.
42. 3s. 4d. : £1 10s. :: 1 yard. Ans. 9 yards.
43. 3 cwt. 2 qrs. : 8 cwt. 1 qr. :: £2. Ans. £4.
44. 10 acres, 3 roods, 20 perches : 21 acres 3 roods :: £60. Ans. £120.
45. 10 tons, 5 cwt., 3 qrs̄., 14 ℔ : 20 tons, 11 cwt., 3 qrs. :: £840. Ans. £1680.

46. What is the price of 31 tuns of wine, at £18 per hhd. *Ans.* £2232.

47. If 1 ounce of spice costs 4*s.*, what will be the price of 16 ℔ ? *Ans.* £51 4*s.*

48. What is the price of 17 tons of butter, at £5 per cwt. ? *Ans.* £1700.

49. If an ounce of silk costs 4*d.*, what will be the price of 15 ℔ ? *Ans.* £4.

50. What will 224 ℔ 6 oz. of spice come to, at 3*s.* per oz. ? *Ans.* £538 10*s.*

51. How much will 12 ℔ 10 oz. of silver come to, at 5*s.* per oz. ? *Ans.* £38 10*s.*

52. What will 156 cwt. 2 qrs. come to, at 7*d.* per ℔ ? *Ans.* £511 4*s.* 8*d.*

53. What will 56 cwt. 2 qrs. cost at 10*s.* 6*d.* per qr. ? *Ans.* £118 13*s.*

54. If 1 yard of cloth costs £1 5*s.*, what will 110 yards, 2 qrs., and 3 nails, come to ? *Ans.* £138 7*s.* 2¼*d.*

55. If 1 cwt. of butter costs £6 6*s.*, how much will 17 cwt., 2 qrs., 7 ℔, cost ? *Ans.* £110 12*s.* 10½*d.*

56. At 15*s.* per cwt., what can I have for £615 15*s.* ? *Ans.* 821 cwt.

57. How much beef can be bought for £760 12*s.*, at 32*s.* per cwt. *Ans.* 475 cwt., 1 qr., 14 ℔.

58. If 12 ℔, 6 oz., 4 dwt., cost £150, what will 3 ℔, 1 oz., 11 dwt., cost ? *Ans.* £37 10*s.*

59. If 10 yards cost 17*s.*, what will 3 yards, 2 qrs. cost ? *Ans.* 5*s.* 11¼*d.*

60. If 12 cwt. 22 ℔ cost £19, what will 2 cwt. 3 qrs. cost ? *Ans.* £4 5*s.* 8¼*d.*

61. If 15 oz., 12 dwt., 16 grs., cost 19*s.*, what will 13 oz. 14 grs. cost ? *Ans.* 15*s.* 10*d.*

38. If the third term consists of more than one denomination—

RULE.—Reduce it to the lowest denomination which it contains, then multiply it by the second, and divide the product by the first term.—The answer will be of that denomination to which the third has been reduced, and may sometimes be changed to a higher [Sec. III. 5].

EXAMPLE 1.—If 3 yards cost 9s. 2¼d., what will 327 yards cost?

The lowest denomination in the third term is farthings.

$$\begin{array}{cccc} \text{yds.} & \text{yds.} & s. & d. \\ 3 & : 327 & :: 9 & 2\frac{1}{4} \end{array} : \frac{327 \times 441}{3} \text{farthings} = 50 \quad 1 \quad 5\frac{1}{4}.$$

$$\begin{array}{c} 12 \\ \hline 110 \text{ pence.} \\ 4 \\ \hline 441 \text{ farthings.} \end{array}$$

EXAMPLE 2.—If 2 yards 3 qrs. cost 11¼d., what will 27 yards, 2 qrs., 2 nails, cost?

The lowest denomination in the first and second is nails, and in the third farthings.

$$\begin{array}{ccccccc} \text{yds.} & \text{qr.} & \text{yds.} & \text{qr.} & \text{n.} & d. \\ 2 & 3 & : 27 & 2 & 2 & :: 11\frac{1}{4} \end{array} : \frac{442 \times 45}{44} \text{ farthings} = 9s.\ 5d.$$

$$\begin{array}{ccc} 4 & 4 & 4 \\ \hline 11 \text{ qr.} & 110 \text{ qr.} & 45 \text{ farthings.} \\ 4 & 4 & \\ \hline 44 \text{ nails.} & 442 \text{ nails.} \end{array}$$

Reducing the third term generally enables us to perform the required multiplication and division with more facility.—It is sometimes, however, unnecessary.

EXAMPLE.—If 3 ℔ cost £3 11s. 4¾d., what will 96 ℔ cost?

$$\begin{array}{ccccccc} \text{℔} & \text{℔} & £ & s. & d. & £ & s. & d. \\ 3 & : 96 & :: 3 & 11 & 4\frac{3}{4} \end{array} : \frac{3 \quad 11 \quad 4\frac{3}{4} \times 96}{3} = 3 \quad 11 \quad 4\frac{3}{4} \times 32 = 114 \quad 4 \quad \dashv$$

EXERCISES.

Find fourth proportionals to

62. 2 tons : 14 tons :: £28 10s.   *Ans.* £199 10s.
63. 1 cwt. : 120 cwt. :: 18s. 6d.   *Ans.* £111.
64. 5 barrels : 100 barrels :: 6s. 7d. *Ans.* £6 11s. 8d.
65. 112 ℔ : 1 ℔ :: £3 10s.   *Ans.* 7½d.
66. 4 ℔ : 112 ℔ :: 5¼d.   *Ans.* 12s. 3d.
67. 7 cwt., 3 qrs., 11 ℔ : 172 cwt., 2 qrs., 18 ℔ :: £3 19s. 4½d.   *Ans.* £87 5s. 4d.

68. 172 cwt., 2 qrs., 18 ℔ : 7 cwt., 3 qrs., 11 ℔ :: £87 6s. 3d.   *Ans.* £3 19s. 4½d.
69. 17 cwt., 2 qrs., 14 ℔ : 2 cwt., 3 qrs., 21 ℔ :: £73. *Ans.* £12 3s. 4d.
70. £87 6s. 3d. : £3 19s. 4½d. :: 172 cwt., 2 qrs., 18 ℔. *Ans.* 7 cwt., 3 qrs., 11 ℔.
71. £3 19s. 4½d. : £87 6s. 3d. :: 7 cwt., 3 qrs., 11 ℔. *Ans.* 172 cwt., 2 qrs., 18 ℔.

---

72. At 18s. 6d. per cwt., what will 120 cwt. cost? *Ans.* £111.
73. At 3¼d. per pound, what will 112 ℔ come to? *Ans.* £1 10s. 4d.
74. What will 120 acres of land come to, at 14s. 6d. per acre?   *Ans.* £87.
75. How much would 324 pieces come to, at 2s. 8½d. per piece?   *Ans.* £43 17s. 6d.
76. What is the price of 132 yards of cloth, at 16s. 4d. per yard?   *Ans.* £107 16s.
77. If 1 ounce of spice costs 3s. 4d., what will 18 ℔ 10 oz. cost?   *Ans.* £49 13s. 4d.
78. If 1 ℔ costs 6s. 8d., what will 2 cwt. 3 qr. come to?   *Ans.* £102 13s. 4d.
79. If £1 2s. be the rent of 1 rood, what will be the rent of 156 acres 3 roods?   *Ans.* £689 14s.
80. At 10s. 6d. per qr., what will 56 cwt. 2 qrs. be worth?   *Ans.* £118 13s.
81. At 15s. 6d. per yard, what will 76 yards 3 qrs. come to?   *Ans.* £59 9s. 7½d.
82. What will 76 cwt. 8 ℔ come to, at 2s. 6d. per ℔?   *Ans.* £1065.
83. At 14s. 4d. per cwt., what will be the cost of 12 cwt. 2 qrs.?   *Ans.* £8 19s. 2d.
84. How much will 17 cwt. 2 qrs. come to, at 19s. 10d. per cwt.   *Ans.* £17 7s. 1d.
85. If 1 cwt. of butter costs £6 6s., what will 17 cwt., 2 qrs., 7 ℔, come to?   *Ans.* £102 12s. 10½d.
86. If 1 qr. 14 ℔ cost £2 15s. 9d., what will be the cost of 50 cwt., 3 qrs., 24 ℔?   *Ans.* £378 16s. 8¼d.

87. If the shilling loaf weigh 3 ℔ 6 oz., when flour sells at £1 13s. 6d. per cwt., what should be its weight when flour sells at £1 7s. 6d.? *Ans.* 4 ℔ 1$\frac{4}{3\frac{3}{}}$ oz.

88. If 100 ℔ of anything cost £25 6s. 3d., what will be the price of 625 ℔? *Ans.* £158 4s. 0$\frac{3}{4}$d.

89. If 1 ℔ of spice cost 10s. 8d., what is half an oz. worth? *Ans.* 4d.

90. Bought 3 hhds. of brandy containing, respectively, 61 gals., 62 gals., and 62 gals. 2 qts., at 6s. 8d. per gallon; what is their cost? *Ans.* £61 16s. 8d.

39. If fractions, or mixed numbers are found in one or more of the terms—

RULE.—Having reduced them to improper fractions, if they are complex fractions, compound fractions, or mixed numbers—multiply the second and third terms together, and divide the product by the first—according to the rules already given [Sec. IV. 36, &c., and 46, &c.] for the management of fractions.

EXAMPLE.—If 12 men build 3$\frac{5}{7}$ yards of a wall in $\frac{3}{4}$ of a week, how long will they require to build 47 yards?

3$\frac{5}{7}$ yards = $\frac{26}{7}$ yards, therefore

$$\frac{26}{7} : 47 :: \frac{3}{4} : \frac{\frac{3}{4} \times 47}{\frac{26}{7}} = 9\frac{1}{2} \text{ weeks, nearly.}$$

40. If all the terms are fractions—

RULE.—Invert the first, and then multiply all the terms together.

EXAMPLE.—If $\frac{3}{4}$ of a regiment consume $\frac{11}{12}$ of 40 tons of flour in $\frac{4}{5}$ of a year, how long will $\frac{8}{9}$ of the same regiment take to consume it?

$$\frac{8}{9} : \frac{3}{4} :: \frac{4}{5} : \frac{3}{4} \times \frac{4}{5} \div \frac{8}{9} = \frac{3}{4} \times \frac{4}{5} \times \frac{9}{8} = \frac{72}{100} = 262 \cdot 8 \text{ days.}$$

This rule follows from that which was given for the division of one fraction by another [Sec. IV. 49].

41. If the first and second, or the first and third terms, are fractions—

RULE.—Reduce them to a common denominator (should they not have one already), and then omit the denominators.

Example.—If $\frac{2}{3}$ of 1 cwt. of rice costs £2, what will $\frac{9}{10}$ of a cwt. cost?

$$\frac{2}{3} : \frac{9}{10} :: 2 : ?$$

Reducing the fractions to a common denominator, we have

$$\frac{20}{30} : \frac{27}{30} :: 2 : ?$$

And omitting the denominator,

$$20 : 27 :: 2 : \frac{27 \times 2}{20} = £2\cdot7 = £2 \; 14s.$$

This is merely multiplying the first and second, or the first and third terms by the common denominator—which [30] does not alter the proportion.

EXERCISES.

91. What will $\frac{3}{4}$ of a yard cost, if 1 yard costs 13s. 6d.?   Ans. 10s. $1\frac{1}{2}d.$

92. If 1 ℔ of spice costs $\frac{3}{4}s.$, what will 1 ℔ 14 oz. cost?   Ans. 1s. $4\frac{1}{8}d.$

93. If 1 oz. of silver costs $5\frac{2}{3}s.$, what will $\frac{3}{4}$ oz. cost? Ans. 4s. 3d.

94. How much will $\frac{1}{4}$ yard come to if $\frac{7}{8}$ cost $\frac{5}{6}s.$? Ans. $\frac{5}{21}s.$

95. If $2\frac{1}{2}$ yards of flannel cost $3\frac{1}{3}s.$, what is the price of $4\frac{3}{4}$ yards?   Ans. 6s. 4d.

96. What will $3\frac{3}{8}$ oz. of silver cost at $6\frac{1}{3}s.$ per oz.? Ans. £1 1s. $4\frac{1}{2}d.$

97. If $\frac{3}{16}$ of a ship costs £273$\frac{1}{3}$, what is $\frac{5}{32}$ of her worth?   Ans. £227 12s. 1d.

98. If 1 ℔ of silk costs $16\frac{2}{3}s.$, how many pounds can I have for $37\frac{1}{2}s.$?   Ans. $2\frac{1}{4}$ ℔.

99. What is the price of $49\frac{3}{11}$ yards of cloth, if $7\frac{5}{8}$ cost £7 18s. 4d.?   Ans. £51 3s. $1\frac{533}{671}d.$

100. If £100 of stock is worth £98$\frac{7}{8}$, what will £362 8s. $7\frac{1}{2}d.$ be worth?   Ans. £358 7s. 1d.

101. If $9\frac{1}{7}s.$ is paid for $4\frac{5}{6}$ yards, how much can be bought for £2$\frac{3}{11}$?   Ans. 24 yards, nearly.

MISCELLANEOUS EXERCISES IN SIMPLE PROPORTION.

102. Sold 4 hhds. of tobacco at $10\frac{1}{2}d.$ per ℔ : No. 1 weighed 5 cwt., 2 qrs.; No. 2, 5 cwt., 1 qr., 14 ℔; No. 3, 5 cwt., 7 ℔; and No. 4, 5 cwt., 1 qr., 21 ℔.   What was their price?   Ans. £104 14s. 9d.

103. Suppose that a bale of merchandise weighs 300 ℔, and costs £15 4s. 9d.; that the duty is 2d. per pound; that the freight is 25s.; and that the porterage home is 1s. 6d.: how much does 1 ℔ stand me in?

```
                       £   s.  d.
                      15   4   9  cost.
                       2  10   0  duty.
                       1   5   0  fieight.
                       0   1   6  porterage.
            ℔   ℔    ───────────
           300 : 1 ::  19   1   3  entire cost.
                      20
                     ─────
                      381
                       12
              300)4575
                  ─────────
                   15¼d.   Answer.
```

104. Received 4 pipes of oil containing 480 gallons, which cost 5s. 5½d. per gallon; paid for freight 4s. per pipe; for duty, 6d. per gallon; for porterage, 1s. per pipe. What did the whole cost; and what does it stand me in per gallon? *Ans.* It cost £144, or 6s. per gallon.

105. Bought three sorts of brandy, and an equal quantity of each sort: one sort at 5s.; another at 6s.; and the third at 7s. What is the cost of the whole— one gallon with another? *Ans.* 6s.

106. Bought three kinds of vinegar, and an equal quantity of each kind: one at 3½d.; another at 4d.; and another at 4½d. per quart. Having mixed them I wish to know what the mixture cost me per quart? *Ans.* 4d.

107. Bought 4 kinds of salt, 100 barrels of each; and the prices were 14s., 16s., 17s., and 19s. per barrel. If I mix them together, what will the mixture have cost me per barrel? *Ans.* 16s. 6d.

108. How many reams of paper at 9s. 9d., and 12s. 3d. per ream shall I have, if I buy £55 worth of both, but an equal quantity of each? *Ans.* 50 reams of each.

109. A vintner paid £171 for three kinds of wine: one kind was £8 10s.; another £9 5s.; and the third

K

£10 15s. per hhd. He had of each an equal quantity, the amount of which is required.

$$\begin{array}{rr} \pounds & s. \\ 8 & 10 \\ 9 & 5 \\ 10 & 15 \\ \hline 28 & 10, \end{array}$$ the price of three hogsheads of each.

$$\underset{28}{\pounds} \quad \underset{10}{s.} \quad : 171 :: 3 : \frac{\pounds 171 \times 3}{\pounds 28 \ 10} = 18 \text{ hhds.}$$

110. Bought three kinds of salt, and of each an equal quantity ; one was 14s., another 16s., and the third 19s. the barrel ; and the whole price was £490. How many barrels had I of each ? *Ans.* 200.

111. A merchant bought certain goods for £1450, with an agreement to deduct £1 per cent for prompt payment. What has he to pay ? *Ans.* £1435 10s.

112. A captain of a ship is provided with 24000 ℔ of bread for 200 men, of which each man gets 4 ℔ per week. How long will it last ? *Ans.* 30 weeks.

113. How long would 3150 ℔ of beef last 25 men, if they get 12 oz. each three times per week ? *Ans.* 56 weeks.

114. A fortress containing 700 men who consume each 10 ℔ per week, is provided with 184000 ℔ of provisions. How long will they last ? *Ans.* 26 weeks and 2 days.

115. In the copy of a work containing 327 pages, a remarkable passage commences at the end of the 156th page. At what page may it be expected to begin in a copy containing 400 pages ? *Ans.* In the 191st page.

116. Suppose 100 cwt., 2 qrs., 14 ℔ of beef for ship's use were to be cut up in pieces of 4 ℔, 3 ℔, 2 ℔, 1 ℔, and ½ ℔—there being an equal number of each. How many pieces would there be in all ? *Ans.* 1073 ; and 3½ ℔ left.

117. Suppose that a greyhound makes 27 springs while a hare makes 25, and that their springs are of equal length. In how many springs will the hare be overtaken, if she is 50 springs before the hound ?

The time taken by the greyhound for one spring is to that required by the hare, as 25 : 27, as 1 : $\frac{27}{25}$, or as 1 : $1\frac{2}{25}$ [12]. The greyhound, therefore, gains $\frac{2}{25}$ of a spring during every spring of the hare. Therefore

$\frac{2}{25}$: 50 :: 1 spring : $50 \div \frac{2}{25} = 675$, the number of springs the hare will make, before it is overtaken.

118. If a ton of tallow costs £35, and is sold at the rate of 10 per cent. profit, what is the selling price? *Ans.* £38 10s.

119. If a ton of tallow costs £37 10s., at what rate must it be sold to gain by 15 tons the price of 1 ton? *Ans.* £40.

120. Bought 45 barrels of beef at 21s. per barrel; among them are 16 barrels, 4 of which would be worth only 3 of the rest. How much must I pay? *Ans.* £43 1s.

121. If 840 eggs are bought at the rate of 10 for a penny, and 240 more at 8 for a penny, do I lose or gain if I sell all at 18 for 2d.? *Ans.* I gain 6d.

122. Suppose that 4 men do as much work as 5 women, and that 27 men reap a quantity of corn in 13 days. In how many days would 21 women do it? *Ans.*

The work of 4 men = that of 5 women. Therefore (dividing each of the equal quantities by 4, they will remain equal), $\dfrac{4 \text{ men's work}}{4}$ (one man's work) $= \dfrac{\text{the work of 5 women}}{4}$. Consequently $1\frac{1}{4}$ times the work of one woman = 1 man's work:—that is, the work of one man, in terms of a woman's work, is $1\frac{1}{4}$; or a woman's work is to a man's work :: 1 : $1\frac{1}{4}$. Hence 27 men's work = $27 \times 1\frac{1}{4}$ women's work; then, in place of saying—

21 women : 27 men :: 13 days : ?

say the work of 21 women : the work of $27 \times 1\frac{1}{4}$ ($= 33\frac{3}{4}$) women :: 13 : $\dfrac{33\frac{3}{4} \times 13}{21} = 20\frac{25}{28}$ days.

123. The ratio of the diameter of a circle to its circumference being that of 1 : 3·14159, what is the circumference of a circle whose diameter is 47·36 feet? *Ans.* 148·78618 feet.

124. If a pound (Troy weight) of silver is worth 66s.,

what is the value of a pound avoirdupoise? *Ans.* £4 0*s.* $2\frac{1}{2}$*d.*

125. A merchant failing, owes £40881871 to his creditors; and has property to the amount of £12577517 10*s.* 11*d.* How much per cent. can he pay? *Ans.* £30 15*s.* $3\frac{3}{4}$*d.*

126. If the digging of an English mile of canal costs £1347 7*s.* 6*d.*, what will be the cost of an Irish mile? *Ans.* £1714 16*s.* $9\frac{3}{4}$*d.*

127. If the rent of 46 acres, 3 roods, and 14 perches, is £100, what will be the rent of 35 acres, 2 roods, and 10 perches? *Ans.* £75 18*s.* $6\frac{3}{4}$*d.*

128. When A has travelled 68 days at the rate of 12 miles a day, B, who had travelled 48 days, overtook him. How many miles a day did B travel, allowing both to have started from the same place? *Ans.* 17.

129. If the value of a pound avoirdupoise weight be £4 0*s.* $2\frac{1}{2}$*d.*, how many shillings may be had for one pound Troy? *Ans.* 66*s.*

130. A landlord abates $\frac{1}{3}$ in a shilling to his tenant; and the whole abatement amounts to £76 3*s.* $4\frac{1}{3}$*d.* What is the rent? *Ans.* £228 10*s.* 1*d.*

131. If the third and tenth of a garden comes to £4 10*s.*, what is the worth of the whole garden? *Ans.* £10 7*s.* $8\frac{1}{4}$*d.*

132. A can prepare a piece of work in $4\frac{1}{2}$ days; B in $6\frac{1}{3}$ days; and C in $8\frac{1}{2}$ days. In what time would all three do it? *Ans.* $2\frac{13}{1447}$.

$4\frac{1}{2}$ days : 1 day :: 1 whole of the work : $\frac{2}{9}$ part of the whole—or what A would do in a day.

$6\frac{1}{3}$ days : 1 day :: 1 whole of the work : $\frac{3}{19}$ part of the whole—or what B would do in a day.

$8\frac{1}{2}$ days : 1 day :: 1 whole of the work : $\frac{2}{17}$ part of the whole—or what C would do in a day.

$\frac{2}{9}+\frac{3}{19}+\frac{2}{17}=\frac{1447}{2907}$=what *all* would do in a day.

Then the $\frac{1447}{2907}$ part of the work : 1 whole of the work :: 1 day (the time all would require to execute $\frac{1447}{2907}$ of the work) : $2\frac{13}{1447}$ days, the time all would take to do the whole of it.

133. A can trench a garden in $8\frac{1}{2}$ days; B in $5\frac{1}{4}$ days; but when A, B, and C work together, it will be finished in $1\frac{1}{3}$ days. In how many days would C be able to do it by himself? *Ans.* $2\frac{166}{631}$ days.

A, B, and C's work in one day=$\frac{3}{4}$ of the whole=$\frac{1233}{1628}$

Subtract- $\begin{cases} \text{A's work in 1 day=}\frac{3}{37} \\ \text{B's work in 1 day=}\frac{4}{31} \end{cases}$ =$\frac{339}{1628}$ of the whole=$\frac{440}{1628}$

C's work in one day remains equal to . . . . $\frac{813}{1628}$

Then $\frac{813}{1628}$ (C's work in one day) : 1 whole of the work :: 1 day : $2\frac{2}{813}$, the time required.

134. A ton of coals yield about 9000 cubic feet of gas ; a street lamp consumes about 5, and an argand burner (one in which the air passes through the centre of the flame) 4 cubic feet in an hour. How many tons of coal would be required to keep 17493 street lamps, and 192724 argand burners in shops, &c., lighted for 1000 hours ? *Ans.* 95373$\frac{4}{9}$.

135. The gas consumed in London requires about 50,000 tons of coal per annum. For how long a time would the gas this quantity may be supposed to produce (at the rate of 9000 cubic feet per ton), keep one argand light (consuming 4 cubic feet per hour) constantly burning ? *Ans.* 12842 years and 170 days.

136. It requires about 14,000 millions of silk worms to produce the silk consumed in the United Kingdom annually. Supposing that every pound requires 3500 worms, and that one-fifth is wasted in throwing, how many pounds of manufactured silk may these worms be supposed to produce ? *Ans.* 1488 tons, 1 cwt., 3 qrs., 17 ℔.

137. If one fibre of silk will sustain 50 grains, how many would be required to support 97 ℔ ? *Ans.* 13580.

138. One fibre of silk a mile long weighs but 12 grains; how many miles would 4 millions of pounds, annually consumed in England, reach ?

Ans. 23333333333$\frac{1}{3}$ miles.

139. A leaden shot of 4$\frac{1}{2}$ inches in diameter weighs 17 ℔; but the size of a shot 4 inches in diameter, is to that of one 4$\frac{1}{2}$ inches in diameter, as 64000 : 91125 : what is the weight of a leaden ball 4 inches in diameter ? *Ans.* 11·9396.

140. The sloth does not advance more than 100 yards in a day. How long would it take to crawl from Dublin to Cork, allowing the distance to be 160 English miles ? *Ans.* 2816 days ; or 8 years, nearly.

141. English race horses have been known to go at the rate of 58 miles an hour. In what time, at this velocity, might the distance from Dublin to Cork be travelled over? *Ans.* 2 hours, 45′ 31″ 2‴

142. An acre of coals 2 feet thick yields 3000 tons; and one 5 feet thick 8000. How many acres of 5 feet thick would give the same quantity as 48 of 2 feet thick? *Ans.* 18.

143. The hair-spring of a watch weighs about the tenth of a grain; and is sold, it is said, for about ten shillings. How much would be the price of a pound of crude iron, costing one halfpenny, made into steel, and then into hair-springs—supposing that, after deducting waste, there are obtained from the iron about 7000 grains of steel? *Ans.* £35000.

COMPOUND PROPORTION.

42. Compound proportion enables us, although two or more proportions are contained in the question, to obtain the required answer by a single stating. In *compound* proportion there are *three* or *more* ratios, one of them imperfect, and the rest perfect.

43. RULE—I. Place the quantity belonging to the imperfect ratio as the third term of the proportion.

II. Put down the terms of each of the other ratios in the first and second places—in such a way that the antecedents may form one column, and the consequents another. In setting down each ratio, consider what effect it has upon the answer—if to increase it, set down the larger term as consequent, and the smaller as antecedent; if to diminish it, set down the smaller term as consequent, and the larger as antecedent.

III. Multiply the quantity in the third term by the product of all the quantities in the second, and divide the result by the product of all those in the first.

44. EXAMPLE 1.—If 5 men build 16 yards of a wall in 20 days, in how many days would 17 men build 37 yards?

The question briefly put down [32], will be as follows:

5 men ⎫
16 yards ⎬ conditions which give 20 days.
20 days ⎭ imperfect ratio.

——————

? days, the number sought.

17 men ⎫
37 yard^s ⎬ conditions which give the *required* number of days.

The imperfect ratio consists of days—therefore we are to put 20, the *given* number of days, in the third place. Two ratios remain to be set down—that of numbers of *men*, and that of numbers of *yards*. Taking the former first, we ask ourselves how it affects the answer, and find that the more men there are, the smaller the *required* number will be—since the greater the number of men, the shorter the time required to do the work. We, therefore, set down 17 as antecedent, and 5 as consequent. Next, considering the ratio consisting of yards, we find that the larger the number of yards, the longer the time, before they are built—therefore increasing their number increases the quantity required. Hence we put 37 as consequent, and 16 as antecedent; and the whole will be as follows :—

$$17 : 5 :: 20 : ?$$
$$16 : 37$$

And $17 : 5$
$16 : 37$ $:: 20 : \dfrac{20 \times 5 \times 37}{17 \times 16} = 13 \cdot 6$ days, nearly

45. The result obtained by the rule is the same as would be found by taking, in succession, the two proportions supposed by the question. Thus

If 5 men would build 16 yards in 20 days, in how many days would they build 37 yards?

$16 : 37 :: 20 : \dfrac{37 \times 20}{16} =$ number of days which 5 men would require, to build 37 yards.

If 5 men would build 37 yards in $\dfrac{20 \times 37}{16}$ days, in how many days would 17 men build them?

$17 : 5 :: \dfrac{20 \times 37}{16} : \dfrac{20 \times 37}{16} \times 5 \div 17 = \dfrac{20 \times 5 \times 37}{17 \times 16}$, the number of days found by the rule.

46. EXAMPLE 2.—If 3 men in 4 days of 12 working hours each build 37 perches, in how many days of 8 working hours ought 22 men to build 970 perches?

$$\begin{array}{l} \text{3 men.} \\ \text{4 days.} \\ \text{12 hours.} \\ \text{37 perches.} \\ \hline \text{? days.} \\ \text{8 hours.} \\ \text{22 men.} \\ \text{970 perches.} \end{array}$$

$$
\begin{array}{l}
22 : 3 \\
8 : 12 \\
37 : 970
\end{array}
:: 4 : \frac{3 \times 12 \times 970 \times 4}{22 \times 8 \times 37} = 21\tfrac{1}{2} \text{ days, nearly.}
$$

The *number* of days is the quantity sought; therefore 4 days constitutes the imperfect ratio, and is put in the third place. The more men the fewer the days necessary to perform the work; therefore, 22 is put first, and 3 second. The smaller the number of working hours in the day, the larger the number of days; hence 8 is put first, and 12 second. The greater the number of perches, the greater the number of days required to build them; consequently 17 is to be put first, and 970 second.

47. The process may often be abbreviated, by dividing one term in the first, and one in the second place; or one in the first, and one in the third place, by the same number.

EXAMPLE 1.—If the carriage of 32 cwt. for 5 miles costs 8*s.*, how much will the carriage of 160 cwt. 20 miles cost?

$$
\begin{array}{l}
32 : 160 \\
5 : 20
\end{array}
:: 8 : \frac{160 \times 20 \times 8}{32 \times 5} = 160
$$

Dividing 32 and 160 by 32 we have 1 and 5 as quotients. Dividing 5 and 20 by 5 we have 1 and 4; and the proportion will be—

$$
\begin{array}{l}
1 : 5 \\
1 : 4
\end{array}
:: 8 : 5 \times 4 \times 8 = 160
$$

48. We are to continue this kind of division as long as possible—that is, so long as any one number will measure a quantity in the first, and another in the second place; or one in the first and another in the third place. This will in some instances change most of the quantities into unity—which of course may be omitted.

EXAMPLE 2.—If 28 loads of stone of 15 cwt. each, build a wall 20 feet long and 7 feet high, how many loads of 19 cwt. will build one 323 feet long and 9 feet high?

$$19 : 15 :: 28 : \frac{15 \times 323 \times 9 \times 28}{19 \times 20 \times 7} = 459.$$
$$20 : 323$$
$$7 : 9$$

Dividing 7 and 28 by 7, we obtain 1 and 4.—Substituting these, we have

$$19 : 15 :: 4 : ?$$
$$20 : 323$$
$$1 : 9$$

Dividing 20 and 15 by 5, the quotients are 4 and 3:

$$19 : 3 :: 4 : ?$$
$$4 : 323$$
$$1 : 9$$

Dividing 4 and 4 by 4, the quotients are 1 and 1:

$$19 : 3 :: 1 : ?$$
$$1 : 323$$
$$1 : 9$$

Dividing 19 and 323 by 19, the quotients are 1 and 17:

$$1 : 3 :: 1 : 3 \times 17 \times 9 = 459$$
$$1 : 17$$
$$1 : 9$$

In this process we merely divide the first and second, or first and third terms, by the same number—which [29] does not alter the proportion. Or we divide the numerator and denominator of the fraction, found as the *fourth term*, by the same number—which [Sec. IV. 15] does not alter the quotient.

EXERCISES IN COMPOUND PROPORTION.

1. If £240 in 16 months gains £64, how much will £60 gain in 6 months? *Ans.* £6.

2. With how many pounds sterling could I gain £5 per annum, if with £450 I gain £30 in 16 months? *Ans.* £100.

3. A merchant agrees with a carrier to bring 15 cwt. of goods 40 miles for 10 crowns. How much ought he to pay, in proportion, to have 6 cwt. carried 32 miles? *Ans.* 16s.

4. If 20 cwt. are carried the distance of 50 miles for £5, how much will 40 cwt. cost, if carried 100 miles? *Ans.* £20.

5. If 200 ℔ of merchandise are carried 40 miles for 3*s.*, how many pounds might be carried 60 miles for £22 14*s.* 6*d.*   *Ans.* 20200 ℔.

6. If 286 ℔ of merchandise aré carried 20 miles for 3*s.*, how many miles might 4 cwt. 3 qrs. be carried for £32 6*s.* 8*d.*?   *Ans.* 2317·627.

7. If a wall of 28 feet high were built in 15 days by 68 men, how many men would build a wall 32 feet high in 8 days?   *Ans.* 146 nearly.

8. If 1 ℔ of thread make 3 yards of linen of $1\frac{1}{4}$ yards wide, how many pounds of thread would be required to make a piece of linen of 45 yards long and 1 yard wide?   *Ans.* 12 ℔.

9. If 3 ℔ of worsted make 10 yards of stuff of $1\frac{1}{2}$ yards broad, how many pounds would make a piece 100 yards long and $1\frac{1}{4}$ broad?   *Ans.* 25 ℔.

10. 80000 cwt. of ammunition are to be removed from a fortress in 9 days; and it is found that in 6 days 18 horses have carried away 4500 cwt. How many horses would be required to carry away the remainder in 3 days?   *Ans.* 604.

11. 3 masters who have each 8 apprentices earn £36 in 5 weeks—each consisting of 6 working days. How much would 5 masters, each having 10 apprentices, earn in 8 weeks, working $5\frac{1}{2}$ days per week—the wages being in both cases the same?   *Ans.* £110.

12. If 6 shoemakers, in 4 weeks, make 36 pair of men's, and 24 pair of women's shoes, how many pair of each kind would 18 shoemakers make in 5 weeks? *Ans.* 135 pair of men's, and 90 pair of women's shoes.

13. A wall is to be built of the height of 27 feet; and 9 feet high of it are built by 12 men in 6 days. How many men must be employed to finish the remainder in 4 days?   *Ans.* 36.

14. If 12 horses in 5 days draw 44 tons of stones, how many horses would draw 132 tons the same distance in 18 days?   *Ans.* 10 horses.

15. If 27*s.* are the wages of 4 men for 7 days,

what will be the wages of 14 men for 10 days? *Ans.* £6 15s.

. 16. If 120 bushels of corn last 14 horses 56 days, how many days will 90 bushels last 6 horses? *Ans.* 98 days.

17. If a footman travels 130 miles in 3 days when the days are 14 hours long, in how many days of 7 hours each will he travel 390 miles? *Ans.* 18.

18. If the price of 10 oz. of bread, when the corn is 4s. 2d. per bushel, be 5d., what must be paid for 3 ℔ 12 oz., when the corn is 5s. 5d. per bushel? *Ans.* 3s. 3d.

19. 5 compositors in 16 days of 14 hours long can compose 20 sheets of 24 pages in each sheet, 50 lines in a page, and 40 letters in a line. In how many days of 7 hours long may 10 compositors compose a volume to be printed in the same letter, containing 40 sheets, 16 pages in a sheet, 60 lines in a page, and 50 letters in a line? *Ans.* 32 days.

20. It has been calculated that a square degree (about 69×69 square miles) of water gives off by evaporation 33 millions of tons of water per day. How much may be supposed to rise from a square mile in a week? *Ans.* 48519·2187 tons.

21. When the mercury in the barometer stands at a height of 30 inches, the pressure of the air on every square inch of surface is 15 ℔. What will be the pressure on the human body—supposing its whole surface to be 14 square feet; and that the barometer stands at 31 inches? *Ans.* 13 tons 19 cwt.

QUESTIONS IN RATIOS AND PROPORTION.

1. What is the rule of proportion; and is it ever called by any other name? [1].

2. What is the difference between simple and compound proportion? [30 and 42].

3. What is a ratio? [7].

4. What are the antecedent and consequent? [7].

5. What is an inverse ratio? [8].

6. What is the difference between an arithmetical and a geometrical ratio? [9].

7. How can we know whether or not an arithmetical or geometrical ratio, is altered in value? [10 and 11].

8. How is one quantity expressed in terms of another? [12].

9. What is a proportion, or analogy? [14].

10. What are means, and extremes? [15].

11. What is the arithmetical, or geometrical mean of two quantities? [19 and 27].

12. How is it known that four quantities are in arithmetical proportion? [16].

13. How is it known that four quantities are in geometrical proportion? [21].

14. How is a fourth proportional to three quantities found? [17 and 22].

15. Mention the principal changes which may be made in a geometrical proportion, without destroying it? [29].

16. How is a question in the simple rule of three to be stated, and solved? [31].

17. Is it necessary, or even correct, to divide the rule of three into the direct, and inverse? [35].

18. How is the question solved, when the first or second terms are not of the same denomination; or one, or both of them contain different denominations? [37].

19. How is a question in the rule of proportion solved, if the third term consists of more than one denomination? [38].

20. How is it solved, if fractions or mixed numbers are found in the first and second, in the first and third, or in all the terms? [39 and 40].

21. How is a question in the rule of compound proportion stated, &c.? [43].

22. Can any of the terms of a question in the rule of compound proportion ever be lessened, or altogether banished? [47 and 48].

ARITHMETIC.

PART II.

SECTION VI.

PRACTICE.

1. Practice is so called from its being the method of calculation *practised* by mercantile men : it is an abridged mode of performing processes dependent on the rule of three—particularly when one of the terms is unity. The statement of a question in practice, in *general terms*, would be, "one quantity of goods is to another, as the price of the former is to the price of the latter."

The simplification of the rule of three by means of *practice*, is principally effected, either by dividing the given *quantity* into "parts," and finding the sum of the prices of these parts ; or by dividing the *price* into "parts," and finding the sum of the prices at each of these parts : in either case, as is evident, we obtain the required price.

2. *Parts* are of two kinds, "aliquot" and "aliquant."

The *aliquant* parts of a number, are those which do *not* measure it—that is, which cannot be multiplied by any integer so as to produce it ; the *aliquot* parts are, as we have seen [Sec. II. 26], those which measure it.

3. To find the aliquot parts of any number—

Rule.—Divide it by its least divisor, and the resulting quotient by *its* least divisor :—proceed thus until the last quotient is unity. All the divisors are the *prime* aliquot parts ; and the product of every two, every three, &c., of them, are the *compound* aliquot parts of the given number.

4. EXAMPLE.—What are the prime, and compound aliquot parts of 84 ?

$$
\begin{array}{r}
2)\overline{84} \\
2)\overline{42} \\
3)\overline{21} \\
7)\overline{7} \\
\hline
1
\end{array}
$$

The prime aliquot parts are 2, 3, and 7 ; and

$$
\left.
\begin{array}{l}
2\times2=\ 4 \\
2\times3=\ 6 \\
2\times7=14 \\
3\times7=21 \\
2\times2\times3=12 \\
2\times2\times7=28 \\
2\times3\times7=42
\end{array}
\right\}
\text{are the compound aliquot parts}
$$

All the aliquot parts, placed in order, are 2, 3, 4, 6, 7, 12, 14, 21, 28, and 42.

5. We may apply this rule to *applicate* numbers.—Let it be required to find the aliquot parts of a pound, in shillings and pence. $240d.=\pounds1.$

$$
\begin{array}{r}
2)\overline{240} \\
2)\overline{120} \\
2)\overline{60} \\
2)\overline{30} \\
3)\overline{15} \\
5)\overline{5} \\
\hline
1
\end{array}
$$

The prime aliquot parts of a pound are, therefore, 2d., 3d., and 5d. : and the compound,

	d.	s.	d.
$2\times2=$	4		
$2\times3=$	6		
$2\times5=$	10		
$2\times2\times2=$	8		
$2\times2\times3=$	12	1	0
$2\times2\times5=$	20	1	8
$2\times3\times5=$	30	2	6
$2\times2\times2\times2=$	16	1	4
$2\times2\times2\times3=$	24	2	0
$2\times2\times2\times5=$	40	3	4
$2\times2\times3\times5=$	60	5	0
$2\times2\times2\times2\times3=$	48	4	0
$2\times2\times2\times2\times5=$	80	6	8
$2\times2\times2\times3\times5=$	120	10	0

And placed in order—

£	d.				£	d.	s.	d.
$\frac{1}{120}$ =	2				$\frac{1}{15}$ =	16 =	1	4
$\frac{1}{80}$ =	3				$\frac{1}{12}$ =	20 =	1	8
$\frac{1}{60}$ =	4				$\frac{1}{10}$ =	24 =	2	0
$\frac{1}{48}$ =	5				$\frac{1}{8}$ =	30 =	2	6
$\frac{1}{40}$ =	6				$\frac{1}{6}$ =	40 =	3	4
$\frac{1}{30}$ =	8				$\frac{1}{5}$ =	48 =	4	0
$\frac{1}{24}$ =	10	s.	d.		$\frac{1}{4}$ =	60 =	5	0
$\frac{1}{20}$ =	12 =	1	0		$\frac{1}{3}$ =	80 =	6	8
					$\frac{1}{2}$ =	120 =	10	0

Aliquot parts of a shilling, obtained in the same way—

s.	d.		s.	d.		s.	d.
$\frac{1}{48}$ =	$\frac{1}{4}$		$\frac{1}{12}$ =	1		$\frac{1}{4}$ =	3
$\frac{1}{24}$ =	$\frac{1}{2}$		$\frac{1}{8}$ =	$1\frac{1}{2}$		$\frac{1}{3}$ =	4
$\frac{1}{15}$ =	$\frac{3}{4}$		$\frac{1}{6}$ =	2		$\frac{1}{2}$ =	6

Aliquot parts of avoirdupoise weight—

Aliquot parts of a ton.			Aliquot parts of a cwt.		Aliquot parts of a quarter.	
ton	cwt.	qr.	cwt.	lb	qr.	lb
$\frac{1}{40}$ =	$\frac{1}{2}$ =	2	$\frac{1}{56}$ =	2	$\frac{1}{14}$ =	2
$\frac{1}{20}$ =	1 =	4	$\frac{1}{28}$ =	4	$\frac{1}{7}$ =	4
$\frac{1}{16}$ =	$1\frac{1}{4}$ =	5	$\frac{1}{16}$ =	7	$\frac{1}{4}$ =	7
$\frac{1}{10}$ =	2 =	8	$\frac{1}{14}$ =	8	$\frac{1}{2}$ =	14
$\frac{1}{8}$ =	$2\frac{1}{2}$ =	10	$\frac{1}{8}$ =	14		
$\frac{1}{5}$ =	4 =	16	$\frac{1}{7}$ =	16		
$\frac{1}{4}$ =	5 =	20	$\frac{1}{4}$ =	28		
$\frac{1}{2}$ =	10 =	40	$\frac{1}{2}$ =	56		

Aliquot parts may, in the same manner, be easily obtained by the pupil from the other tables of weights and measures, page 3, &c.

6. To find the price of a quantity of *one* denomination—the price of a "higher" being given.

RULE.—Divide the price by that number which expresses how many times we must take the lower to make the amount equal to one of the higher denomination.

EXAMPLE.—What is the price of 14 lb of butter at 72s. per cwt.?

We must take 14 lb, or 1 stone 8 times, to make 1 cwt. Therefore the price of 1 cwt. divided by 8, or 72s. ÷ 8 = 9s., is the price of 14 lb.

The table of aliquot parts of avoirdupoise weight shows that 14 lb is the $\frac{1}{8}$ of a cwt. Therefore its price is the $\frac{1}{8}$ of the price of 1 cwt.

EXERCISES.

What is the price of

1. $\frac{1}{4}$ cwt., at 29*s.* 6*d.* per cwt.? *Ans.* 7*s.* $4\frac{1}{2}$*d.*
2. $\frac{1}{2}$ a yard of cloth, at 8*s.* 6*d.* per yard? *Ans.* 4*s.* 3*d.*
3. 14 lb of sugar, at 45*s.* 6*d.* per cwt.? *Ans.* 5*s.* $8\frac{1}{4}$*d.*
4. What is the price of $\frac{3}{4}$ cwt., at 50*s.* per cwt.?

		£	*s.*	*d.*	
	50*s.* =	2	10	0	
qrs. cwt.					£ *s.*
The price of $2 = \frac{1}{2}$ is		1	5	0	$= 2\ 10 \div 2$
" of $1 = \frac{1}{2} \div 2$ is		0	12	6	$= 1\ 5 \div 2$
Therefore the price of $\overline{2+1}$ qrs. ($= \frac{3}{4}$ cwt.) is		1	17	6	

$\frac{3}{4}$ cwt., or 3 qrs. $= \overline{2+1}$ qrs. But 2 qrs. $= \frac{1}{2}$ cwt.; and its price is half that of a cwt. 1 qr. $= \frac{1}{2}$ cwt. $\div 2$; and its price is half the price of 2 qrs. Therefore the price of $\frac{3}{4}$ cwt. is half the price of 1 cwt. plus the half of half the price of one cwt.

What is the price of

5. $\frac{1}{2}$ oz. of cloves, at 9*s.* 4*d.* per lb? *Ans.* $3\frac{1}{2}$*d.*
6. 1 nail of lace, at 15*s.* 4*d.* per yard? *Ans.* $11\frac{1}{2}$*d.*
7. $\frac{1}{2}$ lb, at 23*s.* 4*d.* per cwt? *Ans.* $1\frac{1}{4}$*d.*
8. $\frac{3}{4}$ lb, at 18*s.* 8*d.* per cwt.? *Ans.* $1\frac{1}{2}$*d.*

7. When the price of *more than one* "lower" denomination is required—

RULE.—Find the price of each denomination by the last rule; and the sum of the prices obtained will be the required quantity.

EXAMPLE.—What is the price of 2 qrs. 14 lb of sugar, at 45*s.* per cwt.?

		s.	*d.*	
		45	0	price of 1 cwt.
cwt.				[or $\frac{1}{2}$ of 1 cwt.
2 qrs. $= \frac{1}{2}$	And	22	6	, or 45*s.* $\div 2$, is the price of 2 qrs.,
14 lb $= \frac{1}{8}$, or $\frac{1}{4}$ of 2 qrs.		5	$7\frac{1}{2}$	, or 45*s.* $\div 8 =$ 22*s.* 6*d.* $\div 4$, is the price of 14 lb, the $\frac{1}{8}$ of 1 cwt., or the $\frac{1}{4}$ of 2 qrs.
	And	28	$1\frac{1}{2}$	is the price of 2 qrs. 14 lb.

2 qrs. $= \frac{1}{2}$ of 1 cwt. Therefore 45*s.* (the price of 1 cwt.) $\div 2$, or 26*s.* 6*d.*, is the price of 2 qrs.

14 ℔ is the ⅛ of 1 cwt., or the ¼ of 2 qrs.   Therefore 45s.÷8, or 22s. 6d.÷4=5s. 7½d., is the price of 14 ℔ And 22s. 6d.+5s. 7½d., or the price of 2 qrs. plus the price of 14 ℔, is the price of 2 qrs. 14 ℔.

EXERCISES.

What is the price of
9. 1 qr., 14 ℔ at 46s. 6d. per cwt. ?   *Ans.* 17s. 5¼d.
10. 3 qrs. 2 nails, at 17s. 6d. per yard?   *Ans.* 15s. 3¾d.
11. 5 roods 14 perches, at 3s. 10d. per acre ?   *Ans.* 5s. 1½d.
12. 16 dwt. 14 grs., at £4 4s. 9d. per oz. ?   *Ans.* £3 10s. 3¼d.
13. 14 ℔ 5 oz., at 25s. 4d. per cwt. ?   *Ans.* 3s. 2¾d.

8. When the price of *one* "higher" denomination is required—

RULE.—Find what number of times the lower denomination must be taken, to make a quantity equal to one of the given denomination ; and multiply the price by that number.   (This is the reverse of the rule given above [6]).

EXAMPLE.—What is the price of 2 tons of sugar, at 50s. per cwt. ?
1 cwt. is the $\frac{1}{40}$ of 2 tons ; hence the price of 2 tons will be 40 times the price of 1 cwt.—or 50s.×40=£100.

50s. the price of 1 cwt. multiplied
by 40 the number of hundreds in 2 tons,
gives 2000s.
or £100 as the price of 40 cwt., or 2 tons.

EXERCISES.

What is the price of
.4. 47 cwt., at 1s. 8d. per ℔ ?   *Ans.* £438 13s. 4d.
15. 36 yards, at 4d. per nail ?   *Ans.* £9 12s.
16. 14 acres, at 5s. per perch ?   *Ans.* £560.
17. 12 ℔, at 1¾d. per grain ?   *Ans.* £504.
18. 19 hhds., at 3d. per gallon ?   *Ans.* £14 19s. 3d.

9. When the price of *more than one* "higher" denomination is required—

RULE.—Find the price of each by the last, and add the results together. (This is the reverse of the rule given above [7]).

EXAMPLE.—What is the price of 2 cwt. 1 qr. of flour, at 2s. per stone?

1 stone is the $\frac{1}{16}$ of 2 cwt. Therefore
2s., the price of one stone,
multiplied by 16, the number of stones in 2 cwt.,

gives 32s., the price of 16 stones, or 2 cwt.

There are 2 stones in 1 qr.; therefore 2s. (the price of 1 stone) ×2=4s. is the price of 1 qr. And 32s.+4s.=36s.= £1 16s., is the price of 2 cwt. 1 qr.

EXERCISES.

What is the price of
19. 5 yards, 3 qrs., 4 nails, at 4d. per nail? *Ans.* £1 12s.
20. 6 cwt. 14 ℔, at 3d. per ℔? *Ans.* £8 11s. 6d.
21. 3 ℔ 5 oz., at 2¼d. per oz.? *Ans.* 9s. 11¼d.
22. 9 oz., 3 dwt., 14 grs., at ¾d. per gr.? *Ans.* £13 15s. 4½d.
23. 3 acres, 2 roods, 3 perches, at 5s. per perch? *Ans.* £140 15s.

10. When the price of one denomination is given, to find the price of *any number* of another—
RULE.—Find the price of one of that other denomination, and multiply it by the given number of the latter.

EXAMPLE.—What is the price of 13 stones at 25s. per cwt.?

1 stone=⅛ cwt. Therefore
8)25s., the price of 1 cwt. divided by 8,

gives 3 1½, the price of 1 stone, or ⅛ of 1 cwt.
Multiplying this by 13, the number of stones,

we obtain £2 0 7½ as the price of 13 stones.

1 stone is the ⅛ of 1 cwt. Hence 25s.÷8=3s. 1½d., is the price of one stone; and 3s. 1½d.×13, the price of 13 stones.

EXERCISES.

What is the price of

24. 19 ℔, at 2*d*. per oz. ? *Ans.* £2 10*s*. 8*d*.
25. 13 oz., at 1*s*. 4*d*. per ℔ ? *Ans.* 1*s*. 1*d*.
26. 14 ℔, at 2*s*. 6*d*. per dwt. ? *Ans.* £420
27. 15 acres, at 18*s*. per perch ? *Ans.* £2160.
28. 8 yards, at 4*d*. per nail ? *Ans.* £2 2*s*. 8*d*.
29. 12 hhds., at 5*d*. per pint ? *Ans.* £126.
30. 3 quarts, at 91*s*. per hhd. ? *Ans.* 1*s*. 1*d*.

11. When the price of a given denomination is the aliquot part of a shilling, to find the price of any number of that denomination—

RULE.—Divide the amount of the given denomination by the number expressing what aliquot part the given price is of a shilling, and the quotient will be the required price in shillings, &c.

EXAMPLE.—What is the price of 831 articles at 4*d*. per ?
3)831
277*s*.=£13 17*s*., is the required price.

4*d*. is the ⅓ of a shilling. Hence the price at 4*d*. is ⅓ of what it would be at 1*s*. per article. But the price at 1*s*. per article would be 831*s*. :—therefore the price at 4*d*. is 831*s*. ÷ 3, or 277*s*.

EXERCISES.

What is the price of

31. 379 ℔ of sugar, at 6*d*. per ℔ ? *Ans.* £9 9*s*. 6*d*.
32. 5014 yards of calico, at 3*d*. per yard ? *Ans.* £62 13*s*. 6*d*.
33. 258 yards of tape, at 2*d*. per yard ? *Ans.* £2 3*s*.

12. When the price of a given denomination is the aliquot part of a pound, to find the price of any number of that denomination—

RULE.—Divide the quantity whose price is sought by that number which expresses what aliquot part the given price is of a pound. The quotient will be the required price in pounds, &c.

EXAMPLE.—What is the price of 1732 ℔ of tea, at *5s.* per ℔ ?

5s. is the ¼ of £1 ; therefore the price of 1732 ℔ is the ¼ of what it would be at £1 per ℔. But at £1 per ℔ it would be £1732 ; therefore at *5s.* per ℔ it is £1732 ÷ 4 = £433.

EXERCISES.

What is the price of

34. 47 cwt., at 6s. 8d. per cwt. ? *Ans.* £15 13s. 4d.
35. 13 oz., at 4s. per oz. ? *Ans.* £2 12s.
36. 19 stones, at 2s. 6d. per stone ? *Ans.* £2 7s. 6d.
37. 83 ℔, at 1s. 4d. per ℔ ? *Ans.* £5 10s. 8d.
38. 115 qrs. at 8d. per qr. ? *Ans.* £3 16s. 8d.
39. 976 ℔, at 10s. per ℔ ? *Ans.* £488.
40. 112 ℔, at 5d. per ℔ ? *Ans.* £2 6s. 8d.
41. 563 yards, at 10d. per yard ? *Ans.* £23 9s. 2d.
42. 112 ℔, at 5s. per ℔ ? *Ans.* £28.
43. 795 ℔, at 1s. 8d. per ℔ ? *Ans.* £66 5s.
44. 1000 ℔, at 3s. 4d. per ℔ ? *Ans.* £166 13s. 4d.

13. The *complement* of the price is what it wants of a pound or a shilling.

When the complement of the price *is* the aliquot part or parts of a pound or shilling, but the price *is not*—

RULE.—Find the price at £1, or 1s.—as the case may be—and deduct the price of the quantity calculated at the complement.

EXAMPLE.—What is the price of 1470 yards, at 13s. 4d. per yard ?

6s. 8d. (the complement of 13s. 4d.) is ⅓ of £1.

From £1470, the price at £1 per yard,
subtract 490, the price at 6s. 8d. (the complement) per yard,
and the difference, 980, will be the price at 13s. 4d. per yard.

1470 yards at 13s. 4d., plus 1470 at 6s. 8d., are equal to 1470 at 13s. 4d. + 6s. 8d., or at £1 per yard. Hence the price of 1470 at 13s. 4d. = the price of 1470 at £1, minus the price of 1470 at 6s. 8d. per yard.

EXERCISES.

What is the price of

45. 51 ℔, at 17s. 6d. per ℔? *Ans.* £44 12s. 6d.
46. 39 oz., at 7d. per oz.? *Ans.* £1 2s. 9d.
47. 91 ℔, at 10d. per ℔? *Ans.* £3 15s. 10d.
48. 432 cwt., at 16s. per cwt.? *Ans.* £345 12s.

14. When neither the price nor 'its complement is the aliquot part or parts of a pound or shilling—

RULE 1.—Divide the price into pounds (if there are any), and aliquot parts of a pound or shilling; then find the price at each of these (by preceding rules) :— the sum of the prices will be what is required.

EXAMPLE.—What is the price of 822 ℔, at £5 19s. 3¾d. per ℔? £5 19s. 3¾d.=£5+19s. 3¾d.

$$\text{But } 19s.\ 3\tfrac{3}{4}d.= \begin{cases} \begin{array}{ccl} s. & d. & \pounds \\ 10 & 0 & =\tfrac{1}{2} \\ 6 & 8 & =\tfrac{1}{3} \\ 2 & 6 & =\tfrac{1}{8} \\ 0 & 1\tfrac{1}{2} & =\tfrac{1}{8}\div 20=\tfrac{1}{160},\ \text{or } \tfrac{1}{20} \text{ of the last} \\ 0 & 0\tfrac{1}{4} & =\tfrac{1}{160}\div 6=\tfrac{1}{960},\ \text{or } \tfrac{1}{6} \text{ of the last} \end{array} \end{cases}$$

Hence the price at £5 19s. 3¾d. is equal to

£		£ s. d.		£ s. d.
822×5	=4110 0 0,	the price at	5 0 0 per ℔.
$\frac{822}{2}$	= 411 0 0	„	£⅓ or 0 10 0	„
$\frac{822}{3}$	= 274 0 0	„	£⅓ or 0 6 8	„
$\frac{822}{8}$	= 102 15 0	„	£⅛ or 0 2 6	„
$\frac{822}{160}$ ($\frac{822}{8}\div 20$)=	5 2 9	„	£$\frac{1}{160}$ or 0 0 1½	„
$\frac{822}{960}$ ($\frac{822}{160}\div 6$)=	0 17 1½	„	£$\frac{1}{960}$ or 0 0 0¼	„

And £4903 14 10½ is the price at £5 19 3¾ „

The price at the whole, is evidently equal to the sum of the prices at each of the parts.

If the price were £5 19s. 3¼d. per ℔, we should sub- tract, and not add the price at ¼d. per ℔; and we then would have £4902 0s. 7½d. as the answer.

15. RULE 2.—Find the price at a pound, a shilling, a penny, and a farthing; then multiply each by their

respective numbers, in the given price; and add the
products.   Using the same example—

```
    £   s.  d.                              £    s.   d.
20)822  0  0  (the price at £1)× 5=4110   0   0 the price at £5
12)41   2  0  (the price at 1s.)×19=  780 18   0    ,,    19s.
 4)3    8  6  (the price at 1d.)× 3=   10  5   6    ,,    3d.
        17 1¼ (the price at ¼d.)× 3=    2 11   4½   ;,    ¾d.
```

And the price at £5 19s. 3¾d. is £4903 14 10½

16. RULE 3.—Find the price at the next number of
the highest denomination; and deduct the price at the
difference between the assumed and given price.

Using still the same example—

£6 is next to £5—the highest denomination in the given
price.

```
                           £   s.  d.       £    s.   d.
From the price t £6  0  0  .   .   .   or 4932  0   0
Deduct the price ⎰ the price at 8d.=27 8 0 ⎱ or  28  5  1¼
  at 8¼d.        ⎱    ,,      ¼d.= 0 17 1½ ⎰
```

The difference will be the price at £5 19s. 3¾ or £4903 14 10½

17. RULE 4.—Find the price at the next higher
aliquot part of a pound, or shilling; and deduct the price
at the difference between the assumed, and given price.

EXAMPLE.—What is the price of 84 ℔, at 6s. per ℔?

$$6s.=6s.\ 8d.\ minus\ 8d.=\tfrac{£}{3}\ minus\ \tfrac{£}{3}÷10.$$

Therefore $84÷3=28\quad 0\quad 0$ is the price at 6s. 8d. per ℔.
Deducting ₁₀ of this=2 16   0 the price at 8d.,

we have £25   4   0, the price at 6s.

EXERCISES.

What is the price of

49. 73 ℔, at 13s. per ℔?   *Ans.* £47 9s.

50. 97 cwt., at 15s. 9d. per cwt.?   *Ans.* £76 7s. 9d.

51. 43 ℔, at 3s. 2d. per ℔?   *Ans.* £6 16s. 2d.

52. 13 acres, at £4 5s. 11d. per acre?   *Ans.* £55
16s. 11d.

53. 27 yards, at 7s. 5¾d. per yard?   *Ans.* £10
1s. 11¼d.

18. When the price is an *even* number of shillings,
and less than 20.

RULE.—Multiply the number of articles by half the number of shillings; and consider the tens of the product as pounds, and the units *doubled*, as shillings.

EXAMPLE.—What is the price of 646 ℔, at 16*s*. per ℔?

$$\begin{array}{r} 646 \\ 8 \\ \hline 516|8 \\ |2 \\ \hline £516 \ 16s. \end{array}$$

2*s*. being the tenth of a pound, there are, in the price, half as many tenths as shillings. Therefore half the number of shillings, multiplied by the number of articles, will express the number of tenths of a pound in the price of the entire. The tens of these tenths will be the number of pounds; and the units (being tenths of a pound) will be half the required number of shillings—or, multiplied by 2—the required number of shillings.

In the example, 16*s*., or £·8, is the price of each article. Therefore, since there are 646 articles, 646 × £·8 = £516·8 is the price of them. But 8 tenths of a pound (the *units* in the product obtained), are twice as many shillings; and hence we are to multiply the units in the product by 2.

EXERCISES.

What is the price of
54. 3215 ells, at 6*s*. per ell? *Ans.* £964 10*s*.
55. 7563 ℔, at 8*s*. per ℔? *Ans.* £3025 4*s*.
56. 269 cwt., at 16*s*. per cwt.? *Ans.* £215 4*s*.
57. 27 oz., at 4*s*. per oz.? *Ans.* £5 8*s*.
58. 84 gallons, at 14*s*. per gallon? *Ans.* £58 16*s*.

19. When the price is an *odd* number of shillings, and less than 20—

RULE.—Find the amount at the next lower even number of shillings; and add the price at one shilling.

EXAMPLE.—What is the price of 275 ℔, at 17*s*. per ℔?

$$\begin{array}{r} 275 \\ 8 \\ \hline \end{array}$$

The price at 16*s*. (by the last rule) is   220   0
The price at 1*s*. is 275*s*.=   .   .   13   15

Hence the price at 16*s*.+1*s*., or 17*s*., is £233 15*s*.

The price at 17s. is equal to the price at 16s., plus the price at one shilling.

59. 86 oz., at 5s. per oz. ?　*Ans.* £21 10s.
60. 62 cwt., at 19s. per cwt. ?　*Ans.* £58 18s.
61. 14 yards, at 17s. per yard ?　*Ans.* £11 18s.
62. 439 tons, at 11s. per ton ?　*Ans.* £241 9s.
63. 96 gallons, at 7s. per gallon ?　*Ans.* £33 12s.

20. When the quantity is represented by a mixed number—

RULE.—Find the price of the integral part. Then multiply the given price by the numerator of the fraction, and divide the product by its denominator—the quotient will be the price of the fractional part. The sum of these prices will be the price of the whole quantity.

EXAMPLE.—What is the price of 8¾ ℔ of tea, at 5s. per ℔ ?

$$\begin{array}{ccc} & \pounds & s. \quad d. \\ \text{The price of 8 ℔ is } 8\times 5s.= & 2 & 0 \ \text{-}0 \\ \text{The price of } \tfrac{3}{4} \text{ ℔ is } \dfrac{3\times 5s.}{4}= & 0 & 3 \quad 9 \\ \hline \text{And the price of 8¾ ℔ is } \ . \quad . & 2 & 3 \quad 9 \end{array}$$

The price of ¾ of a pound, is evidently ¾ of the price of a pound.

What is the price of
64. 5½ dozen, at 3s. 3d. per dozen ?　*Ans.* 17s. 10½d.
65. 273¼ ℔, at 2s. 6d. per ℔ ?　*Ans.* £34 3s. 1½d.
66. 530¾ ℔, at 14s. per ℔ ?　*Ans.* £371 10s. 6d.
67. 178⅜ cwt., at 17s. per cwt. ?　*Ans.* £151 12s. 4½d.
68. 762⅖ cwt , at £1 12s. 6d. per cwt. ?　*Ans.* £1239 4s. 6d.
69. 817 5/10 cwt., at £3 7s. 4d. per cwt. ?　*Ans.* £2751 11s. 6¼d.

21. The rules for finding the price of several denominations, that of one being given [7 and 9], may be abbreviated by those which follow—

Avoirdupoise Weight.—Given the price per cwt., to find the price of hundreds, quarters, &c.—

RULE.—Having brought the tons, if any, to cwt., multiply 1 by the number of hundreds, and consider the product as pounds sterling ; 5 by the number of quarters, and consider the product as shillings ; 2⅐, the number of pounds, and consider the product as pence :—the sum of all the products will be the price at £1 per cwt. From this find the price, at the given number of pounds, shillings, &c.

EXAMPLE.—What is the price of 472 cwt., 3 qrs., 16 ·℔, at £5 9s. 6d. per cwt. ?

$$
\begin{array}{rrl}
£ & s. & d. \\
1 & 5 & 2⅐ \\
\text{Multipliers } 472 & 3 & 16 \\
\hline
472 & 17 & 10¼ \text{ is the price at £1 per cwt.} \\
 & & 5 \\
\hline
2364 & 9 & 3¼ \text{ the price, at £5 per cwt.} \\
212 & 16 & 0¾ \text{ the price, at 9s. } (£\tfrac{1}{20}×9.) \\
11 & 16 & 5¼ \text{ the price, at 6d. } (£\tfrac{1}{20}÷2.) \\
\hline
2589 & 1 & 9¼ \text{ the price, at £5 9s. 6d.} \\
\end{array}
$$

At £1 per cwt., there will be £1 for every cwt. We multiply the qrs. by 5, for shillings ; because, if one cwt. costs £1, the fourth of 1 cwt., or one quarter, will cost the fourth of a pound, or 5s.—and there will be as many times 5s. as there are quarters. The pounds are multiplied by 2⅐ ; because if the quarter costs 5s., the 28th part of a quarter, or 1 ℔, must cost the 28th part of 5s., or 2⅐d.—and there will be as many times 2⅐d. as there are pounds.

EXERCISES.

What is the price of
70. 499 cwt., 3 qrs., 25 ℔, at 25s. 11d. per cwt. ? *Ans.* £647 17s. 7½d.
71. 106 cwt., 3 qrs., 14 ℔, at 18s. 9d. per cwt. ? *Ans.* £100 3s. 10¾d.

72. 2061 cwt., 2 qrs., 7 ℔, at 16*s.* 6*d.*, per cwt.? *Ans.* £1700 15*s.* 9¼*d.*

73. 106 cwt., 3 qrs., 14 ℔, at 9*s.* 4*d.* per cwt.? *Ans.* £49 17*s.* 6*d.*

74. 26 cwt., 3 qrs., 7 ℔, at 15*s.* 9*d.* per cwt.? *Ans.* £21 2*s.* 3¼*d.*

75. 432 cwt., 2 qrs., 22 ℔, at 18*s.* 6*d.* per cwt.? *Ans.* £400 4*s.* 10½*d.*

76. 109 cwt., 0 qrs., 15 ℔, at 19*s.* 9*d.* per cwt.? *Ans.* £107 15*s.* 4¾*d.*

77. 753 cwt., 1 qr., 25 ℔, at 15*s.* 2*d.* per cwt.? *Ans.* £571 7*s.* 8*d.*

78. 19 tons, 19 cwt., 3 qrs., 27½℔, at £19 19*s.* 11¾*d.* per ton? *Ans.* £399 19*s.* 6*d.*

22. To find the price of cwt., qrs., &c., the price of a pound being given—

RULE.—Having reduced the tons, if any, to cwt., multiply 9*s.* 4*d.* by the number of pence contained in the price of one pound :—this will be the price of one cwt. Divide the price of one cwt. by 4, and the quotient will be the price of one quarter, &c.

Multiply the price of 1 cwt. by the number of cwt.; the price of a quarter by the number of quarters; the price of a pound by the number of pounds; and the sum of the products will be the price of the given quantity.

EXAMPLE.—What is the price of 4 cwt., 3 qrs., 7 ℔, at 8*d.* per ℔.?

```
        s.  d.
        9   4
            8                          s.  d.
    4)74   8 the price of 1 cwt. ×4, will give 298  8 the price of 4 cwt.
   28)18   8 the price of 1 qr.  ×3, will give  56  0 the price of 3 qrs.
        8   8 the price of 1 ℔   ×7, will give   4  8 the price of 7 ℔.
                                         20)359  4
```
And the price of the whole will be £17 19 4

At 1*d.* per ℔ the price of 1 cwt. would be 112*d.* or 9*s.* 4*d.* :—therefore the price per cwt. will be as many times 9*s.* 4*d.* as there are pence in the price of a pound. The price of a quarter is ¼ the price of 1 cwt.; and there will be as many times the price of a quarter, as there are quarters, &c.

EXERCISES.

What is the price of ·

79. 1 cwt., at 6*d*. per ℔ ? *Ans.* £2 16*s*.
80. 3 cwt., 2 qrs., 5 ℔, at 4*d*. per ℔ ? *Ans.* £6 12*s*. 4*d*.
81. 51 cwt., 3 qrs., 21 ℔, at 9*d*. per ℔ ? *Ans.* £218 2*s*. 9*d*.
82. 42 cwt., 0 qrs., 5 ℔, at 25*d*. per ℔ ? *Ans.* £490 10*s*. 5*d*.
83. 10 cwt., 3 qrs., 27 ℔, at 51*d*.-per ℔ ? *Ans.* £261 11*s*. 9*d*.

23. Given the price of a pound, to find that of a ton—
RULE.—Multiply £9 6*s*. 8*d*. by the number of pence contained in the price of a pound.

EXAMPLE.—What is the price of a ton, at 7*d*. per ℔ ?

$$\begin{array}{ccc} £ & s. & d. \\ 9 & 6 & 8 \\ & & 7 \\ \hline 65 & 6 & 8 \end{array}$$ is the price of 1 ton.

If one pound costs 1*d*., a ton will cost 2240*d*., or £9 6*s*. 8*d*. Hence there will be as many times £9 6*s*. 8*d*. in the price of a ton, as there are pence in the price of a pound.

EXERCISES.

What is the price of

84. 1 ton, at 3*d*. per ℔ ? *Ans.* £28.
85. 1 ton, at 9*d*. per ℔ ? *Ans.* £84.
86. 1 ton, at 10*d*. per ℔ ? *Ans.* £93 6*s*. 8*d*.
87. 1 ton, at 4*d*. per ℔ ? *Ans.* £37 6*s*. 8*d*.

The price of any *number* of tons will be found, if we multiply the price of 1 ton by that number.

24. *Troy Weight.*—Given the price of an ounce—to find that of ounces, pennyweights, &c.—
RULE.—Having reduced the pounds, if any, to ounces, set down the ounces as pounds sterling ; the dwt. as shillings ; and the grs. as halfpence :—this will give the price at £1 per ounce. Take the same part, or parts, &c., of this, as the price per ounce is of a pound.

EXAMPLE 1.—What is the price of 538 oz., 18 dwt., 14 grs., at 11s. 6d. per oz. ?

$$11s.\ 6d.=\frac{£1}{2}+\frac{£\frac{1}{2}}{10}+\frac{£\frac{1}{2}}{10}\div 2.$$

£	s.	d.	
2)538	18	7	is the price, at £1 per ounce.
10)269	9	3½	is the price, at 10s. per ounce.
2) 26	18	11¼	is the price, at 1s. per ounce.
13	9	5¾	is the price, at 6d. per ounce.

And 309 17 8¼ is the price, at 11s. 6d. per ounce. 14 halfpence are set down as 7 pence.

If one ounce, or 20 dwt. costs £1, 1 dwt. or the 20th part of an ounce will cost the 20th part of £1—or 1s. ; and the 24th part of 1 dwt., or 1 gr. will cost the 24th part oɟ 1s.—or ½d.

EXAMPLE 2.—What is the price of 8 oz. 20 grs., at £3 2s. 6d. per oz. ?

£	s.	d.	
8	0	10	is the price, at £1 per ounce.
		3	

	£	s.	d.	
	24	2	6	is the price, at £3 per ounce.
Price at £1 ÷ 10 =	0	16	1	is the price, at 2s. per ounce.
Price at 2s. ÷ 4 =	0	4	0¼	is the price, at 6d. per ounce.

And £25 2 7¼ is the price, at £3 2s. 6d. per oz.

EXERCISES.

What is the price of

88. 147 oz., 14 dwt., 14 grs., at 7s. 6d. per oz. ? Ans. £55 7s. 11½d.

89. 194 oz., 13 dwt., 16 grs., at 11s. 6d. per oz. ? Ans. £111 18s. 10¼d.

90. 214 oz., 14 dwt., 16 grs., at 12s. 6d. per oz. ? Ans. £134 4s. 2d.

91. 11 ℔, 10 oz., 10 dwt., 20 grs., at 10s. per oz. ? Ans. £71 5s. 5d.

92. 19 ℔, 4 oz., 3 grs., at £2 5s. 2d. per oz. ? Ans. £523 18s. 11½d.

93. 3 oz., 5 dwt., 12 grs., at £1 6s. 8d. per oz. ? Ans. £4 7s. 3¾d.

25. *Cloth Measure.*—Given the price per yard—to find the price of yards, quarters, &c.—

RULE.—Multiply £1 by the number of yards; 5s. by the number of quarters; 1s. 3d. by the number of nails; and add these together for the price of the quantity at £1 per yard. Take the same part, or parts, &c., of this, as the price is of £1.

EXAMPLE 1.—What is the price of 97 yards, 3 qrs., 2 nails, at 8s. per yard?

```
                £1   5s.  1s. 3d.
   Multipliers  97    3    2
            2)97  17    6  is the price, at £1 per yard.
            5)48  18    9  is the price, at 10s. per yard.
From this subtract 9  15   9  the price, at 2s. per yard.
And the remainder 39  3    0  is the price at 8s. (10s.—2s.)
```

If a yard costs £1, a quarter of a yard must cost 5s.; and a nail, or the 4th of a yard, will cost the 4th part of 5s. or 1s. 3d.

EXAMPLE 2.—What is the price of 17 yards, 3 qrs., 2 nails, at £2 5s. 9d. per yard?

```
                £1   5s.  1s. 3d.
   Multipliers  17    3    2
                17   17    6  is the price, at £1 per yard.
                         2
                35   15    0  is the price, at £2 per yard.
```
The price at £1 ÷ 4 = 4 9 4½ is the price, at 5s.
The price at 5s. ÷ 10 = 0 8 11¼ is the price, at 6d.
The price at 6d. ÷ 2 = 0 4 5¼ is the price, at 3d.

And £40 17 9¼ is the price, at £2 5s. 9d.

EXERCISES.

What is the price of

94. 176 yards, 2 qrs., 2 nails, at 15s. per yard? *Ans.* £132 9s. 4½d.

95. 37 yards, 3 qrs., at £1 5s. per yard? *Ans.* £47 3s. 9d.

96. 49 yards, 3 qrs., 2 nails, at £1 10s. per yard? *Ans.* £74 16s. 3d.

97. 98 yards, 3 qrs., 1 nail, at £1 15s. per yard? *Ans.* £172 18s. 5¼d.

98. 3 yards, 1 qr., at 17s. 6d. per yard? *Ans.* £2 16s. 10½d.

99. 4 yards, 2 qrs., 3 nails, at £1 2s. 4d. per yard? *Ans.* £5 4s. 8¼d.

26. *Land Measure.*—RULE.—Multiply £1 by the number of acres; 5s. by the number of roods; and 1½d. by the number of perches:—the sum of the products will be the price at £1 per acre. From this find the price, at the given sum.

EXAMPLE.—What is the rent of 7 acres, 3 roods, 16 perches, at £3 8s. per acre?

```
                        £.   s.  d.
                        1    5   1¼
    Multipliers         7    3   16
                        ─────────────
Sum of the products     7    17  0, or the price at £1 per acre.
                             3
                        ─────────────
                        23   11  0  the price at £3 per acre.
                        3    18  6  the price at 10s. per acre.
                        ─────────────
                        27   9   6  the price at £3 10s. per acre.
    Subtract            0    15  8¼ the price at 2s. per acre.
                        ─────────────
    And                 26   13  9¼ is the price at £3 8s.
```

If one acre costs £1, a quarter of an acre, or one rood, must cost 5s.; and the 40th part of a quarter, or one perch, must cost the 40th part of 5s.—or 1¼d.

What is the rent of

100. 176 acres, 2 roods, 17 perches, at £5 6s. per acre? *Ans.* £936 0s. 3d.

101. 256 acres, 3 roods, 16 perches, at £6 6s. 6d. per acre? *Ans.* £1624 11s. 6¼d.

102. 144 acres, 1 rood, 14 perches, at £5 6s. 8d. per acre? *Ans.* £769 16s.

103. 344 acres, 3 roods, 15 perches, at £4 1s. 1d. per acre? *Ans.* £1398 1s. 1d.

27· *Wine Measure.*—To find the price of a hogshead, when the price of a quart is given—

RULE.—For each hogshead, reckon as many pounds, and shillings as there are pence per quart.

EXAMPLE.—What is the price of a hogshead at 9*d*. per ,juart? *Ans.* £9 9s.

One hogshead at 1*d*. per quart would be 63×4, since there are 4 quarts in one gallon, and 63 gallons in one hhd. But $63 \times 4d. = 252d. = £1\ 1s.$; and, therefore, the price, at 9*d*. per quart, will be nine times as much—or $9 \times £1\ 1s. = £9\ 9s.$

EXERCISES.

What is the price of
104. 1 hhd. at 18*d*. per quart? *Ans.* £18 18s.
105. 1 hhd. at 19*d*. per quart? *Ans.* £19 19s.
106. 1 hhd. at 20*d*. per quart? *Ans.* £21.
107. 1 hhd. at 2*s*. per quart? *Ans.* £25 4s.
108. 1 hhd. at 2*s*. 6*d*. per quart? *Ans.* £31 10s.

When the price of a pint is given, of course we know that of a quart.

28. Given the price of a quart, to find that of a tun—
RULE.—Take 4 times as many pounds, and 4 times as many shillings as there are pence per quart.

EXAMPLE.—What is the price of a tun at 11*d*. per quart?

$$
\begin{array}{cc}
£ & s. \\
11 & 11 \\
\underline{4} & \\
46 & 4
\end{array}
$$
is the price of a tun.

Since a tun contains 4 hogsheads, its price must be 4 times the price of a hhd.; that is, 4 times as many pounds and shillings, as pence per quart [27].

EXERCISES.

What is the price of
109. 1 tun, at 19*d*. per quart? *Ans.* £79 16s.
110. 1 tun, at 20*d*. per quart? *Ans.* £84.
111. 1 tun, at 2*s*. per quart? *Ans.* £100 16s.
112. 1 tun, at 2*s*. 6*d*. per quart? *Ans.* £126.
113. 1 tun, at 2*s*. 8*d*. per quart? *Ans.* £134 8s.

29. *A number of Articles.*—Given the price of 1 article in pence, to find that of any number—
RULE.—Divide the number by 12, for shillings and

pence; and multiply the quotient by the number of pence in the price.

EXAMPLE.—What is the price of 438 articles, at 7*d.* each?

12)438

    36*s.* 6*d.*, the price at 1*d.* each.

       7

20)255 6

£12 15 .6 the price at 7*d.* each.

438 articles at 1*d.* each will cost 438*d.*=36*s.* 6*d.* At 7*d.* each, they will cost 7 times as much—or 7×36*s.* 6*d.*=255*s.* 6*d.* = £12 15*s.* 6*d.*

EXERCISES.

What is the price of

114. 176 ℔, at 3*d.* per ℔?   *Ans.* £2 4*s.*
115. 146 yards, at 9*d.* per yard?   *Ans.* £5 9*s.* 6*d.*
116. 180 yards, at 10½*d.* per yard?   *Ans.* £7 17*s.* 6*d.*
117. 192 yards, at 7½*d.* per yard?   *Ans.* £6.
118. 240 yards, at 8½*d.* per yard?   *Ans.* £8 10*s.*

30. *Wages.*—Having the wages per day, to find their amount per year—

RULE.—Take so many pounds, half pounds, and 5 pennies sterling, as there are pence per day.

EXAMPLE.—What are the yearly wages, at 5*d.* per day?

£ *s.* *d.*

1 10 5

     5 the number of pence per day.

7. 12 1 the wages per year.

One penny per day is equal to 365*d.*=240*d.*+120*d.*+5*d.*= £1+10*s.*+5*d.* Therefore any number of pence per day, must be equal to £1 10*s.* 5*d.* multiplied by that number.

What is the amount per year, at

119. 3*d.* per day?   *Ans.* £4 11*s.* 3*d.*
120. 7*d.* per day?   *Ans.* £10 12*s.* 11*d.*
121. 9*d.* per day?   *Ans.* £13 13*s.* 9*d.*
122. 14*d.* per day?   *Ans.* £21 5*s.* 10*d.*
123. 2*s.* 3*d.* per day?   *Ans.* £41 1*s.* 3*d.*
124. 8½*d.* per day?   *Ans.* £12 18*s.* 6¼*d.*

BILLS OF PARCELS.

Dublin, 16th April, **1844.**

Mr. John Day,

Bought of Richard Jones.

	s.	d.		£	s.	d.
15 yards of fine broadcloth, at	13	6	per yard	10	2	6
24 yards of superfine ditto, at	18	9	„	22	10	0
27 yards of yard wide ditto, at	8	4	„	11	5	0
16 yards of drugget, at . .	6	3	„	5	0	0
12 yards of serge, at . .	2	10	„	1	14	0
32 yards of shalloon, at . .	1	8	„	2	13	4

Ans. £53 4 10

Dublin, 6th May, 1844.

Mr. James Paul,

Bought of Thomas Norton.

	s.	d.	
9 pair of worsted stockings, at	4	6	per pair
6 pair of silk ditto, at . .	15	9	„
17 pair of thread ditto, at .	5	4	„
23 pair of cotton ditto, at .	4	10	„
14 pair of yarn ditto, at .	2	4	„
18 pair of women's silk gloves, at	4	2	.„
19 yards of flannel, at . .	1	7½	per yard

Ans. £23 15 4¼

Dublin, 17th May, 1844.

Mr. James Gorman,

Bought of John Walsh & Co.

	s.	d.	
40 ells of dowlas, at .	1	6	per ell
34 ells of diaper, at . .	1	4¼	„
31 ells of Holland, at .	5	8	„
29 yards of Irish cloth, at	2	4	per yard
17½ yards of muslin, at .	7	2½	„
13¾ yards of cambric, at .	10	6	„
54 yards of printed calico, at	1	2½	„

Ans. £34 5 10¼

L 2

Dublin, 20*th May*, 1844.

Lady Denny,

Bought of Richard Mercer

	s.	d.	
$9\frac{1}{2}$ yards of silk, at . .	12	9	per yard
13 yards of flowered do., at	15	6	"
$11\frac{3}{4}$ yards of lustring, at .	6	10	"
14 yards of brocade, at .	11	3	"
$12\frac{1}{4}$ yards of satin, at .	10	8	"
$11\frac{3}{8}$ yards of velvet, at .	18	0	"

Ans. £44 15 10

Dublin, 21*st May*, 1844.

Mr. Jonas Darling,

Bought of William Roper.

	s.	d.	
$15\frac{1}{2}$ lb of currants, at . .	0	4	per lb
$17\frac{1}{4}$ lb of Malaga raisins, at .	0	$5\frac{1}{2}$	"
$19\frac{3}{4}$ lb of raisins of the sun, at .	0	6	"
17 lb of rice, at . . .	0	$3\frac{1}{2}$	"
$8\frac{1}{4}$ lb of pepper, at . .	1	6	"
3 loaves of sugar, weight $32\frac{1}{2}$ lb, at	0	$8\frac{1}{2}$	"
13 oz. of cloves, at . . .	0	9	per oz.

Ans. £3 13 $0\frac{1}{2}$

Dublin, 27*th June*, 1844.

Mr. Thomas Wright,

Bought of Stephen Brown & Co.

	s.	d.	
252 gallons of prime whiskey, at	6	4	per gallon
252 gallons of old malt, at .	6	8	"
252 gallons of old malt, at .	8	0	"

Ans. £264 12 0

MISCELLANEOUS EXERCISES.

What is the price of

1. 4715 yards of tape, at $\frac{1}{4}$*d.* per yard? *Ans.* £4 18*s.* $2\frac{3}{4}$*d.*

2. 354 lb, at $1\frac{1}{4}$*d.* per lb? *Ans.* £1 16*s.* $10\frac{1}{2}$*d.*

3. 4756 lb of sugar, at $12\frac{1}{4}$*d.* per lb? *Ans.* £242 15*s.* 1*d.*

4. 425 pair of silk stockings, at 6*s.* per pair? *Ans.* £127 10*s.*

5. 3754 pair of gloves, at 2*s.* 6*d.* ? *Ans.* £469 5*s.*

6. 3520 pair of gloves, at 3*s.* 6*d.* ? *Ans.* £616.

7. 7341 cwt., at £2 6*s.* per cwt. ? £16884 6*s.*

8. 435 cwt., at £2 7*s.* per cwt. ? *Ans.* £1022 5*s.*

9. 4514 cwt., at £2 17*s.* 7½*d.* per cwt. ? *Ans.* £13005 19*s.* 3*d.*

10. 3749⅜ cwt., at £3 15*s.* 6*d.* per cwt. ? *Ans.* £14153 17*s.* 9¾*d.*

11. 17 cwt., 1 qr., 17 ℔, at £1 4*s.* 9*d.* per cwt. ? £21 10*s.* 8¼*d.*

12. 78 cwt., 3 qrs., 12 ℔, at £2 17*s.* 9*d.* per cwt. ? *Ans.* £227 14*s.*

13. 5 oz., 6 dwt., 17 grs., at 5*s.* 10*d.* per oz. ? *Ans.* £1 11*s.* 1½*d.*

14. 4 yards, 2 qrs., 3 nails, at £1 2*s.* 4*d.* per yard? *Ans.* £5 4*s.* 8¼*d.*

15. 32 acres, 1 rood, 14 perches, at £1 16*s.* per acre ? *Ans.* £58 4*s.* 1¾*d.*

16. 3 gallons, 5 pints, at 7*s.* 6*d.* per gallon ? *Ans.* £1 7*s.* 2¼*d.*

17. 20 tons, 19 cwt., 3 qrs., 27½ ℔, at £10 10*s.* per ton ? *Ans.* £220 9*s.* 11½*d.* nearly

18. 219 tons, 16 cwt., 3 qrs., at £11 7*s.* 6*d.* per ton ? *Ans.* £2500 13*s.* 0½*d.*

QUESTIONS IN PRACTICE.

1. What is practice ? [1].

2. Why is it so called ? [1].

3. What is the difference between aliquot, and aliquant parts ? [2].

4. How are the aliquot parts of abstract, and of applicate numbers found ? [3].

5. What is the difference between prime, and compound aliquot parts ? [3].

6. How is the price of any denomination found, that of another being given ? [6 and 8].

7. How is the price of two or more denominations found, that of one being given ? [7 and 9].

8. The price of one denomination being given, how do we find that of any number of another ? [10].

9. When the price of any denomination is the aliquot part of a shilling, how is the price of any number of that denomination found? [11].

10. When the price of any denomination is the aliquot part of a pound, how is the price of any number of that denomination found? [12].

11. What is meant by the complement of the price? [13].

12. When the complement of the price of any denomination is the aliquot part of a pound or shilling, but the price is not so, how is the price of any number of that denomination found? [13].

13. When neither the price of a given denomination, nor its complement, is the aliquot part of a pound or shilling, how do we find the price of any number of that denomination? [14, 15, 16, and 17].

14. How do we find the price of any number of articles, when the price of each is an even or odd number of shillings, and less than 20? [18 and 19].

15. How is the price of a quantity, represented by a mixed number, found? [20].

16. How do we find the price of cwt., qrs., and ℔, when the price of 1 cwt. is given? [21].

17. How do we find the price of cwt., qrs., and ℔, when the price of 1 ℔ is given? [22].

18. How is the price of a ton found, when the price of 1 ℔ is given? [23].

19. How do we find the price of oz., dwt., and grs., when the price of an ounce is given? [24].

20. How do we find the price of yards, qrs., and nails, when the price of a yard is given? [25].

21. How do we find the price of acres, roods, and perches? [26].

22. How may the price of a hhd. or a tun be found, when the price of a quart is given? [27 and 28].

23. How may the price of any number of articles be found, the price of each in pence being given? [29].

How are wages per year found, those per day being given? [30].

TARE AND TRET.

31. The *gross* weight is the weight both of the goods, and of the bag, &c., in which they are.

Tare is an allowance for the bag, &c., which contains the article.

Suttle is the weight which remains, after deducting the tare.

Tret is, usually, an allowance of 4 ℔ in every 104 ℔, or $\frac{1}{26}$ of the weight of goods liable to waste, after the tare has been deducted. _

Cloff is an allowance of 2 ℔ in every 3 cwt., after both tare and tret have been deducted.

What remains after making all deductions is called the *net,* or *neat* weight.

Different allowances are made in different places, and for different goods; but the mode of proceeding is in all cases very simple, and may be understood from the following—

EXERCISES.

1. Bought 100 carcases of beef at 18*s.* 6*d.* per cwt.; gross weight 450 cwt., 2 qrs., 23 ℔; tret 8 ℔ per carcase. What is to be paid for them?

	cwt.	qrs.	℔.	
				100 carcasses.
Gross	450	2	23	8 ℔ per carcass.
Tret	7	0	16	—— cwt. qrs. ℔.

Tret, on the entire, 800 ℔ = 7   0   16

443   2   7 at 18*s.* 6*d.* per cwt. = £410 5*s.* 10$\frac{7}{8}$*d.*

2. What is the price of 400 raw hides, at 19*s.* 10*d.* per cwt.; the gross weight being 306 cwt., 3 qrs., 15 ℔; and the tret 4 ℔ per hide? *Ans.* £290 3*s.* 2$\frac{3}{8}$*d.*

3. If 1 cwt. of butter cost £3, what will be the price of 250 firkins; gross weight 127 cwt., 2 qrs., 21 ℔; tare 11 ℔ per firkin? *Ans.* £309 8*s.* 0$\frac{7}{8}$*d.*

4. What is the price of 8 cwt., 3 qrs., 11 ℔, at 15*s.* 6*d.* per cwt., allowing the usual tret? *Ans.* £6 11*s.* 10$\frac{3}{4}$*d.*

5. What is the price of 8 cwt. 21 ℔, at 18s. 4½d. per cwt., allowing the usual tret ? *Ans.* £7 4s. 8½d.

6. Bought 2 hhds. of tallow; No. 1 weighing 10 cwt., 1 qr., 11 ℔, tare 3 qrs., 20 ℔ ; and No. 2, 11 cwt., 0 qr., 17 ℔, tare 3 qrs., 14 ℔ ; tret 1 ℔ per cwt. What do they come to, at 30s. per cwt. ?

	cwt.	qrs.	℔.			cwt.	qrs.	℔.
Gross weight of No. 1,	10	1	11	.	Tare 0	3	20	
Gross weight of No. 2,	11	0	17	.	Tare 0	3	14	
Gross weight, .	.	21	2	0.		1	3	6
Tare, . .	.	1	3	6				
Suttle, . .	.	19	2	22				
Tret 1 ℔. per cwt.	.	0	0	19$\frac{3}{5}\frac{2}{8}$				

Net weight, 19 2 2$\frac{1}{5}\frac{7}{8}$. The price, at 30s. per cwt., is £29 5s. 7$\frac{3}{7}\frac{1}{8}\frac{7}{4}$d.

It is evident that the tret may be found by the following proportion—

cwt. cwt. qrs. ℔. ℔. ℔.
1 : 19 2 22 :: 1 : 19$\frac{3}{5}\frac{2}{8}$.

7. What is the price of 4 hhds. of copperas; No. 1 weighing gross 10 cwt., 2 qrs., 4 ℔, tare 3 qrs. 4 ℔ ; No. 2, 11 cwt., 0 qr., 10 ℔, tare 3 qrs. 10 ℔ ; No. 3, 12 cwt. 1 qr., tare 3 qrs. 14 ℔ ; No. 4, 11 cwt., 2 qrs., 14 ℔, tare 3 qrs. 18 ℔ ; the tret being 1 ℔ per cwt. ; and the price 10s. per cwt. ? *Ans.* £20 17s. 1$\frac{4}{7}\frac{8}{8}\frac{5}{4}$d.

8. What will 2 bags of merchandise come to ; No. 1, weighing gross 2 cwt., 3 qrs., 10 ℔ ; No. 2, 3 cwt., 3 qrs., 10 ℔ ; tare, 16 ℔ per bag ; tret 1 ℔ per cwt. ; and at 1s. 8d. per ℔ ? *Ans.* £59 2s. 8$\frac{1}{4}$d.

9. A merchant has sold 3 bags of pepper; No. 1, weighing gross 3 cwt. 2 qrs. ; No. 2, 4 cwt., 1 qr., 7 ℔; No. 3, 3 cwt., 3 qrs., 21 ℔ ; tare 40 ℔ per bag ; tret 1 ℔ per cwt. ; and the price being 15d. per ℔. What do they come to ? *Ans.* £74 1s. 7$\frac{2}{2}\frac{3}{8}$d.

10. Bought 3 packs of wool, weighing, No. 1, 3 cwt., 1 qr., 12 ℔ ; No. 2, 3 cwt., 3 qrs., 7 ℔ ; No. 3, 3 cwt., 2 qrs., 15 ℔ ; tare 30 ℔ per pack ; tret 8 ℔ for every 20 stone ; and at 10s. 6d. per stone. What do they amount to ?

	cwt.	qrs.	℔.		℔.
No. 1,	3	1	12	Tare	30
No. 2,	3	3	7	Tare	30 ·
No. 3,	3	2	15	Tare	30
Gross,	10	3	6		90=3 qrs. 6 ℔.
Tare,	0	3	6		
Suttle,	10	0	0=70 stones.		

st.		st.		℔.		℔.	st.	℔.
20	:	70	::	8	:	28 = 1		12

	st.	℔.
Suttle,	70	0
Tret,	1	12

Net weight, 68. 4, at 10s. 6d. per stone=£35 16s. 7½d.

11. Sold 4 packs of wool at 9s. 9d. per stone; weighing, No. 1, 3 cwt., 3 qrs., 27 ℔.; No. 2, 3 cwt., 2 qrs.,
16 ℔.; No. 3, 4 cwt., 1 qr., 10 ℔.; No. 4, 4 cwt., 0
qr., 6 ℔ : tare 30 ℔ per pack, and tret 8 ℔ for every
20 stone. What is the price? $Ans.$ £49 15s. $2\frac{29}{128}d$.

12. Bought 5 packs of wool; weighing, No. 1, 4 cwt.,
2 qrs., 15 ℔; No. 2, 4 cwt. 2 qrs.; No. 3, 3 cwt.,
3 qrs., 21 ℔; No. 4, 3 cwt., 3 qrs., 14 ℔; No. 5, 4
cwt., 0 qr., 14 ℔ : tare 28 ℔ per pack; tret 8 ℔ for
every 20 stone; and at 11s. 6d. per stone. What is
the price? $Ans.$ £77 15s. $8\frac{13}{16}d$.

13. Sold 3 packs of wool; weighing gross, No. 1, 3
cwt., 1 qr., 27 ℔; No. 2, 3 cwt., 2 qrs., 16 ℔; No. 3,
4 cwt., 0 qr., 21 ℔ : tare 29 ℔ per pack; tret 8 ℔ for
every 20 stone; and at 11s. 7d. per stone. What is the
price? $Ans.$ £41 13s. $7\frac{281}{640}d$.

14. Bought 50 casks of butter, weighing gross, 202
cwt., 3 qrs., 14 ℔; tare 20 ℔ per cwt. What is the
net weight?

	cwt.	qrs.	℔.		cwt.	qrs.	℔.
	202	3	14	Gross weight,	202	3	14
	20			Tare, . .	36	0	25¼
qrs. cwt.	4040 ℔.			Net weight,	166	2	16¼

qrs. cwt.		
2 = ⅐	10	
1 = ¼	5 = ½ of the last, ⎫ =the tare on 3 qr. 14 ℔
14 = ⅛	2½ = ½ of the last, ⎭

Tare, 4057½ ℔ = 36 cwt., 0 qr., 25½ ℔.

15. The gross weight of ten hhds. of tallow is 104 cwt., 2 qrs., 25 ℔; and the tare 14 ℔ per cwt. What is the net weight? *Ans.* 91 cwt., 2 qrs., 14⅞ ℔.

16. The gross weight of six butts of currants is 58 cwt., 1 qr., 18 ℔; and the tare 16 ℔ per cwt. What is the net weight? *Ans.* 50 cwt., 0 qr., 7⅗ ℔.

17. What is the net weight of 39 cwt., 3 qrs., 21 ℔; the tare being 18 ℔ per cwt.; the tret 4 ℔ for 104 ℔; and the cloff 2 ℔ for every 3 cwt.?

	cwt.	qrs.	℔
Gross weight,	39	3	21
Tare, . .	6	1	13
Suttle, . .	33	2	2
Tret=$\frac{1}{26}$th, or	1	1	4
	32	0	26
	0	0	22
Net weight,	32	0	4

℔.	℔. cwt.	cwt.	qrs.	℔
$18 = \begin{cases} 16=\frac{1}{7} \\ 2=\frac{1}{7}\div 8 \end{cases}$		39	3	21
		5	2	23
		0	2	24
Tare,		6	1	13

2 ℔ in 3 cwt. is the $\frac{1}{168}$th part of 3 cwt.
Hence the cloff of 32 cwt. 26 ℔ is its $\frac{1}{168}$th part, or 0 0 ·22

18. What is the net weight of 45 hhds. of tobacco; weighing gross, 224 cwt., 3 qrs., 20 ℔; tare 25 cwt. 3 qrs.; tret 4 ℔ per 104; cloff 2 ℔ for every 3 cwt.? *Ans.* 190 cwt., 1 qr., 14$\frac{9}{28}$ ℔.

19. What is the net weight of 7 hhds. of sugar, weighing gross, 47 cwt., 2 qrs., 4 ℔; tare in the whole, 10 cwt., 2 qrs., 14 ℔; and tret 4 ℔ per 104 ℔? *Ans.* 35 cwt., 1 qr., 27 ℔.

20. In 17 cwt., 0 qr., 17 ℔, gross weight of galls, how much net; allowing 18 ℔ per cwt. tare; 4 ℔ per 104 ℔ tret; and 2 ℔ per 3 cwt. cloff? *Ans.* 13 cwt., 3 qrs., 1 ℔ nearly.

QUESTIONS.

1. What is the gross weight? [31].
2. What is tare? [31].
3. What is suttle? [31].
4. What is tret? [31].
5. What is cloff? [31].
6. What is the net weight? [31].
7. Are the allowances made, always the same? [31].

SECTION VII.

INTEREST, &c.

1. Interest is the price which is allowed for the use of money; it depends on the plenty or scarcity of the latter, and the risk which is run in lending it.

Interest is either simple or compound. It is *simple* when the interest due is not added to the sum lent, so as to bear interest.

It is *compound* when, after certain periods, it is made to bear interest—being added to the sum, and considered as a part of it.

The money lent is called the *principal.* The sum allowed for each hundred pounds " per annum" (for a year) is called the " *rate* per cent."—(per £100.) The *amount* is the sum of the principal and the interest due.

SIMPLE INTEREST.

2. To find the interest, at any rate per cent., on any sum, for one year—

RULE I.—Multiply the sum by the rate per cent., and divide the product by 100.

EXAMPLE.—What is the interest of £672 14*s.* 3*d.* for one year, at 6 per cent. (£6 for every £100.)

```
  £    s.   d.
 672  14   3
            6
 ───────────────
 40·36  5   6
   20
 ───────────────
  7·25  The quotient, £40 7s. 3d., is the interest required.
   12
 ─────
  3·06
```

We have divided by 100, by merely altering the decimal point [Sec. I. 34].

If the interest were 1 per cent., it would be the hundredth part of the principal—or the principal multiplied by $\frac{1}{100}$; but being 6 per cent., it is 6 times as much—or the principal multiplied by $\frac{6}{100}$.

3. RULE II.—Divide the interest into parts of £100; and take corresponding parts of the principal.

EXAMPLE.—What is the interest of £32 4s. 2d., at 6 per cent.?

£6=£5+£1, or $£\frac{100}{20}$ plus $£\frac{100}{20} \div 5$. Therefore the interest is the $\frac{1}{20}$ of the principal, plus the $\frac{1}{5}$ of the $\frac{1}{20}$.

```
            £    s.   d.
      20)32    4    2
      ────────────────
      5)1   12    2¼  is the interest, at 5 per cent.
        0    6    5¼  is the interest, at 1 per cent.
      ────────────────
  And 1   18    7¾  is the interest, at 6 (5 + 1) per cent.
```

EXERCISES.

1. What is the interest of £344 17s. 6d. for one year, at 6 per cent.? *Ans.* £20 13s. 10½d.

2. What is the interest of £600 for one year, at 5 per cent.? *Ans.* £30.

3. What is the interest of £480 15s. for one year, at 7 per cent.? *Ans.* £33 13s. 0⅖d.

4. What is the interest of £240 10s. for one year, at 4 per cent.? *Ans.* £9 12s. 4⅘d.

4. To find the interest when the rate per cent. consists of more than one denomination—

RULE.—Find the interest at the highest denomination; and take parts of this, for those which are lower. The sum of the results will be the interest, at the given rate.

EXAMPLE.—What is the interest of £97 8s. 4d., for one year, at £5 10s. per annum?

$$£5=£\frac{100}{20}; \text{ and } 10s.=£\frac{5}{10}.$$

```
        £    s.   d.
  20)97    8    4
  ────────────────
  10)4   17    5  is the interest, at 5 per cent.
     0    9    9  is the interest, at 10s. per cent.
  ────────────────
  And 5   7    2  is the interest, at £5 + 10s. per cent.
```

At 5 per cent. the interest is the $\frac{1}{20}$ of the principal; at 10s. per cent. it is the $\frac{1}{10}$ of what it is at 5 per cent. Therefore, at £5 10s. per cent., it is the sum of both.

EXERCISES.

5. What is the interest of £371 19s. $7\frac{1}{2}d.$ for one year, at £3 15s. per cent.? *Ans.* £13 18s. $11\frac{3}{4}d.$

6. What is the interest of £84 11s. $10\frac{1}{2}d.$ for one year, at £4 5s. per cent.? *Ans.* £3 11s. $10\frac{3}{4}d.$

7. What is the interest of £91 0s. $3\frac{3}{4}d.$ for one year, at £6 12s. 9d. per cent.? *Ans.* £6 0s. $10\frac{1}{4}d.$

8. What is the interest of £968 5s. for one year, at £5 14s. 6d. per cent.? *Ans.* £55 8s. 8d.

5. To find the interest of any sum, for *several* years—

RULE.—Multiply the interest of one year by the number of years.

EXAMPLE.—What is the interest of £32 14s. 2d. for 7 years, at 5 per cent.?

```
       £ . s.  d.
  20)32 14   2
       1 12  8¼ is the interest for one year, at 5 per cent.
             7
```

And 11 8 $1\frac{1}{4}$ is the interest for 7 years, at 5 per cent. This rule requires no explanation.

EXERCISES.

9. What is the interest of £14 2s. for 3 years, at 6 per cent? *Ans.* £2 10s. 9d.

10. What is the interest of £72 for 13 years, at £6 10s. per cent.? *Ans.* £60 16s. $9\frac{1}{2}d.$

11. What is the interest of £853 0s. $6\frac{1}{2}d.$ for 11 years, at £4 12s. per cent.? *Ans.* £431 12s. $7\frac{3}{4}d.$

6. To find the interest of a given sum for years, months, &c.—

RULE.—Having found the interest for the years, as already directed [2, &c.], take *parts* of the interest of one year, for that of the months, &c.; and then add the results.

EXAMPLE.—What is the interest of £86 8s. 4d. for 7 years and 5 months, at 5 per cent.?

```
              £    s.   d.
        20)86    8    4
             ─────────────
               4    6    5   is the interest for 1 year, at 5 per cent.
                         7
             ─────────────
£  s.  d.         30    4  11   is the interest for 7 years.
4  6  5 ÷3=  1    8   9¾  is the interest for 4 months.
1  8  9¾÷4=  0    7   2½  is the interest for 1 month.
```

And 32   0   11¼ is the required interest.

EXERCISES.

12. What is the interest of £211 5s. for 1 year and 6 months, at 6 per cent.? *Ans.* £19 0s. 3d.

13. What is the interest of £514 for 1 year and 7½ months, at 8 per cent.? *Ans.* £66 16s. 4⅘d.

14. What is the interest of £1090 for 1 year and 5 months, at 6 per cent.? *Ans.* £92 13s.

15. What is the interest of £175 10s. 6d. for 1 year and 7 months, at 6 per cent.? *Ans.* £16 13s. 5$\frac{27}{100}$d.

16. What is the interest of £571 15s. for 4 years and 8 months, at 6 per cent.? *Ans.* £160 1s. 9⅗d.

17. What is the interest of £500 for 2 years and 10 months, at 7 per cent.? *Ans.* £99 3s. 4d.

18. What is the interest of £93 17s. 4d. for 7 years and 11 months, at 6 per cent.? *Ans.* £44 11s. 7½d.

19. What is the interest of £84 9s. 2d. for 8 years and 8 months, at 5 per cent.? *Ans.* £36 11s. 11¼d.

7. To find the interest of any sum, for any time, at 5, or 6, &c., per cent.

At 5 per cent.—

RULE.—Consider the years as shillings, and the months as pence; and find what aliquot part or parts of a pound these are. Then take the same part or parts of the principal.

To find the interest at 6 per cent., find the interest at 5 per cent., and to it *add* its fifth part, &c.

The interest at 4 per cent. will be the interest at 5 per cent., *minus* its fifth part, &c.

8. EXAMPLE 1.—What is-the interest of £427 5s. 9d. for 6 years and 4 months, at 5 per cent. ?

6 years and 4 months are represented by 6s. 4d.; but 6s. 4d.=5s.+1s.+4d.=$\frac{1}{4}$+$\frac{1}{20}$ of a pound + the $\frac{1}{3}$ of the $\frac{1}{20}$.

```
        £    s.   d.
  4)427   5   9
  5)106  16   5¼  is the ¼ of principal.
  3)21    7   3½  is the ¹⁄₂₀ (⅕ of ¼) of principal.
      7   2   5   is the ¹⁄₆₀ (⅓ of ¹⁄₂₀) of principal.
```

And 135 6 1¾ is the required interest.

The interest of £1 for 1 year, at 5 per cent., would be 1s.; for 1 month 1d.; for any number of years, the same number of shillings; for any number of months, the same number of pence; and for years *and* months, a corresponding number of shillings *and* pence. But whatever part, or parts, these shillings, and pence are of a pound, the interest of any other sum, for the same time and rate, must be the same part or parts of that other sum—since the interest of any sum is proportional to the interest of £1.

EXAMPLE 2.—What is the interest of £14 2s. 2d. for 6 years and 8 months, at 6 per cent. ?

6s. 8d. is the $\frac{1}{3}$ of a pound.

```
        £   s.   d.
  3)14   2   2
  5)4   14   0¾  is the interest, at 5 per cent.
    0   18   9¾  is the interest, at 1 per cent.

    5   12  10¼  is the interest, at 6 (5+1) per cent.
```

EXERCISES.

20. Find the interest of £1090 17s. 6d. for 1 year and 8 months, at 5 per cent. *Ans.* £90 18s. 1½d.

21. Find the interest of £976 14s. 7d. for 2 years and 6 months, at 5 per cent.? *Ans.* £122 1s. 9⅞d.

22. Find the interest of £780 17s. 6d. for 3 years and 4 months, at 6 per cent.? *Ans.* £156 3s. 6d.

23. What is the interest of £197 11s. for 2 years and 6 months, at 5 per cent.? £24 13s. 10½d.

24. What is the interest of £279 11s. for 7½ months, at 4 per cent.? *Ans.* £6 19s. 9$\frac{5}{10}$d.

25. What is the interest of £790 16s. for 6 years and 8 months, at 5 per cent.? *Ans.* £263 12s.

26. What is the interest of £124 2*s*. 9*d*. for 3 years and 3 months, at 5 per cent.? *Ans*. £20 3*s*. 5$\frac{3}{4}$*d*.

27. What is the interest of £1837 4*s*. 2*d*. for 3 years and 10 months, at 8 per cent.? *Ans*. £563 8*s*. 3*d*.

9. When the *rate*, or number of years, or both of them, are expressed by a mixed number—

RULE.—Find the interest for 1 year, at 1 per cent., and multiply this by the number of pounds and the fraction of a pound (if there is one) per cent.; the sum of these products, or one of them, if there is but one, will give the interest for one year. Multiply this by the number of years, and by the fraction of a year (if there is one); and the sum of these products, or one of them, if there is but one, will be the required interest.

EXAMPLE 1.—Find the interest of £21 2*s*. 6*d*. for 3$\frac{3}{4}$ years at 5 per cent.?

£21 2*s*. 6*d*. ÷ 100 = 4*s*. 2$\frac{3}{4}$*d*. Therefore

£	*s*.	*d*.	
0	4	2$\frac{3}{4}$	is the interest for 1 year, at 1 per cent.
		5	
1	1	1$\frac{3}{4}$	is the interest for 1 year, at 5 per cent.
		3	
3	3	5$\frac{1}{4}$	is the interest for 3 years, at do.
0	15	10$\frac{1}{4}$	is the interest for $\frac{3}{4}$ of a year (£1 1*s*. 1$\frac{3}{4}$*d*. × $\frac{3}{4}$), at do.
3	19	3$\frac{1}{2}$	is the interest for 3$\frac{3}{4}$ years, at do.

EXAMPLE 2.—What is the interest of £300 for 5$\frac{3}{4}$ years, at 3$\frac{3}{4}$ per cent?

	£	*s*.	*d*.	
£300 ÷ 100 =	3	0	0	is the interest for 1 year, at 1 per cent.
			3	
	9	0	0	is the interest for 1 year, at 3 per cent.
	2	5	0	is the interest for 1 year, at £$\frac{3}{4}$ (£3 × $\frac{3}{4}$)
	11	5	0	is the interest for 1 year, at 3$\frac{3}{4}$ per cent
			5	
	56	5	0	is the interest for 5 years, at 3$\frac{3}{4}$ per cent
	5	12	6	is the interest for $\frac{1}{2}$ year (£11 5*s*. ÷ 2)
	2	16	3	is the do. for $\frac{1}{4}$ year (£5 12*s*. 6$\frac{3}{4}$*d*. ÷ 2)
And	64	13	9	is the interest for 5$\frac{3}{4}$ years, at 3$\frac{3}{4}$ do.

28. What is the interest of £379 2s. 6d. for 4⅓ years, at 5⅝ per cent.? *Ans.* £91 5s. 5d.

29. What is the interest of £640 10s. 6d. for 2½ years, at 4½ per cent.? *Ans.* £72 1s. 2$\frac{7}{40}$d.

30. What is the interest of £600 10s. 6d. for 3⅓ years, at 5¾ per cent.? *Ans.* £115 2s. 0$\frac{3}{20}$d.

31. What is the interest of £212 8s. 1½d. for 6⅔ years, at 5¾ per cent.? *Ans.* £81 8s. 5⅝d.

10. To find the interest for *days*, at 5 per cent—

RULE.—Multiply the principal by the number of days, and divide the product by 7300.

EXAMPLE.—What is the interest of £26 4s. 2d. for 8 days?

$$
\begin{array}{rrr}
£ & s. & d. \\
26 & 4 & 2 \\
 & & 8 \\
\hline
209 & 13 & 4 \\
20 & & \\
\hline
4193 & & \\
12 & & \\
\hline
\end{array}
$$

7300)50320($6\frac{3520}{7365}$d.
43800
———
6520

The required interest is $6\frac{3520}{7365}$, or *7d.*—since the remainder is *greater* than half the divisor.

The interest of £1 for 1 year is $£\frac{1}{20}$, and for 1 day $\frac{1}{20} \div 365 = \frac{1}{20 \times 365} = 7300$; that is, the 7300th part of the principal. Therefore the interest of any other sum for one day is the 7300th part of that sum; and for any number of days, it is that number, multiplied by the 7300th part of the principal— or, which is the same thing, the principal multiplied by the number of days, and divided by 7300.

32. Find the intérest of £140 10s. for 76 days, at 5 per cent. *Ans.* £1 9s. 3$\frac{21}{365}$d.

33. Find the interest of £300 for 91 days, at 5 per cent. *Ans.* £3 14s. 9$\frac{53}{73}$d.

34. What is the interest of £800 for 61 days, at 5 per cent.? *Ans.* £6 13s. 8$\frac{28}{73}$d.

11. To find the interest for *days*, at *any other* rate—
RULE.—Find the interest at 5 per cent., and take parts of this for the remainder.

EXAMPLE.—What is the interest of £3324 6s. 2d. for 11 days, at £6 10s. per cent. ?

£3324 6s. 2d. × 11 ÷ 7300 = £5 0s. 2¼d.   Therefore

	£	s.	d.	
5)	5	0	2¼	is the interest for 11 days, at 5 per cent.
2)	1	0	0⅜	is the interest for 11 days, at 1 per cent.
	0	10	0	is the interest for 11 days, at 10s. per cent

And   6 10   2⅔ is the interest for 11 days, at £6 10s. (£5 +
£1 + 10s.)
This rule requires no explanation.

EXERCISES.

35. What is the interest of £200 from the 7th May to the 26th September, at 8 per cent. ?   *Ans.* £6 4s. 5⁶⁷⁄₇₃d.

36. What is the interest of £150 15s. 6d. for 53 days, at 7 per cent. ?   *Ans.* £1 10s. 7¾d.

37. What is the interest of £371 for 1 year and 213 days, at 6 per cent. ?   *Ans.* £35 5s. 0d.

38. What is the interest of £240 for 1 year and 135 days, at 7 per cent. ?   *Ans.* £23 0s. 3²⁄₇₃d.

Sometimes the number of days is the aliquot part of a year; in which case the process is rendered more easy.

EXAMPLE.—What is the interest of £175 for 1 year and 73 days, at 8 per cent. ?

1 year and 73 days = 1⅕ year. Hence the required interest is the interest for 1 year + its fifth part. But the interest of £175 for 1 year, at the given rate is £14. Therefore its interest for the given time is £14 + £¹⁴⁄₅ = £14 + £2 16s. = £16 16s.

12. To find the interest for *months*, at 6 per cent—
RULE.—If the number expressing the months is *even*, multiply the principal by half *the number* of months and divide by 100. But if it is *odd*, multiply by the half of *one less than the number* of months; divide the result by 100; and add to the quotient what will be obtained if we divide it by one less than the number of months.

EXAMPLE 1.—What is the interest of £72 6s. 4d. for 8 months at 6 per cent. ?

$$
\begin{array}{rrr}
£ & s. & d. \\
72 & 6 & 4 \\
 & & 4 \\
\hline
£2\cdot89 & 5 & 4 \\
20 & & \\
\hline
\end{array}
$$

17·85s.   The required interest is £2 17s. 10¼d.
12
____

10·24d.
4
____

0·96=¼d. nearly.

Solving the question by the rule of three, we shall have—

£100 : £72 6s. 4d. :: £6 : $\dfrac{£72 \ 6s. \ 4d. \times 8 \times 6}{100 \times 12}=$ (dividing both numerator and denominator by 6 [Sec. IV. 4]).

$\dfrac{£72 \ 6s. \ 4d. \times 8 \times 6 \div 6}{100 \times 12 \div 6} = \dfrac{£72 \ 6s. \ 4d. \times 8}{100 \times 2}=$ (dividing both numerator and denominator by 2) $\dfrac{£72 \ 6s. \ 4d. \times 8 \div 2}{100 \times 2 \div 2}=$

$$\dfrac{£72 \ 6s. \ 4d. \times 4}{100}$$

—that is, the required interest is equal to the given sum, multiplied by half the number which expresses the months, and divided by 100.

EXAMPLE 2.—What is the interest of £84 6s. 2d. for 11 months, at 6 per cent. ?   11=10+1   ∴   10÷2=5.

$$
\begin{array}{rrr}
£ & s. & d. \\
84 & 6 & 2 \\
 & & 5 \\
\hline
£4\cdot21 & 10 & 10 \\
20 & & \\
\end{array}
$$

One less than the given number of months=10.

4·30s.   10)4 4 3¾ is the interest for 10 months, at 6 per cent.
12   0 8 5¼ is the interest for 1 month, at same rate.
____

3·70d. And 4 12 9 is the interest for 11 (10+1) months, at 6 do
4
____

2·80f.=¾d. nearly.

The interest for 11 months is evidently the interest of $\overline{11-1}$ month, plus the interest of $\overline{11-1}$ month $\div \overline{11-1}$.

M

EXERCISES.

39. What is the interest of £250 17s. 6d. for 8 months, at 6 per cent. ? *Ans.* £10 0s. 8⅔d. '

40. What is the interest of £571 15s. for 8 months, at 6 per cent. ? *Ans.* £22 17s. 4⅘d.

41. What is the interest of £840 for 6 months, at 6 per cent. ? *Ans.* £25 4s.

42. What is the interest of £3790 for 4 months, at 6 per cent. ? *Ans.* £75 16s.

43. What is the interest of £900 for 10 months, at 6 per cent. ? *Ans.* £45.

44. What is the interest of £43 2s. 2d· for 9 months, at 6 per cent. ? *Ans.* £1 18s. 9½d.

13. To find the interest of money, left after one or more payments—

RULE.—If the interest is paid by *days*, multiply the sum by the number of days which have elapsed before any payment was made. Subtract the first payment, and multiply the remainder by the number of days which passed between the first and second payments. Subtract the second payment, and multiply this remainder by the number of days which passed between the second and third payments. Subtract the third payment, &c. Add all the products together, and find the interest of their sum, for 1 day.

If the interest is to be paid by the *week* or *month*, substitute *weeks* or *months* for *days*, in the above rule.

EXAMPLE.—A person borrows £117 for 94 days, at 8 per cent., promising the principal in parts at his convenience, and interest corresponding to the money left unpaid, up to the different periods. In 6 days he pays £17; in 7 days more £20; in 15 more £32; and at the end of the 94 days, all the money then due. What does the interest come to ?

$$
\left.
\begin{array}{l}
117 \times 6 = 702 \times 1 \\
100 \times 7 = 700 \times 1 \\
80 \times 15 = 1200 \times 1 \\
48 \times 66 = 3168 \times 1
\end{array}
\right\} = £5770
$$

The interest on 5770 for 1 day, at 5 per cent., is 15s. 9¾d. Therefore

```
 £   s.   d.
5)0  15   9¾ is the interest, at 5 per cent.
 0   3    2  is the interest, at 1 per cent.
3)0  18  11¾ is the interest, at 6 per cent.
 0   6    4  is the interest, at 2 per cent.
```

And 1 5 3¾ is the interest, at 8 per cent., for the given sums and times.

If the entire sum were 6 days unpaid, the interest would be the same as that of 6 times as much, for 1 day. Next, £100 due for 7 days, should produce as much as £700, for 1 day, &c. And all the sums due for the different periods should produce as much as the sum of their equivalents, in 1 day.

EXERCISES.

45. A merchant borrows £250 at 8 per cent. for 2 years, with condition to pay before that time as much of the principal as he pleases. At the expiration of 9 months he pays £80, and 6 months after £70—leaving the remainder for the entire term of 2 years. How much interest and principal has he to pay, at the end of that time? *Ans.* £127 16s.

46. I borrow £300 at 6 per cent. for 18 months, with condition to pay as much of the principal before the time as I please. In 3 months I pay £60; 4 months after £100; and 5 months after that £75. How much principal and interest am I to pay, at the end of 18 months? *Ans.* £79 15s.

47. A gives to B at interest on the 1st November, 1804, £6000, at 4½ per cent. B is to repay him with interest, at the expiration of 2 years—having liberty to pay before that time as much of the principal as he pleases. Now B pays

			£
The 16th December, 1804,	.		900
The 11th March, 1805,	.	.	1260
The 30th March,	.	.	600
The 17th August,	.	.	800
The 12 February, 1806	.	.	1048

How much principal and interest is he to pay on the 1st November, 1806? *Ans.* £1642 9s. $2\frac{1545}{1823}d.$

48. Lent at interest £600 the 13th May, 1833, for

1 year, at 5 per cent.—with condition that the receiver may discharge as much of the principal before the time as he pleases. Now he pays the 9th July £200; and the 17th September £150. How much principal and interest is he to pay at the expiration of the year? *Ans.* £266 13s. 5$\frac{7}{73}$d.

14. It is hoped that the pupil, from what he has learned of the properties of proportion, will easily understand the modes in which the following rules are proved to be correct.

Of the principal, amount, time, and rate—given any three, to find the fourth.

Given the amount, rate of interest, and time; to find the principal—

RULE.—Say as £100, plus the interest of it, for the given time, and at the given rate, is to £100; so is the given amount to the principal sought.

EXAMPLE.—What will produce £862 in 8 years, at 5 per cent. ?

£40 (=£5×8) is the interest for £100 in 8 years at the given rate. Therefore

$$£140 : £100 :: £862 : \frac{862 \times 100}{140} = £615 \; 14s. \; 3\frac{1}{2}d.$$

When the time and rate are given— .

£100 : any other sum :: interest of £100 : interest of that other sum.

By alternation [Sec. V. 29], this becomes—

£100 : interest of £100 :: any other sum : interest of that sum.

And, saying " the first + the second : the second," &c. [Sec. V. 29] we have—

£100 + its interest : £100 :: any other sum + its interest : that sum—which is exactly the rule.

EXERCISES.

49. What principal put to interest for 5 years will amount to £402 10s., at 3 per cent. per annum ? *Ans.* £350.

50. What principal put to interest for 9 years, at 4 per cent., will amount to £734 8s. ? *Ans.* £540.

51. The amount of a certain principal, bearing interest for 7 years, at 5 per cent., is £334 16s. What is the principal? *Ans.* £248.

15. Given the time, rate of interest, and principal—to find the amount—

RULE.—Say, as £100 is to £100 plus its interest for the given time, and at the given rate, so is the given sum to the amount required.

EXAMPLE.—What will £272 come to, in 5 years, at 5 per cent.?

£125 (=£100+£5×5) is the principal and interest of £100 for 5 years; then—

£100 : £125 :: £272 : $\dfrac{272 \times 125}{100}$ =£340, the required amount.

We found by the last rule that

£100+its interest : £100 :: any other sum+its interest : that sum.

Inversion [Sec. V. 29] changes this into,

£100 : £100+its interest :: any other sum : that other sum+its interest—which is the present rule.

EXERCISES.

52. What will £350 amount to, in 5 years, at 3 per cent. per annum? *Ans.* £402 10s.

53. What will £540 amount to, in 9 years, at 4 per cent. per annum? *Ans.* £734 8s.

54. What will £248 amount to, in 7 years, at 5 per cent. per annum? *Ans.* £334 16s.

55. What will £973 4s. 2d. amount to, in 4 years and 8 months, at 6 per cent.? *Ans.* £1245 14s. 1¾d.

56. What will £42 3s. 9½d. amount to, in 5 years and 3 months, at 7 per cent.? *Ans.* £57 13s. 10½d.

16. Given the amount, principal, and rate—to find the time—

RULE.—Say, as the interest of the given sum for 1 year is to the given interest, so is 1 year to the required time.

EXAMPLE.—When would £281 13s. 4d. become £338, at 5 per cent. ?

£14 1s. 8d. (the interest of £281 13s. 4d. for 1 year [2]) : £56 6s. 8d. (the given interest) :: 1 : $\dfrac{£56\ 6s.\ 8d.}{£14\ 1s.\ 8d.} = 4$, the required number of years.

17. Hence, briefly, to find the time—Divide the interest of the given principal for 1 year, into the entire interest, and the quotient will be the time.

It is evident, the principal, and rate being given, the interest is proportional to the time ; the longer the time, the more the interest, and the reverse. That is—
The interest for one time : the interest for another :: the former time : the latter.
Hence, the interest of the given sum for one year (the interest for *one* time) : the given interest (the interest or the same sum for *another* time) :: 1 year (the time which produced the former) : the time sought (that which produced the latter)—which is the rule.

EXERCISES.

57. In what time would £300 amount to £372, at 6 per cent. ? *Ans.* 4 years.
58. In what time would £211 5s. amount to £230 5s. 3d., at 6 per cent. ? *Ans.* In 1 year and 6 months.
59. When would £561 15s. become £719 0s. 9¾d., at 6 per cent. ? *Ans.* In 4 years and 8 months.
60. When would £500 become £599 3s. 4d., at 7 per cent. ? *Ans.* In 2 years and 10 months.
61. When will £436 9s. 4d. become £571 8s. 1¼d., at 7 per cent. ? *Ans.* In 4 years and 5 months.

18. Given the amount, principal, and time—to find the rate—
RULE.—Say, as the principal is to £100, so is the given interest, to the interest of £100—which will give the interest of £100, at the same rate, and for the same time. Divide this by the time, and the quotient will be the rate.

EXAMPLE.—At what rate will £350 amount to £402 10s in 5 years?

$$£350 : £100 :: £52 10s. : \frac{£52 \ 10s. \times 100}{350} = £15,$$ the interest of £100 for the same time, and at the same rate. Then $\frac{15}{5} = 3$, is the required number of years.

We have seen [14] that the time and rate being the same,

£100 : any other sum :: the interest of £100 : interest of the other sum.

This becomes, by inversion [Sec. **V**. 29]—

Any sum : £100 :: interest of the former : interest of 100 (for same number of years).

But the interest of £100 divided by the number of years which produced it, gives the interest of £100 for 1 year—or, in other words, the *rate*.

EXERCISES.

62. At what rate will £300 amount in 4 years to £372? *Ans.* 6 per cent.

63. At what rate will £248 amount in 7 years to £334 16s.? *Ans.* 5 per cent.

64. At what rate will £976 14s. 7d. amount in 2 years and 6 months to £1098 16s. 4¾d.? *Ans.* 5 per cent.

Deducting the 5th part of the interest, will give the interest of £976 14s. 7d. for 2 years.

65. At what rate will £780 17s. 6d. become £937 1s. in 3 years and 4 months? *Ans.* 6 per cent.

66. At what rate will £843 5s. 9d. become £1047 1s. 7¾d., in 4 years and 10 months? *Ans.* At 5 per cent.

67. At what rate will £43 2s. 4¼d. become £60 7s. 4½d., in 6 years and 8 months? *Ans.* At 6 per cent.

68. At what rate will £473 become £900 13s. 6¼d. in 12 years and 11 months? *Ans.* At 7 per cent.

COMPOUND INTEREST.

19. Given the principal, rate, and time—to find the amount and interest—

RULE I.—Find the interest due at the first time of payment, and add it to the principal. Find the interest of that sum, considered as a new principal, and add it to what it would produce at the next payment. Consider that new sum as a principal, and proceed as before. Continue this process through all the times of payment.

Example.—What is the compound interest of £97, for 4 years, at 4 per cent. half-yearly?

	£	s.	d.	
	97	0	0	
	3	17	7¼	is the interest, at the end of 1st half-year.
	100	17	7¼	is the amount, at end of 1st half-year.
	4	0	8½	is the interest, at the end of 1st year.
	104	18	3¾	is the amount, at the end of 1st year.
	4	3	11¼	is the interest, at the end of 3rd half-year.
	109	2	3	is the amount, at the end of 3rd half-year.
	4	7	3½	is the interest, at the end of 2nd year.
	113	9	6½	is the amount, at the end of 2ud year.
	4	10	9½	is the interest, at the end of 5th half-year.
	118	0	4	is the amount, at the end of 5th half-year.
	4	14	5	is the interest, at the end of 3rd year.
	122	14	9	is the amount, at the end of 3rd year.
	4	18	2¼	is the interest, at the end of 7th half-year.
	127	12	11¼	is the amount, at the end of 7th half-year.
	5	2	1½	is the interest, at the end of 4th year.
	132	15	0¾	*is the amount, at the end of 4th year.*
	97	0	0	is the principal.
And	35	15	0¾	is the compound interest of £97, in 4 years.

20. This is a tedious mode of proceeding, particularly when the times of payment are numerous; it is, therefore, better to use the following rules, which will be found to produce the same result—

Rule II.—Find the interest of £1 for one of the payments at the given rate. Find the product of so many factors (each of them £1 + its interest for one payment) as there are times of payment; multiply this product by the given principal; and the result will be the principal, plus its compound interest for the given time. From this subtract the principal, and the remainder will be its compound interest.

EXAMPLE 1.—What is the compound interest of £237 for 3 years, at 6 per cent. ?

£·06 is the interest of £1 for 1 year, at the given rate; and there are 3 payments. Therefore £1·06 (£1 + £0·6) is to be taken 3 times to form a product. Hence 1·06 × 1·06 × 1·06 × £237 is the *amount* at the end of three years; and 1·06 × 1·06 × 1·06 × £237 − £237 is the compound interest.

The following is the process in full—

$$
\begin{array}{l}
£ \\
1·06 \text{ the amount of £1, in one year.} \\
1·06 \text{ the multiplier.} \\
\overline{} \\
1·1236 \text{ the amount of £1, in two years.} \\
1·06 \text{ the multiplier.} \\
\overline{} \\
1·191016 \text{ the amount of £1, in three years.}
\end{array}
$$

Multiplying by 237, the principal,

£ s. d.

we find that 282·270792 = 282 5 5 is the amount;
and subtracting 237 0 0, the principal,

we obtain 45 5 5 as the compound interest.

EXAMPLE 2.—What are the amount and compound interest of £79 for 6 years, at 5 per cent. ?

The amount of £1 for 1 year, at this rate would be £1·05. Therefore £1·05 × 1·05 × 1·05 × 1·05 × 1·05 × 1·05 × 79 is the amount, &c. And the process in full will be—

$$
\begin{array}{l}
£ \\
1·05 \\
1·05 \\
\overline{} \\
1·1025 \text{ the amount of £1, in two years.} \\
1·1025 \\
\overline{} \\
1·21551 \text{ the amount of £1, in four years.} \\
1·1025 \\
\overline{} \\
1·34010 \text{ the amount of £1, in six years.} \\
79
\end{array}
$$

£ s. d.

£105·86790 = 105 17 4¼ is the required amount
  79 0 0

And 26 17 4¼ is the required interest

M 2

EXAMPLE 3.—What are the amount, and compound interest of £27, for 4 years, at £2 10s. per cent. half-yearly.

The amount of £1 for one payment is £1·025. Therefore £1·025 × 1·025 × 1·025 × 1·025 × 1·025 × 1·025 × 1·025 × 1·025 × 27 is the amount, &c. And the process in full will be

£
1·025
1·025
——————
1·05063 the amount of £1, in one year.
1·05063
——————
1·10382 the amount of £1, in two years.
1·10382
——————
1·21842 the amount of £1, in four years.
27

£ s. d.
£32·89734 = 32 17 11¼ is the required amount.
27 0 0

And 5 17 11¼ is the required interest.

21. RULE III.—Find by the interest table (at the end of the treatise) the amount of £1 at the given rate, and for the given number of payments; multiply this by the given principal, and the product will be the required amount. From this product subtract the principal, and the remainder will be the required compound interest.

EXAMPLE.—What is the amount and compound interest of £47 10s. for 6 years, at 3 per cent., half-yearly ?

£47 10s. = £47·5.

We find by the table that
£1·42576 is the amount of £1, for the given time and rate.
47·5 is the multiplier.

£ s. d.
67·7236 = 67 14 5¾ is the required amount.
47 10 0

And 20 4 5¾ is the required interest.

22. Rule I. requires no explanation.

REASON OF RULE II.—When the time and rate are the same, two *principals* are proportional to their corresponding amounts. Therefore

£1 (one principal) : £1·06 (its corresponding amount) :: £1·06 (another principal) : £1·06 × 1·06 (its corresponding amount).

Hence the amount of £1 for two years, is £1·06 × 1·06—or the product of two factors, each of them the amount of £1 for one year.

Again, for similar reasons,

£1 : £1·06 :: £1·06 × 1·06 : £1·06 × 1·06 × 1·06·

Hence the amount of £1 for three years, is £1·06 × 1·06 × 1·06—or the product of three factors, each of them the amount of £1 for one year.

The same reasoning would answer for any number of payments.

The amount of any principal will be as much greater than the amount of £1, at the same rate, and for the same time, as the principal itself is greater than £1. Hence we multiply the amount of £1, by the given principal.

Rule III. requires no explanation.

23. When the decimals become numerous, we may proceed as already directed [Sec. II. 58].

We may also shorten the process, in many cases, if we remember that the product of two of the factors multiplied by itself, is equal to the product of four of them; that the product of four multiplied by the product of two is equal to the product of six; and that the product of four multiplied by the product of four, is equal to the product of eight, &c. Thus, in example 2,

$$1·1025 \; (= 1·05 \times 1·05) \times 1·1025 = 1·05 \times 1·05 \times 1·05 \times 1·05$$

EXERCISES.

1. What are the amount and compound interest of £91 for 7 years, at 5 per cent. per annum? *Ans.* £128 0s. 11d. is the amount; and £37 0s. 11d., the compound interest.

2. What are the amount and compound interest of £142 for 8 years, at 3 per cent. half-yearly? *Ans.* £227 17s. 4½d. is the amount; and £85 17s. 4½d., the compound interest.

3. What are the amount and compound interest of £63 5s. for 9 years, at 4 per cent. per annum? *Ans.* £90 0s. 5¾d. is the amount; and £26 15s. 5¾d., the compound interest.

4. What are the amount and compound interest of £44 5s. 9d. for 11 years, at 6 per cent. per annum?

Ans. £84 1*s.* 5*d.* is the amount; and £39 15*s.* 8*d.*, the compound interest.

5. What are the amount and compound interest of £32 4*s.* 9¾*d.* for 3 years, at £2 10*s.* per cent. half-yearly? *Ans.* £37 7*s.* 8½*d.* is the amount; and £5 2*s.* 10½*d.*, the compound interest.

6. What are the amount and compound interest of £971 0*s.* 2¼*d.* for 13 years, at 4 per cent. per annum? *Ans.* £1616 15*s.* 11¾*d.* is the amount; and £645 15*s.* 9½*d.*, the compound interest.

24. Given the amount, time, and rate—to find the principal; that is, to find the *present worth* of any sum to be due hereafter—a certain rate of interest being allowed for the money now paid.

RULE.—Find the product of as many factors as there are times of payment—each of the factors being the *amount* of £1 for a single payment; and divide this product into the given amount.

EXAMPLE.—What sum would produce £834 in 5 years, at 5 per cent. compound interest?

The amount of £1 for 1 year at the given rate is £1·05; and the product of this taken 5 times as a factor 1·05 × 1·05 × 1·05 × 1·05 × 1·05, which (according to the table) is 1·27628. Then

£834 ÷ 1·27628 = £653 9*s.* 2¾*d.*, the required principal.

25. REASON OF THE RULE.—We have seen [21] that the *amount* of any sum is equal to the amount of £1 (for the same time, and at the same rate) multiplied by the principal; that is,

The amount of the given principal = the given principal × the amount of £1.

If we divide each of these equal quantities by the same number [Sec. V. 6], the quotients will be equal. Therefore—

The amount of the given principal ÷ the amount of £1 = the given principal × the amount of £1 ÷ the amount of £1. That is, the amount of the given principal (the given amount) divided by the amount of £1, is equal to the principal, or quantity required—which is the rule.

EXERCISES.

7. What ready money ought to be paid for a debt of £629 17*s.* 1¹¹⁄₂₃*d.*, to be due 3 years hence, allowing 8 per cent. compound interest? *Ans.* £500.

8. What principal, put to interest for 6 years, would amount to £268 0s. 4½d., at 5 per cent. per annum? *Ans.* £200.

9. What sum would produce £742 19s. 11½d. in 14 years, at 6 per cent. per annum? *Ans.* £328 12s. 7d.

10. What is £495 19s. 11¾d., to be due in 18 years, at 3 per cent. half-yearly, worth at present? *Ans.* £171 2s. 8¾d.

26. Given the principal, rate, and amount—to find the time—

RULE I.—Divide the amount by the principal; and into the quotient divide the amount of £1 for one payment (at the given rate) as often as possible—the number of times the amount of £1 has been used as a divisor, will be the required number of payments.

EXAMPLE.—In what time will £92 amount to £106 13s. 0¾d., at 3 per cent. half-yearly?

£106 13s. 0¾d. ÷ £92 = 1·15927. The amount of £1 for one payment is £1·03. But 1·15927 ÷ 1·03 = 1·1255; 1·1255 ÷ 1·03 = 1·09272; 1·09272 ÷ 1·03 = 1·0609; and 1·0609 ÷ 1·03 = 1·03; 1·03 ÷ 1·03 = 1. We have used 1·03 as a divisor 5 times; therefore the time is 5 payments, or 2½ years. Sometimes there will be a remainder after dividing by 1·03, &c., as often as possible.

In explaining the method of finding the powers and roots of a given quantity, we shall, hereafter, notice a shorter method of ascertaining how often the amount of one pound can be used as a divisor.

27. RULE II.—Divide the given principal by the given amount, and ascertain by the interest table in how many payments £1 would be equal to a quantity nearest to the quotient—considered as pounds: this will be the required time.

EXAMPLE.—In what time will £50 become £100, at 6 per cent. per annum compound interest?

£100 ÷ 50 = 2.

We find by the tables that in 11 years £1 will become £1·8983, which is less; and in 12 years that it will become £2·0122, which is greater than 2. The answer nearest to the truth, therefore, is 12 years.

28. REASON OF RULE I.—The given amount is [20] equal to the given principal, multiplied by a product which contains as many factors as there are times of payment—each factor being the amount of £1, for one payment. Hence it is evident, that if we divide the given amount by the given principal, we must have the product of these factors; and that, if we divide this product, and the successive quotients by one of the factors, we shall ascertain their *number*.

REASON OF RULE II.—We can find the required number of factors (each the amount of £1), by ascertaining how often the amount of £1 may be considered as a factor, without forming a product *much* greater or less than the quotient obtained when we divide the given amount by the given principal. Instead, however, of calculating *for ourselves*, we may have recourse to tables constructed by those who have already made the necessary multiplications—which saves much trouble.

29. When the quotient [27] is greater than any amount of £1, at the given rate, in the table, divide it by the greatest found in the table; and, if necessary, divide the resulting quotient in the same way. Continue the process until the quotient obtained is not greater than the largest *amount* in the table. Ascertain what *number of payments* corresponds to the last quotient, and add to it so many times the largest *number of payments* in the table, as the largest *amount* in the table has been used for a divisor.

EXAMPLE.—When would £22 become £535 12s. 0¾d., at 3 per cent. per annum?

£535 12s. 0¾d. ÷ 22 = 24·34560, which is greater than any amount of £1, at the given rate, contained in the table. 24·34560 ÷ 4·3839 (the greatest amount of £1, at 3 per cent., found in the table) = 5·55339; but this latter, also, is greater than any amount of £1 at the given rate in the tables. 5·55339 ÷ 4·3839 = 1·26677, which is found to be the amount of £1, at 3 per cent. per payment, in 8 payments. We have divided by the highest amount for £1 in the tables, or that corresponding to fifty payments, *twice*. Therefore, the required time, is 50 + 50 + 8 payments, or 108 years.

EXERCISES.

11. When would £14 6s. 8d. amount to £18 2s. 8¾d., at 4 per cent. per annum, compound interest? *Ans.* In 6 years.

12. When would £54 2s. 8d. amount to £76 3s. 5d., at 5 per cent. per annum, compound interest? *Ans.* In 7 years.

13. In what time would £793 0s. 2¼d. become £1034 13s. 10½d., at 3 per cent. half-yearly, compound interest? *Ans.* In 4½ years.

14. When would £100 become £1639 7s. 9d., at 6 per cent. half-yearly, compound interest? *Ans.* In 24 years.

QUESTIONS.

1. What is interest? [1].

2. What is the difference between simple and compound interest? [1].

3. What are the principal, rate, and amount? [1].

4. How is the simple interest of any sum, for 1 year, found? [2 &c.].

5. How is the simple interest of any sum, for several years, found? [5].

6. How is the interest found, when the rate consists of more than one denomination? [4].

7. How is the simple interest of any sum, for years, months, &c., found? [6].

8. How is the simple interest of any sum, for any time, at 5 or 6, &c. per cent. found? [7].

9. How is the simple interest found, when the rate, number of years, or both are expressed by a mixed number? [9].

10. How is the simple interest for days, at 5 per cent., found? [10].

11. How is the simple interest for days, at any other rate, found? [11].

12. How is the simple interest of any sum, for months at 6 per cent., found? [12].

13. How is the interest of money, left after one or more payments, found? [13].

14. How is the principal found, when the amount, rate, and time are given? [14].

15. How is the amount found, when the time, rate, and principal are given? [15].

16. How is the time found, when the amount, prin-cipal, and rate are given ? [16].

17. How is the rate found, when the amount, princi-pal, and time are given ? [18].

---

18. How are the amount, and compound interest found, when the principal, rate, and time are given ? [19].

19. How is the present worth of any sum, at com-pound interest for any time, at any rate, found ? [24].

20. How is the time found, when the principal, rate of compound interest, and amount are given ? [26].

DISCOUNT.

30. Discount is money allowed for a sum paid before it is due, and should be such as would be produced by what is paid, were it put to interest from the time the payment *is*, until the time it *ought to be* made.

The *present worth* of any sum, is that which would, at the rate allowed as discount, produce it, if put to interest until the sum becomes due.

31. A bill is not payable until three days after the time mentioned in it; these are called *days of grace*. Thus, if the time expires on the 11th of the month, the bill will not be payable until the 14th—except the latter falls on a Sunday, in which case it becomes payable on the preceding Saturday. A bill at 91 days will not be due until the 94th day after date.

32. When goods are purchased, a certain discount is often allowed for *prompt* (immediate) payment.

The discount generally taken is larger than is sup-posed. Thus, let what is allowed for paying money one year before it is due be 5 per cent. ; in ordinary circumstances £95 would be the payment for £100. But £95 would not in one year, at 5 per cent., produce more than £99 15s., which is less than £100 ; the error, however, is inconsiderable when the time or sum is small. Hence to find the discount and present worth at any rate, we may *generally* use the following—

33. RULE.—Find the interest for the sum to be paid, at the discount allowed; consider this as discount, and deduct it from what is due; the remainder will be the required present worth.

EXAMPLE.—£62 will be due in 3 months; what should be allowed on immediate payment, the discount being at the rate of 6 per cent. per annum?

The interest on £62 for 1 year at 6 per cent. per annum is £3 14s. 4¾d.; and for 3 months it is 18s. 7¼d. Therefore £62 minus 18s. 7¼d.=£61 1s. 4¾d., is the required present worth.

34. To find the present worth *accurately*—

RULE.—Say, as £100 plus its interest for the given time, is to £100, so is the given sum to the required present worth.

EXAMPLE.—What would, at present, pay a debt of £142 to be due in 6 months, 5 per cent. per annum discount being allowed?

$$
\overset{£}{102.5} \; \overset{£ \quad £ \; s.}{(100+2\ 10)} \; : \; \overset{£}{100} \; :: \; \overset{£}{142} : \frac{\overset{£}{100 \times 142}}{102.5} = \overset{£ \quad s. \quad d.}{138 \; 10 \; 8}
$$

This is merely a question in a rule already given [14].

EXERCISES.

1. What is the present worth of £850 15s., payable in one year, at 6 per cent. discount? *Ans.* £802 11s. 10¾d.

2. What is the present worth of £240 10s., payable in one year, at 4 per cent. discount? *Ans.* £231 5s.

3. What is the present worth of £550 10s., payable in 5 years and 9 months, at 6 per cent. per an. discount? *Ans.* £409 5s. 10½d.

4. A debt of £1090 will be due in 1 year and 5 months, what is its present worth, allowing 6 per cent. per an. discount? *Ans.* £1004 12s. 2d.

5. What sum will discharge a debt of £250 17s. 6d., to be due in 8 months, allowing 6 per cent. per an. discount? *Ans.* £241 4s. 6¼d.

6. What sum will discharge a debt of £840, to be due in 6 months, allowing 6 per cent. per an. discount? *Ans.* £815 10s. 8¼d.

7. What ready money now will pay a debt of £200, to be due 127 days hence, discounting at 6 per cent. per an.? *Ans.* £195 18s. 2½d.

8. What ready money now will pay for £1000, to be due in 130 days, allowing 6 per cent. per an. discount? *Ans.* £979 1s. 7d.

9. A bill of £150 10s. will become due in 70 days, what ready money will now pay it, allowing 5 per cent. per an. discount? *Ans.* £149 1s. 5d.

10. A bill of £140 10s. will be due in 76 days, what ready money will now pay it, allowing 5 per cent. per an. discount? *Ans.* £139 1s. 0½d.

11. A bill of £300 will be due in 91 days, what will now pay it, allowing 5 per cent. per an. discount? *Ans.* £296 6s. 1½d.

12. A bill of £39 5s. will become due on the first of September, what ready money will pay it on the preceding 3rd of July, allowing 6 per cent. per an.? *Ans.* £38 18s. 7¼d.

13. A bill of £218 3s. 8¼d. is drawn of the 14th August at 4 months, and discounted on the 3rd of Oct.; what is then its worth, allowing 4 per cent. per an. discount? *Ans.* £216 8s. 1½d.

14. A bill of £486 18s. 8d. is drawn of the 25th March at 10 months, and discounted on the 19th June, what then is its worth, allowing 5 per cent. per an. discount? *Ans.* £472 9s. 11¾d.

15. What is the present worth of £700, to be due in 9 months, discount being 5 per cent. per an.? *Ans.* £674 13s. 11½d.

16. What is the present worth of £315 12s. 4⅕d., payable in 4 years, at 6 per cent. per an. discount? *Ans.* £254 10s. 7¼d.

17. What is the present worth and discount of £550 10s. for 9 months, at 5 per cent. per an.? *Ans.* £530 12s. 0½d. is the present worth; and £19 17s. 11¼d. is the discount.

18. Bought goods to the value of £35 13s. 8d. to be paid in 294 days; what ready money are they now worth, 6 per cent. per an. discount being allowed? *Ans.* £34 0s. 9¼d.

19. If a legacy of £600 is left to me on the 3rd of May, to be paid on Christmas day following, what must I receive as present payment, allowing 5 per cent. per an. discount ?   *Ans.* £581 4s. 2¼d.

20. What is the discount of £756, the one half payable in 6, and the remainder in 12 months, 7 per cent. per an. being allowed ?   *Ans.* £37 14s. 2½d.

21. A merchant owes £110, payable in 20 months, and £224, payable in 24 mouths ; the first he pays in 5 months, and the second in one month after that. What did he pay, allowing 8 per cent. per an. ?   *Ans.* £300.

QUESTIONS FOR THE PUPIL.

1. What is discount ? [30].
2. What is the *present worth* of any sum ? [30].
3. What are *days of grace* ? [31].
4. How is discount *ordinarily* calculated ? [33].
5. How is it *accurately* calculated ? [34].

COMMISSION, &c.

35. *Commission* is an allowance per cent. made to a person called an *agent,* who is employed to sell goods.

Insurance is so much per cent. paid to a person who undertakes that if certain goods are injured or destroyed, he will give a stated sum of money to the owner.

Brokerage is a small allowance, made to a kind of agent called a *broker,* for assisting in the disposal of goods, negotiating bills, &c.

36. To compute commission, &c.—

RULE.—Say, as £100 is to the rate of commission, so is the given sum to the corresponding commission.

EXAMPLE.—What will be the commission on goods worth £437 5s. 2d., at 4 per cent. ?

$$£100 : £4 :: £437 \ 5s. \ 2d. : \frac{4 \times £437 \ 5s. \ 2d.}{100} = £17 \ 9s.$$

9¾d., the required commission.

37. To find what insurance must be paid so that, if the goods are lost, both their value and the insurance paid may be recovered—

RULE.—Say, as £100 minus the rate per cent. is to £100, so is the value of the goods insured, to the required insurance.

EXAMPLE.—What sum must I insure that if goods worth £400 are lost, I may receive both their value and the insurance paid, the latter being at the rate of 5 per cent.?

$$£95 : £100 :: £400 : \frac{£100 \times 400}{95} = £421 \ 1s. \ 0\frac{3}{4}d.$$

If £100 were insured, only £95 would be actually received, since £5 was paid for the £100. In the example, £421 1s. 0¾d. are received; but deducting £21 1s. 0¾d., the insurance, £400 remains.

EXERCISES.

1. What premium must be paid for insuring goods to the amount of £900 15s., at 2½ per cent.? *Ans.* £22 10s. 4½d.

2. What premium must be paid for insuring goods to the amount of £7000, at 5 per cent.? *Ans.* £350.

3. What is the brokerage on £976 17s. 6d., at 5s. per cent.? *Ans.* £2 8s. 10⅛d.

4. What is the premium of insurance on goods worth £2000, at 7½ per cent.? *Ans.* £150.

5. What is the commission on £767 14s. 7d., at 2½ per cent.? *Ans.* £19 3s. 10⅝d.

6. How much is the commission on goods worth £971 14s. 7d., at 5s. per cent.? *Ans.* £2 8s. 7$\frac{3}{20}$d.

7. What is the brokerage on £3000, at 2s. 6d. per cent.? *Ans.* £3 15s.

8. How much is to be insured at 5 per cent. on goods worth £900, so that, in case of loss, not only the value of the goods, but the premium of insurance also, may be repaid? *Ans.* £947 7s. 4$\frac{8}{19}$d.

9. Shipped off for Trinidad goods worth £2000, how much must be insured on them at 10 per cent., that in case of loss the premium of insurance, as well as their value, may be recovered? *Ans.* £2222 4s. 5⅓d.

QUESTIONS FOR THE PUPIL.

1. What is commission? [35].
2. What is insurance? [35].
3. What is brokerage? [35].

4. How are commission, insurance, &c., calculated ? [36].

5. How is insurance calculated, so that both the insurance and value of the goods may be received, if the latter are lost ? [37].

PURCHASE OF STOCK.

38. Stock is money borrowed by Government from individuals, or contributed by merchants, &c., for the purpose of trade, and bearing interest at a fixed, or variable rate. It is transferable either entirely, or in part, according to the pleasure of the owner.

If the price per cent. is more than £100, the stock in question is said to be *above*, if less than £100, *below* "par."

Sometimes the shares of trading companies are only gradually paid up; and in many cases the whole price of the share is not demanded at all—they may be £50, £100, &c., shares, while only £5, £10, &c., may have been paid on each. One person may have many shares. When the interest per cent. on the money paid is considerable, stock often sells for more than what it originally cost; on the other hand, when money becomes more valuable, or the trade for which the stock was contributed is not prosperous, it sells for less.

39. To find the value of any amount of stock, at any rate per cent.—

RULE.—Multiply the amount by the value per cent., and divide the product by 100.

EXAMPLE.—When £69⅓ will purchase £100 of stock, what will purchase £642 ?

$$\frac{£642 \times 69\frac{1}{3}}{100} = £443 \; 15s. \; 7\frac{3}{4}d.$$

It is evident that £100 of stock is to any other amount of it, as the price of the former is to that of the latter. Thus

$$£100 : £642 :: £69\tfrac{1}{3} : \frac{£642 \times 69\frac{1}{3}}{100}$$

EXERCISES.

1. What must be given for £750 16s. in the 3 per cent. annuities, when £64⅛ will purchase £100 ? *Ans.* £481 9s. $0\frac{3}{25}d$.

2. What must be given for £1756 7s. 6d. India stock, when £196¼ will purchase £100 ? *Ans.* £3446 17s. 8⅝d.

3. What is the purchase of £9757 bank stock, at £125⅝ per cent. ? *Ans.* £12257 4s. 7½d.

QUESTIONS.

1. What is stock ? [38].

2. When is it *above,* and when *below* "par" ? [38].

3. How is the value of any amount of stock, at any rate per cent., found ? [39].

EQUATION OF PAYMENTS.

40. This is a process by which we discover a time, when several debts to be due at *different* periods may be paid, *at once,* without loss either to debtor or creditor.

RULE.—Multiply each payment by the time which should elapse before it would become due ; then, add the products together, and divide their sum by the sum of the debts.

EXAMPLE 1.—A person owes another £20, payable in 6 months ; £50, payable in 8 months ; and £90, payable in 12 months. At what time may all be paid together, without loss or gain to either party ?

$$
\begin{array}{l}
\pounds \qquad \pounds \\
20\times 6 = \ \ 120 \\
50\times 8 = \ \ 400 \\
90\times 12 = 1080 \\
\hline
160 \quad 160)\overline{1600}(10 \text{ the required number of months.} \\
\qquad\quad 160
\end{array}
$$

EXAMPLE 2.—A debt of £450 is to be paid thus : £100 immediately, £300 in four, and the rest in six months. When should it be paid altogether ?

$$
\begin{array}{l}
\pounds \qquad\quad \pounds \\
100\ \times\ 0 = \ \ \ \ 0 \\
300\ \times\ 4 = 1200 \\
\ \ 50\ \times\ 6 = \ \ 300 \\
\hline
450 \quad\ 450)\overline{1500}(3\tfrac{1}{3} \text{ months.} \\
\qquad\qquad 1350 \\
\qquad\qquad \dfrac{150}{450} = \tfrac{1}{3}
\end{array}
$$

41. We have (according to a principle formerly used [13]) reduced each debt to a sum which would bring the same interest, in one month. For 6 times £20, to be due in 1 month, should evidently produce the same as £20, to be due in 6 months—and so of the other debts. And the interest of £1600 for the smaller time, will just be equal to the interest of the smaller sum for the larger time.

EXERCISES.

1. A owes B £600, of which £200 is payable in 3 months, £150 in 4 months, and the rest in 6 months; but it is agreed that the whole sum shall be paid at once. When should the payment be made? *Ans.* In $4\frac{1}{2}$ months.

2. A debt is to be discharged in the following manner: $\frac{1}{4}$ at present, and $\frac{1}{4}$ every three months after until all is paid. What is the equated time? *Ans.* $4\frac{1}{2}$ months.

3. A debt of £120 will be due as follows: £50 in 2 months, £40 in 5, and the rest in 7 months. When may the whole be paid together? *Ans.* In $4\frac{1}{4}$ months.

4. A owes B £110, of which £50 is to be paid at the end of 2 years, £40 at the end of $3\frac{1}{2}$, and £20 at the end of $4\frac{1}{2}$ years. When should B receive all at once? *Ans.* In 3 years.

5. A debt is to be discharged by paying $\frac{1}{2}$ in 3 months, $\frac{1}{3}$ in 5 months, and the rest in 6 months. What is the equated time for the whole? *Ans.* $4\frac{1}{6}$ months.

QUESTIONS.

1. What is meant by the equation of payments? 40].

2.—What is the rule for discovering when money, tc be due at different times, may be paid at once? [40].

SECTION VIII.

EXCHANGE, &c.

1 Exchange enables us to find what amount of the money of one country is equal to a given amount of the money of another.

Money is of two kinds, *real*—or coin, and *imaginary*—or money of exchange, for which there is no coin ; as, for example, " one *pound* sterling." ·

The *par* of exchange is that amount of the money of one country *actually* equal to a given sum of the money of another ; taking into account the value of the metals they contain. The *course* of exchange is that sum which, in point of fact, would be allowed for it.

2. When the course of exchange with any place is *above* " păr," the balance of trade is *against* that place. Thus if Hamburgh receives merchandise from London to the amount of £100,000, and ships off, in return, goods to the amount of but £50,000, it can pay only half what it owes by bills of exchange, and for the remainder must obtain bills of exchange from some place else, giving for them a premium—which is so much lost. But the exchange cannot be much above par, since, if the premium to be .paid for bills of exchange is high, the merchant will export goods at less profit; or he will pay the expense of transmitting and insuring coin, or bullion.

3. The nominal value of commodities in these countries was from four to fourteen times less formerly than at present ; that is, the same amount of money would then buy much more than now. We may estimate the value of money, at any particular period, from the amount of corn it would purchase at that time. The value of money fluctuates from the nature of the crops, the state of trade, &c.

In exchange, a variable is given for a fixed sum; thus London receives different values for £1 from different countries.

Agio is the difference which there is in some places between the *current* or *cash* money, and the *exchange* or *bank money*—which is finer.

The following tables of foreign coins are to be made familiar to the pupil.

FOREIGN MONEY.

MONEY OF AMSTERDAM.

Flemish Money.

Pennings				
8	.	.	.	. make 1 grote or penny.
16 or	grotes 2	.	.	1 stiver.
320	40 or	stivers 20	.	1 florin or guilder.
800	100	50 or	guilders 2¼	1 rixdollar.
1920	240	120 or	6	1 pound.

MONEY OF HAMBURGH.

Flemish Money.

Pfennings			
6	.	.	make 1 grote or penny.
72 or	grotes 12	.	1 skilling.
1440	240 or	skillings. 20	1 pound.

Hamburgh Money.

Pfennings	Pence			
12 or	2	.	.	make 1 schilling, equal to 1 stiver.
192	32 or	schillings 16	.	1 mark.
384	64	32 or	marks 2	1 dollar of exchange.
576	96	48 or	3	1 rixdollar.

We find that 6 schillings=1 skilling.

Hamburgh money is distinguished by the word "Hambro." "Lub," from Lubec, where it was coined, was formerly used for this purpose; thus, "one mark Lub."

We exchange with Holland and Flanders by the pound sterling.

N

FRENCH MONEY.

Accounts were formerly kept in livres, &c.

Derniers.				
12				make 1 sou.
	sous			
240 or	20			1 livre.
		livres		
720	60 or	3		1 ecu or crown.

Accounts are now kept in francs and centimes.

Centimes			
10			make 1 decime.
	decimes		
100 or	10		1 franc.

81 livres=80 francs.

PORTUGUESE MONEY.

Accounts are kept in milrees and rees.

Rees			
400			make 1 crusado.
	crusados		
1000 or	2½		1 milree.
4800	12		1 moidere.

SPANISH MONEY.

Spanish money is of two kinds, *plate* and *vellon* ; the latter being to the former as 32 is to 17. *Plate* is used in exchange with us. Accounts are kept in piastres, and maravedi.

Maravedies			
34			make 1 real.
	reals		
272 or	8		1 piastre or piece of eight.
		piastres	
1088	32 or	4	1 pistole of exchange.
375			1 ducat.

AMERICAN MONEY.

In some parts of the United States accounts are kept in dollars, dimes, and cents.

Cents		
10		make 1 dime.
	dimes	
100 or	10	1 dollar.

In other parts accounts are kept in pounds, shillings, and pence. These are called *currency*, but they are of much less value than with us, paper money being used.

DANISH MONEY.

Pfennings
12	.	.	.	.	make 1 skilling.

skillings
192 or	16	.	.	.	1 mark.

marks
1152	96 or	6	.	.	.	1 rixdollar

6 Danish=3 Hamburgh marks.

VENETIAN MONEY.

Denari (the plural of denaro)
12	.	.	.	make 1 soldo.

soldi
240 or	20	.	.	.	1 lira.

lire soldi
1488	124 or	6	4	.	.	1 ducat current.
1920	160	8		.	.	1 ducat effective

AUSTRIAN MONEY.

Pfennings
4	.	.	.	.	make 1 creutzer.

creutzers
240 or	60	.	.	.	1 florin.

florins
360	90 or	1½	.	.	.	1 rixdollar.

NEAPOLITAN MONEY.

Grains
10	.	.	.	make 1 carlin.

carlins
100 or	10	.	.	.	1 ducat regno.

MONEY OF GENOA.

Lire   soldi
4 and 12 make 1 scudo di cambio, or crown of exchange
10 and 14   1 scudo d'oro, or gold crown.

OF GENOA AND LEGHORN.

Denari di pezza
12	.	.	.	make 1 soldo di pezza.

soldi di pezza
240 or	20	1 pezza of 8 reals.

Denari di lira
12	.	.	. make 1 soldo di lira.

soldi di lira
240 or	20	1 lira.
1380	115 or	5¾	.	.	1 pezza of 8 reals.

SWEDISH MONEY.

Fennings, or oers
12	.		.	make 1 skilling.

shillings
576 or	48	.	1 rixdollar.

RUSSIAN MONEY.

Copecs
100   .   .   .   make 1 ruble.

EAST INDIAN MONEY.

Cowries
2560   .   .   .   make 1 rupee.

Rupees
100,000   .   .   1 lac.
10,000,000   .   .   1 crore.

The cowrie is a small shell found at the Maldives, and near
Angola: in Africa about 5000 of them pass for a pound.

The rupee has different values: at Calcutta it is 1s. 11¾d.;
the Sicca rupee is 2s. 0¾d.; and the current rupee 2s.—if we
divide any number of these by 10, we change them to pounds
of our money; the Bombay rupee is 2s. 3d., &c. A sum of
Indian money is expressed as follows; 5·38220, which means
5 lacs and 38220 rupees.

---

4. To reduce bank to current money—

RULE.—Say, as £100 is to £100 + the agio, so is
the given amount of bank to the required amount of
current money.

EXAMPLE.—How many guilders, current money, are equal
to 463 guilders, 3 stivers, and $13\frac{64}{65}$ pennings banco, agio
being $4\frac{5}{7}$?

100	:	104$\frac{5}{7}$	::	463 g. 3 st. 13$\frac{64}{65}$ p. : ?
7		7		20

     700   733   9263 stivers.
     65   16

45500   148221 pennings.

Multiplying by 65, and adding 64 to the product,

will give 9634429
Multiplying by 733

and dividing by 45500)7062036457

will give 155209 pennings.

16)155209

20)9700   9

And 485 g. 0 st. 9$\frac{9}{4}\frac{9}{5}\frac{5}{5}\frac{7}{0}$ p. is the amount sought.

5. We multiply the first and second terms by 7, and add the
numerator of the fraction to one of the products. This is the
same thing as reducing these terms to fractions having 7 for
their denominator, and then multiplying them by 7 [Sec. V. 29].

For the same reason, and in the same way, we multiply the
first and third terms by 65, to banish the fraction, without
destroying the proportion.

The remainder of the process is according to the rule of proportion [Sec. V. 31]. We reduce the answer to pennings, stivers, and guilders.

EXERCISES.

1. Reduce 374 guilders, 12 stivers, bank money, to current money, agio being $4\frac{5}{7}$ per cent.? *Ans.* 392 g., 5 st., $3\frac{19}{175}$ p.

2. Reduce 4378 guilders, 8 stivers, bank money, to current money, agio being $4\frac{5}{9}$ per cent.? *Ans.* 4577 g., 17 st., $3\frac{77}{225}$ p.

3. Reduce 873 guilders, 11 stivers, bank money, to current money, agio being $4\frac{7}{8}$ per cent.? *Ans.* 916 g., 2 st., $11\frac{19}{50}$ p.

4. Reduce 1642 guilders, bank money, to current money, agio being $4\frac{11}{12}$ per cent.? *Ans.* 1722 g., 14 st., $10\frac{2}{15}$ p.

6. To reduce current to bank money—

RULE—Say, as £100+the agio is to £100, so is the given amount of current to the required amount of bank money.

EXAMPLE.—How much bank money is there in 485 guilders and $9\frac{24985}{45500}$ pennings, agio being $4\frac{5}{7}$?

$$\begin{array}{ccccccc} & & & & \text{g.} & \text{st.} & \text{p.} \\ 104\frac{5}{7} & : & 100 & :: & 485 & 0 & 9\frac{24985}{45500} \; : \; ? \\ 7 & & 7 & & 20 & & \end{array}$$

$$\begin{array}{ccc} \overline{733} & \overline{700} & \overline{9700} \\ 45500 & & 16 \end{array}$$

$$\overline{33351500} \qquad\qquad \overline{155209}$$

Multiplying by 45500 the denominator,

7062009500

and adding 25957 the numerator,

we get 7062035457

700

33351500)4943424819900

Quotient 148221$\frac{64}{65}$

16)148221$\frac{64}{65}$

20)9263

463   3   13$\frac{64}{65}$ is the amount sought.

EXERCISES.

5. Reduce 58734 guilders, 9 stivers, 11 pennings, current money, to bank money, agio being $4\frac{5}{6}$ per cent. ? *Ans.* 56026 g., 10 st., $11\frac{161}{629}$ p.

6. Reduce 4326 guilders, 15 pennings, current money, to bank money, agio being $4\frac{6}{7}$ per cent. ? *Ans.* 4125 g., 13 st., $2\frac{189}{367}$ p.

7. Reduce 1186 guilders, 4 stivers, 8 pennings, current, to bank money, agio being $4\frac{5}{8}$ per cent. ? *Ans.* 1136 g., 10 st, $0\frac{69}{187}$ p.

8. Reduce 8560 guilders, 8 stivers, 10 pennings, current, to bank money, agio being $4\frac{3}{5}$ per cent. ? *Ans.* 8183 g., 19 st., $5\frac{513}{523}$ p.

7. To reduce foreign money to British, &c.—

RULE.—Put the amount of British money considered in the rate of exchange as third term of the proportion, its value in foreign money as first, and the foreign money to be reduced as second term.

EXAMPLE 1.—*Flemish Money.*—How much British money is equal to 1054 guilders, 7 stivers, the exchange being 33s. 4d. Flemish to £1 British ?

$$33s. 4d. \quad : \quad 1054g. \ 7st. \ :: \ £1 \ : \ ?$$
$$12 \qquad\qquad 20$$

$$\overline{400 \text{ pence.}} \qquad \overline{21087 \text{ stivers.}}$$
$$\qquad\qquad\qquad \frac{.\ 2}{\quad}$$
$$400)\overline{42174} \text{ Flemish pence.}$$
$$£105{\cdot}435 = £105 \ 8s. \ 8\frac{1}{2}d.$$

£1, the amount of British money considered in the rate, is put in the third term; 33s. 4d., its value in foreign money, in the first; and 1054 g. 7 st., the money to be reduced, in the second.

9. How many pounds sterling in 1680 guilders, at 33s. 3d. Flemish per pound sterling ? *Ans.* £168 8s. $5\frac{7}{133}d.$

10. Reduce 6048 guilders, to British money, at 33s. 11d. Flemish per pound British ? *Ans.* £594 7s. $11\frac{215}{407}d.$

11. Reduce 2048 guilders, 15 stivers, to British money, at 34s. 5d. Flemish per pound sterling ? *Ans.* £198 8s. $6\frac{114}{413}d.$

12. How many pounds sterling in 1000 guilders, 10 stivers, exchange being at 33*s*. 4*d*. per pound sterling? *Ans.* £100 1*s*.

EXAMPLE 2.—*Hamburgh Money.*—How much British money is equivalent to 476 marks, 9 skillings, the exchange being 33*s*. 6*d*. Flemish per pound British?

$$
\begin{array}{ccc}
s. & d. & m. & s. \\
33 & 6 & : & 476 & 9\tfrac{2}{3} & :: £1 & : ? \\
12 & & & 32 & 2 \\
\hline
\end{array}
$$

402 grotes.　　15232+19$\tfrac{1}{3}$=15251$\tfrac{1}{3}$ grotes.

402)15251$\tfrac{1}{3}$

£37·9386=£37 18*s*. 9*d*.

Multiplying the schillings by 2, and the marks by 32, reduces both to pence.

13. How much British money is equivalent to 3083 marks, 12$\tfrac{2}{3}$ schillings Hambro', at 32*s*. 4*d*. Flemish per pound sterling? *Ans.* £254 6*s*. 8*d*.

14. How much English money is equal to 5127 marks, 5 schillings, Hambro' exchange, at 36*s*. 2*d*. Flemish per pound sterling? *Ans.* £378 1*s*.

15. How many pounds sterling in 2443 marks, 9$\tfrac{1}{2}$ schillings, Hambro', at 32*s*. 6*d*. Flemish per pound sterling? *Ans.* £200 10*s*.

16. Reduce 7854 marks, 7 schillings Hambro', to British money, exchange at 34*s*. 11*d*. Flemish per pound sterling, and agio at 21 per cent.? *Ans.* £495 15*s*. 0$\tfrac{3}{4}$*d*.

EXAMPLE 3.—*French Money.*—Reduce 8654 francs, 42 centimes, to British money, the exchange being 23f., 50c., per £1 British.

$$
\begin{array}{ccccc}
f. & c. & f. & c. \\
23 & 50 & : 8654 & 42 & :: 1 & : \dfrac{8654\cdot42}{23\cdot50} = £368 \; 5s. \; 5\tfrac{1}{2}d.
\end{array}
$$

42 centimes are 0·42 of a franc, since 100 centimes make 1 franc.

17. Reduce 17969 francs 85 centimes, to British money, at 23 francs 49 centimes per pound sterling? *Ans.* £765.

18. Reduce 7672 francs, 50 centimes, to British money, at 23 francs, 25 centimes per pound sterling? *Ans.* £330.

19. Reduce 15647 francs 36 centimes, to British money, at 23 francs 15 centimes per pound sterling ? *Ans.* £675 18s. 2¾d.

20. Reduce 450 francs 58½ centimes, to British money, at 25 francs 5 centimes per pound sterling ? *Ans.* £176 14s.

EXAMPLE 4.—*Portuguese Money.*—How much British money is equal to 540 milrees, 420 rees, exchange being at 5s. 6d. per milree ?

m.　m.　r.　s.　d.
1 : 540·420 :: 5 ·6 : 540·420×5s. 6d.=£148 12s. 3¾d.

In this case the British money is the variable quantity, and 5s. 6d. is that amount of it which is considered in the rate.

The rees are changed into the decimal of a milree by putting them to the right hand side of the decimal point, since one ree is the thousandth of a milree.

21. In 850 milrees, 500 rees, how much British money, at 5s. 4d. per milree ?　*Ans.* £226 16s.

22. Reduce 2060 milrees, 380 rees, to English money, at 5s. 6¾d. per milree ?　*Ans.* £573 0s. 10¼d.

23. In 1785 milrees, 581 rees, how many pounds sterling, exchange at 64½ per milree ?　*Ans.* £479 17s. 6d.

24. In 2000 milrees, at 5s. 8½d. per milree, how many pounds sterling ?　*Ans.* £570 16s. 8d.

EXAMPLE 5.—*Spanish Money.*—Reduce 84 piastres, 6 reals, 19 maravedi, to British money, the exchange being 49d. the piastre.

p.　　　p.　r.　m.　d.
1　:　84　6　19 :: 49 : ?
8　　　8
——　　——
8　　　678 reals.
34　　　34
——　　——
272　　23052 maravedi.
　　　·49
　　——————
272)1129548
　　4152·7, &c.=£17 6s. 0¾d.

25. Reduce 2448 piastres to British money, exchange at 50*d.* sterling per piastre ? *Ans.* £510.

26. Reduce 30000 piastres to British money, at 40*d* per piastre ? *Ans.* £5000.

27. Reduce 1025 piastres, 6 reals, $22\frac{159}{137}$ maravedi, to British money, at $39\frac{1}{4}d.$ per piastre? *Ans.* £167 15*s.* 4*d.*

EXAMPLE 6.—*American Money.*—Reduce 3765 dollars to British money, at 4*s.* per dollar. $4s.=£\frac{1}{5}$; therefore

5)3765   dol. dol. *s.* £
753 is the required sum. Or 1 : 3765 :: 4 : 753

28. Reduce £292 3*s.* $2\frac{2}{5}d.$ American, to British money, at 66 per cent. ? *Ans.* £176.

29. Reduce 5611 dollars, 42 cents., to British money, at 4*s.* $5\frac{1}{2}d.$ per dollar ? *Ans.* £1250 17*s.* 7*d.*

30. Reduce 2746 dollars, 30 cents., to British money, at 4*s.* $3\frac{1}{2}d.$ per dollar ? *Ans.* £589 6*s.* $2\frac{1}{2}d.$

From these examples the pupil will very easily understand how any other kind of foreign, may be changed to British money.

8. To reduce British to foreign money—

RULE.—Put that amount of foreign money which is considered in the rate of exchange as the third term, its value in British money as the first, and the British money to be reduced as the second term.

EXAMPLE 1.—*Flemish Money.*—How many guilders, &c., in £236 14*s.* 2*d.* British, the exchange being 34*s.* 2*d.* Flemish to £1 British ?

```
  £            £   s.  d.       s.  d.
  1    :     236 14   2  ::    34   2 : ?
 20          20               12
 ──          ────             ──
 20          4734            410  pence.
 12          12
───          ─────
240          56810d.
             410
       240)23292100
         12)97050·4, &c.
          20)8087   6
            £404   7   6½  Flemish.
```

We might take parts for the 34s. 2d.—

$$34s.\ 2d. = £1 + 10s. + 4s. + 2d.$$

	£	£	s.	d.
£1 =	1	236	14	2
10s. =	½	118	7	1
4s. =	⅕	47	6	10
2d. =	$\frac{1}{120}$ ($\frac{1}{24}$ of ⅕)	1	19	5½

£404 7 6½ Flemish.

EXERCISES.

31. In £100 1s., how much Flemish money, exchange at 33s. 4d. per pound sterling? *Ans.* 1000 guilders, 10 stivers.

32. Reduce £168 8s. $5\frac{7}{133}d$. British into Flemish, exchange being 33s. 3d. Flemish per pound sterling? *Ans.* 1680 guilders.

33. In £199 11s. $10\frac{22}{139}d$. British, how much Flemish money, exchange 34s. 9d. per pound sterling? *Ans.* 2080 guilders, 15 stivers.

34. Reduce £198 8s. $6\frac{114}{413}d$. British to Flemish money, exchange being 34s. 5d. Flemish per pound sterling? *Ans.* 2048 guilders, 15 stivers.

EXAMPLE 2.—*Hamburgh Money.*—How many marks, &c., in £24 6s. British, exchange being 33s. 2d. per £1 British?

£1 : £24 6s. :: 33s. 2d. : ?

20	20	12
20	486	398 grotes.
	398	

20)193428

2)9671 8 pence.

16)4835 schillings, 1 penny.

302 marks, 3 schillings, 1 penny.

35. Reduce £254 6s. 8d. English to Hamburgh money, at 32s. 4d. per pound sterling? *Ans.* 3083 marks, $12\frac{2}{3}$ stivers.

36. Reduce £378 1s. to Hamburgh money, at 36s. 2d. Flemish per pound sterling? *Ans.* 5127 marks, 5 schillings.

37. Reduce £536 to Hamburgh money, at 36s. 4d. per pound sterling? *Ans.* 7303 marks.

38. Reduce £495 15*s*. 0¾*d*. to Hamburg currency, at 34*s*. 11*d*. per pound sterling ; agio at 21 per cent. ? *Ans*. 7854 marks 7 schillings.

EXAMPLE 3.—*French Money*.—How much French money is equal in value to £83 2*s*. 2*d*., exchange being 23 francs 25 centimes per £1 British ?

```
 £        £   s.   d.        f.
 1    :  83   2    2  ::  23·25  :  ?
 20       20
 ──       ──
 20     1662
 12       12
───     ─────
240    19946
        23·25
      ──────────
240)463744·50
```

19322·7, or 19322f. 70c. is the required sum.

39. Reduce £274 5*s*. 9*d*. British to francs, &c., exchange at 23 francs 57 centimes per pound sterling ? *Ans*. 6464 francs 96 centimes.

40. In £765, how many francs, &c., at 23 francs 49 centimes per pound sterling ? *Ans*. 17969 francs 85 centimes.

41. Reduce £330 to francs, &c., at 23 francs 25 centimes per pound sterling ? *Ans*. 7672 francs 50 cents.

42. Reduce £734 4*s*. to French money, at 24 francs 1 centime per pound sterling ? *Ans*. 1769 francs 42½ centimes.

EXAMPLE 4.—*Portuguese Money*.—How many milrees and rees in £32 6*s*. British, exchange being 5*s*. 9*d*. British per milree ?

```
 s.  d.      £    s.
 5   9   :  32    6  ::  1000  :  ?
 12         20
 ──         ──
 69        646
            12
         ──────
           7752
           1000
        ──────────
69)7752000
```

112348 rees=112 milrees 348 rees, is the required sum.

43. Reduce £226 16*s.* to milrees, &c., at 5*s.* 4*d.* per milree?_ *Ans.* 850 milrees 500 rees.

44. Reduce £479 17*s.* 6*d.* to milrees, &c., at 64½*d.* per milree? *Ans.* 1785 milrees 581 rees.

45. Reduce £570 16*s.* 8*d.* to milrees, &c., at 5*s.* 8½*d.* per milree? *Ans.* 2000 milrees.

46. Reduce £715 to milrees, &c., at 5*s.* 8*d.* per milree? *Ans.* 2523 milrees 529$\frac{7}{17}$ rees.

EXAMPLE 5.—*Spanish Money.*—How many piastres, &c., in £62 British, exchange being 50*d.* per piastre?

$$
\begin{array}{cccccc}
d. & \pounds \\
50 & : & 62 & :: & 1 & : & ? \\
& & 20 \\
\hline
& & 1240 & & p. & r. & m. \\
& & 12 & & 297 & 0 & 32\tfrac{1}{2}\tfrac{6}{5}, \text{ is the required sum.}\\
\hline
50)& & 14880 \\
\hline
& & 297\text{·}6 & \text{piastres.} \\
& & 8 \\
\hline
& & 48 & \text{reals.} \\
& & 34 \\
\hline
50)& & 1632 \\
\hline
& & 32\tfrac{1}{2}\tfrac{6}{5} & \text{maravedis.}
\end{array}
$$

47. How many piastres, &c., shall I receive for £510 sterling, exchange at 50*d.* sterling per piastre? *Ans.* 2448 piastres.

48. Reduce £5000 to piastres, at 40*d.* per piastre? *Ans.* 30000 piastres.

49. Reduce £167 15*s.* 4*d.* to piastres, &c., at 39¼*d.* per piastre? *Ans.* 1025 piastres, 6 reals, 22$\frac{1}{1}\frac{5}{3}\frac{0}{7}$ maravedis.

50. Reduce £809 9*s.* 8*d.* to piastres, &c., at 40¾*d.* per piastre? *Ans.* 4767 piastres, 4 reals, 2$\frac{8}{1}\frac{2}{6}\frac{}{3}$ maravedis.

EXAMPLE 6.—*American Money.*—Reduce £176 British to American currency, at 66 per cent.

$$
\begin{array}{cccccc}
\pounds & & \pounds & & \pounds \\
100 & : & 176 & :: & 166 & : & : \\
& & 166 \\
\hline
100)& & 29216 \\
\hline
\end{array}
$$

£292 3*s.* 2½*d.*, is the required sum.

EXERCISES.

51. Reduce £753 to dollars, at 4*s.* per dollar ? *Ans.* 3765 dollars.

52. Reduce £532 4*s.* 8*d.* British to American money, at 64 per cent. ? *Ans.* £872 17*s.* 3*d.*

53. Reduce £1250 17*s.* 7*d.* sterling to dollars, at 4*s.* 5½*d.* per dollar ? *Ans.* 5611 dollars 42 cents.

54. Reduce £589 6*s.* 2$\frac{9}{20}$*d.* to dollars, at 4*s.* 3½*d.* per dollar ? *Ans.* 2746 dollars 30 cents.

55. Reduce £437 British to American money, at 78 per cent. ? *Ans.* £777 17*s.* 2½*d.*

9. To reduce florins, &c., to pounds, &c., Flemish—
RULE —Divide the florins by 6 for pounds, and— adding the remainder (reduced to stivers) to the stivers —divide the sum by 6, for skillings, and double the remainder, for grotes.

EXAMPLE.—How many pounds, skillings, and grotes, in 165 florins 19 stivers ?

$$\begin{array}{cc} \text{f.} & \text{st.} \\ 6)165 & 19 \\ \hline \end{array}$$

£27 13*s.* 2*d.*, the required sum.

6 will go into 165, 27 times—leaving 3 florins, or 60 stivers, which, with 19, make 79 stivers; 6 will go into 79, 13 times— leaving 1 ; twice 1 are 2.

10. REASON OF THE RULE.—There are 6 times as many florins as pounds; for we find by the table that 240 grotes make £1, and that 40 ($^{2\frac{4}{0}}$) grotes make 1 florin. There are 6 times as many stivers as skillings; since 96 pennings make 1 skilling, and 16 ($\frac{96}{6}$) pfennings make one stiver. Also, since 2 grotes make one stiver, the remaining stivers are equal to twice as many grotes.

Multiplying by 20 and 2 would reduce the florins to grotes; and dividing the grotes by 12 and 20 would reduce them to pounds. Thus, using the same example—

$$\begin{array}{cc} \text{f.} & \text{st.} \\ 165 & 19 \\ 20 & \\ \hline 3319 & \\ 2 & \\ \hline 12)6638 & \\ \hline 20)553 & 2 \\ \hline \end{array}$$

£27 13*s.* 2*d.*, as before, is the result.

56. In 142 florins 17 stivers, how many pounds, &c.,
Ans. £23 16*s.* 2*d.*

57. In 72 florins 14 stivers, how many pounds, &c.,
Ans. £12 2*s.* 4*d.*

58. In 180 florins, how many pounds, &c.? *Ans.* £30.

11. To reduce pounds, &c., to florins, &c.—

RULE.—Multiply the stivers by 6; add to the product
half the number of grotes, then for every 20 contained
in the sum carry 1, and set down what remains above
the twenties as stivers. Multiply the pounds by 6, and,
adding to the product what is to be carried from the
stivers, consider the sum as florins.

EXAMPLE.—How many florins and stivers in 27 pounds,
13 skillings, and 2 grotes?

$$£\quad s.\quad d.$$
$$27\quad 13\quad 2$$
$$6$$

165fl. 19st., the required sum.

6 times 13 are 78, which, with half the number ($\frac{2}{2}$) of
grotes, make 79 stivers—or 3 florins and 19 stivers (3 twenties,
and 19); putting down 19 we carry 3. 6 times 27 are 162,
and the 3 to be carried are 165 florins.

This rule is merely the converse of the last. It is evident
that multiplying by 20 and 12, and dividing the product by 2
and 20, would give the same result. Thus

$$£\quad s.\quad d.$$
$$27\quad 13\quad 2$$
$$20$$

$$553$$
$$12$$

$$2)6638$$
$$20)3319$$

165fl. 19st., the same result as before.

59. How many florins and stivers in 30 pounds, 12
skillings, and 1 grote? *Ans.* 183 fl., 12 st., 1 g.

60. How many florins, &c., in 129 pounds 7 skillings? *Ans.* 776 fl. 2 st.

61. In 97 pounds, 8 skillings, 2 grotes, how many
florins, &c.? *Ans.* 584 fl. 9 st.

QUESTIONS.

1. What is exchange? [1].
2. What is the difference between real and imaginary money? [1].
3. What are the *par* and *course* of exchange? [1].
4. What is *agio*? [3].
5. What is the difference between current or cash money and exchange or bank money? [3].
6. How is bank reduced to current money? [4].
7. How is current reduced to bank money? [6].
8. How is foreign reduced to British money? [7].
9. How is British reduced to foreign money? [8].
10. How are florins, &c., reduced to pounds Flemish, &c.? [9].
11. How are pounds Flemish, &c., reduced to florins, &c.? [11].

ARBITRATION OF EXCHANGES.

12. In the rule of *exchange* only *two* places are concerned; it may sometimes, however, be more beneficial to the merchant to draw *through* one or more other places. The mode of estimating the value of the money of any place, not drawn directly, but through one or more other places, is called the *arbitration of exchanges*, and is either *simple* or *compound*. It is " simple" when there is only *one* intermediate place, " compound" when there are *more* than one.

All questions in this rule may be solved by one or more proportions.

13. *Simple Arbitration of Exchanges.*—Given the *course* of exchange between each of two places and a third, to find the *par* of exchange between the former.

RULE.—Make the given sums of money belonging to the third place the first and second terms of the proportion ; and put, as third term, the equivalent of what is in the first. The fourth proportional will be the value of what is in the second term, in the kind of money contained in the third term.

EXAMPLE.—If London exchanges with Paris at 10*d*. per franc, and with Amsterdam at 34*s*. 8*d*. per £1 sterling, what ought to be the course of exchange, between Paris and Amsterdam, that a merchant may without loss remit from London to Amsterdam through Paris ?

£1 : 10*d*. :: 34*s*. 8*d*. (the equivalent, in Flemish money, of £1) : ? the equivalent of 10*d*. British (or of a franc) in Flemish money.

Or 240 : 10 :: 34*s*. 8*d*. : $\dfrac{34s.\ 8d. \times 10}{240} = 17\frac{1}{3}d.$, the required value of 10*d*. British, or of a franc, in Flemish money.

£1 and 10*d*. are the two given sums of English money, or that which belongs to the *third* place ; and 34*s*. 8*d*. is the given equivalent of £1.

It is evident that, 17⅓*d*. (Flemish) being the value of 10*d*., the equivalent in British money of a franc, when *more* than 17⅓*d*. Flemish is given for a franc, the merchant will gain if he remits through Paris, since he will thus indirectly receive more than 17⅓*d*. for 10*d*. sterling—that is, more than its equivalent, in Flemish money, at the given course of exchange between London and Amsterdam. On the other hand, if less than 17⅓*d*. Flemish is allowed for a franc, he will lose by remitting through Paris; since he will receive a franc for 10*d*. (British); but he will not receive 17⅓*d*. for the franc:—while, had he remitted 10*d*., the value of the franc, to Amsterdam *directly*, he would have been allowed 17⅓*d*.

EXERCISES.

1. If the exchange between London and Amsterdam is 33*s*. 9*d*. per pound sterling, and the exchange between London and Paris 9½*d*. per franc, what is the *par* of exchange between Amsterdam and Paris ? *Ans.* Nearly 16*d*. Flemish per franc.

2. London is indebted to Petersburgh 5000 rubles ; while the exchange between Petersburgh and London is at 50*d*. per ruble, but between Petersburgh and Holland it is at 90*d*. Flemish per ruble, and Holland and England at 36*s*. 4*d*. Flemish per pound sterling. Which will be the more advantageous method for London to be drawn upon—the direct or the indirect ? *Ans.* London will gain £9 11*s*. 1$\frac{63}{109}$*d*. if it makes payments by way of Holland.

5000 rubles=£1041 13*s*. 4*d*. British, or £1875 Flemish ; but £1875 Flemish=£1032 2*s*. 2$\frac{46}{109}$*d*. British.

14. *Compound Arbitration of Exchanges.*—To find what should be the course of exchange between two places, through two or more others, that it may be on a par with the course of exchange between the same two places, *directly*—

RULE.—Having reduced monies of the same kind to the same denomination, consider each course of exchange as a ratio; set down the different ratios in a vertical column, so that the antecedent of the second shall be of the same kind as the consequent of the first, and the antecedent of the third, of the same kind as the consequent of the second—putting down a note of interrogation for the unknown term of the imperfect ratio. Then divide the product of the consequents by the product of the antecedents, and the quotient will be the value of the given sum if remitted through the intermediate places.

Compare with this its value as remitted by the *direct* exchange.

15. EXAMPLE.—£824 Flemish being due to me at Amsterdam, it is remitted to France at 16*d*. Flemish per franc; from France to Venice at 300 francs per 60 ducats; from Venice to Hamburgh at 100*d*. per ducat; from Hamburgh to Lisbon at 50*d*. per 400 rees; and from Lisbon to England at 5*s*. 8*d*. sterling per milree. Shall I gain or lose, and how much, the exchange between England and Amsterdam being 34*s*. 4*d*. per £1 sterling?

$$16d. \ : \ 1 \text{ franc.}$$
$$300 \text{ francs} \ : \ 60 \text{ ducats.}$$
$$1 \text{ ducat} \ : \ 100 \text{ pence Flemish.}$$
$$50 \text{ pence Flemish} \ : \ 400 \text{ rees.}$$
$$1000 \text{ rees} \ : \ 68 \text{ pence British.}$$
$$? \ : \ £824 \text{ Flemish.}$$

$$\frac{1\times60\times100\times400\times68\times824}{16\times300\times1\times50\times1000} = \text{(if we reduce the terms}$$

[Sec. V. 47]) $\dfrac{17\times824}{25} = £560 \ 6s. \ 4\tfrac{4}{5}d.$

But the exchange between England and Amsterdam for £824 Flemish is £480 sterling.

Since 34*s*. 4*d*. : £824 :: £1 ∷ $\dfrac{£824}{34s. \ 4d.} = £480.$

I gain therefore by the circular exchange £560 6*s*. 4⅘*d*. minus £480 = £80 6*s*. 4⅘*d*.

If commission is charged in any of the places, it must be deducted from the value of the sum which can be obtained in that place.

The process given for the compound arbitration of exchange may be proved to be correct, by putting down the different proportions, and solving them in succession. Thus, if 1·6d. are equal to 1 franc, what will 300 francs (=60 ducats) be worth. If the quantity last found is the value of 60 ducats, what will be that of one ducat (=100d.), &c. ?

EXERCISES.

3. If London would remit £1000 sterling to Spain, the direct exchange being 42½d. per piastre of 272 maravedis ; it is asked whether it will be more profitable to remit directly, or to remit first to Holland at 35s. per pound ; thence to France at 19⅓d. per franc ; thence to Venice at 300 francs per 60 ducats ; and thence to Spain at 360 maravedis per ducat ? *Ans.* The circular exchange is more advantageous by 103 piastres, 3 reals, 19$\frac{2}{20}$ maravedis.

4. A merchant at London has credit for 680 piastres at Leghorn, for which he can draw directly at 50d. per piastre ; but choosing to try the circular way, they are by his orders remitted first to Venice at 94 piastres per 100 ducats ; thence to Cadiz at 320 maravedis per ducat ; thence to Lisbon at 630 rees per piastre of 272 maravedis ; thence to Amsterdam at 50d. per crusade of 400 rees ; thence to Paris at 18⅔d. per franc ; and thence to London at 10½d. per franc ; how much is the circular remittance better than the direct draft, reckoning ½ per cent. for commission ? *Ans.* £14 12s. 7¼d.

16. To estimate the gain or loss per cent.—

RULE.—Say, as the par of exchange is to the course of exchange, so is £100 to a fourth proportional. From this subtract £100.

EXAMPLE.—The par of exchange is found to be 18⅕d. Flemish, but the course of exchange is 19d. per franc ; what is the gain per cent. ?

$$18\tfrac{1}{5}\,d. : 19d. :: £100 : \frac{£19 \times 100}{18\tfrac{1}{5}} = £104 \; 7s. \; 11d.$$

Thus the gain per cent. =£104 7s. 11d. minus £100 = £4 7s. 11d. if the merchant remits through Paris.

If in remitting through Paris commission must be paid, it is to be deducted from the gain.

EXERCISES.

5. The par of exchange is found to be 18¾d. Flemish, but the course of exchange is 19⅓d., what is the gain per cent.? *Ans.* £4 18s. 2¼d.

6. The par of exchange is 17⅝d. Flemish, but the course is 18⅔d., what is the gain per cent.? *Ans.* £4 6s. 11½d.

7. The par of exchange is 18⅐d. Flemish, but the course of exchange is 17⅔⅞d., what is the loss per cent.? *Ans.* £1 16s. 2d.

QUESTIONS.

1. What is meant by arbitration of exchanges? [12].
2. What is the difference between simple and compound arbitration? [12].
3. What is the rule for simple arbitration? [13].
4. What is the rule for compound arbitration? [14].
5. How are we to act if commission is charged in any place? [15].
6. How is the gain or loss per cent. estimated? [16].

PROFIT AND LOSS.

17. This rule enables us to discover how much we gain or lose in mercantile transactions, when we sell at certain prices.

Given the prime cost and selling price, to find the gain or loss in a certain quantity.

RULE.—Find the price of the goods at prime cost and at the selling price; the difference will be the gain or loss on a given quantity.

EXAMPLE.—What do I gain, if I buy 460 ℔ of butter at 6d., and sell it at 7d. per ℔?

The total prime cost is 460d. × 6 = 2760d.
The total selling price is 460d. × 7 = 3220d.
The total gain is 3220d. minus 2760d. = 460d. = £1 18s. 4d

EXERCISES.

1. Bought 140 ℔ of butter, at 10*d*. per ℔, and sold it at 14*d*. per ℔ ; what was gained? *Ans.* £2 6*s*. 8*d*.

2. Bought 5 cwt., 3 qrs., 14 ℔ of cheese, at £2 12*s*. per cwt., and sold it for £2 18*s*. per cwt. What was the gain upon the whole? *Ans.* £1 15*s*. 3*d*.

3. Bought 5 cwt., 3 qrs., 14 ℔ of bacon, at 34*s*· per cwt., and sold it at 36*s*. 4*d*. per cwt. What was the gain on the whole? *Ans.* 13*s*. 8½*d*.

4. If a chest of tea, containing 144 ℔ is bought for 6*s*. 8*d*. per ℔, what is the gain, the price received for the whole being £57 10*s*.? *Ans.* £9 10*s*.

18. To find the gain or loss per cent.—

RULE.—Say, as the cost is to the selling price, so is £100 to the required sum. The fourth proportional minus £100 will be the gain per cent.

EXAMPLE 1.—What do I gain per cent. if I buy 1460 ℔ of beef at 3*d*., and sell it at 3½*d*. per ℔?

$$3d. \times 1460 = 4380d., \text{ is the cost price.}$$
$$\text{And } 3\tfrac{1}{2}d. \times 1460 = 5110d., \text{ is the selling price.}$$

Then $4380 : 5110 :: 100 : \dfrac{5110 \times 100}{4380} = £116 \ 13s. \ 4d.$

Ans. £116 13*s*. 4*d*. minus £100 (=£16 13*s*. 4*d*.) is the gain per cent.

REASON OF THE RULE.—The price is to the price plus the gain in one case, as the price (£100) is to the price plus the gain (£100+the gain on £100) in another.

Or, the price is to the price plus the gain, as any multiple or part of the former (£100 for instance) is to an equimultiple of the latter (£100+the gain on £100).

EXAMPLE 2.—A person sells a horse for £40, and loses 9 per cent., while he should have made 20 per cent. What is his entire loss?

£100 minus the loss, per cent., is to £100 as £40 (what the horse cost, minus what he lost by it) is to what it cost.

$91 : 100 :: 40 : \dfrac{100 \times 40}{91} = £43 \ 19s. \ 1\tfrac{1}{2}d.,$ what the horse cost.

But the person should have gained 20 per cent., or ⅕ of the price ; therefore his profit should have been $\dfrac{£43 \ 19s. \ 1\tfrac{1}{2}d.}{5} = £8 \ 15s. \ 9\tfrac{3}{4}d.$ And

£ s. d.

3 19 1¼ is the difference between cost and selling price.

8 15 9¾ is what he should have received above cost.

12 14 11¼ is his total loss.

EXERCISES.

5. Bought beef at 6*d.* per ℔, and sold it at 8*d.* What was the gain per cent.? *Ans.* 33⅓.

6. Bought tea for 5*s.* per ℔, and sold it for 3*s.* What was the loss per cent.? *Ans.* 40.

7. If a pound of tea is bought for 6*s.* 6*d.*, and sold for 7*s.* 4*d.*, what is the gain per cent.? *Ans.* 12 $\frac{32}{39}$.

8. If 5 cwt., 3 qrs., 26 ℔, are bought for £9 8*s.*, and sold for £11 18*s.* 11*d.*, how much is gained per cent.? *Ans.* 27 $\frac{47}{564}$.

9. When wine is bought at 17*s.* 6*d.* per gallon, and sold for 27*s.* 6*d.*, what is the gain per cent.? *Ans.* 57⅐.

10. Bought a quantity of goods for £60, and sold them for £75; what was the gain per cent.? *Ans.* 25.

11. Bought a tun of wine for £50, ready money, and sold it for £54 10*s.*, payable in 8 months. How much per cent. per annum is gained by that rate? *Ans.* 13½.

12. Having sold 2 yards of cloth for 11*s.* 6*d.*, I gained at the rate of 15 per cent. What would I have gained if I had sold it for 12*s.*? *Ans.* 20 per cent.

13. If when I sell cloth at 7*s.* per yard, I gain 10 per cent.; what will I gain per cent. when it is sold for 8*s.* 6*d.*? *Ans.* £33 11*s.* 5⅐*d.*

7*s.* : 8*s.* 6*d.* :: £110 : £133 11*s.* 5⅐*d.* And £133 11*s.* 5⅐*d.* −£100=£33 11*s.* 5⅐*d.*, is the required gain.

19. Given the cost price and gain, to find the selling price—

RULE.—Say, as £100 is to £100 plus the gain per cent., so is the cost price to the required selling price.

EXAMPLE.—At what price per yard must I sell 427 yards of cloth which I bought for 19*s.* per yard, so that I may gain 8 per cent.?

$$100 : 108 :: 19s. : \frac{108 \times 19s.}{100} = £1\ 0s.\ 6\tfrac{1}{4}d.$$

This result may be proved by the last rule.

EXERCISES.

14. Bought velvet at 4s. 8d. per yard; at what price must I sell it, so as to gain $12\frac{1}{2}$ per cent.? *Ans.* At 5s. 3d.

15. Bought muslin at 5s. per yard; how must it be sold, that I may lose 10 per cent.? *Ans.* At 4s. 6d.

16. If a tun of brandy costs £40, how must it be sold, to gain $6\frac{1}{4}$ per cent.? *Ans.* For £42 10s.

17. Bought hops at £4 16s. per cwt.; at what rate must they be sold, to lose 15 per cent.? *Ans.* For £4 1s. $7\frac{1}{2}d$.

18. A merchant receives 180 casks of raisins, which stand him in 16s. each, and trucks them against other merchandize at 28s. per cwt., by which he finds he has gained 25 per cent.; for what, on an average, did he sell each cask? *Ans.* 80 ℔, nearly.

20. Given the gain, or loss per cent., and the selling price, to find the cost price—

RULE.—Say, as £100 plus the gain (or as £100 minus the loss) is to £100, so is the selling to the cost price.

EXAMPLE 1.—If I sell 72 ℔ of tea at 6s. per ℔, and gain 9 per cent., what did it cost per ℔?

$$109 : 100 :: 6 : \frac{£100 \times 6}{109} = 5s.\ 6d.$$

What produces £109 cost £100; therefore what produces 6s. must, at the same rate, cost 5s. 6d.

EXAMPLE 2.—A merchant buys 97 casks of butter at 30s. each, and selling the butter at £4 per cwt., makes 20 per cent.; for how much did he buy it per cwt.?

$30s. \times 97 = 2910s.$, is the total price.

Then $100 : 120 :: 2910 : \dfrac{2910s. \times 120}{100} = 3492s.$, the selling price. And $\dfrac{3492s.}{80s.} \left(= \dfrac{3492s.}{£4} \right) = 43{\cdot}65$, is the number of cwt.; and $\dfrac{43{\cdot}65}{97} = 50\frac{194}{485}$ ℔, is the average weight of each cask.

$$\overset{\text{℔}}{\text{Then } 50\tfrac{194}{485}} \cdot \overset{\text{℔}}{112} :: \overset{s.}{30} : \frac{\overset{s.}{112 \times 30}}{50\tfrac{194}{485}} = 66s.\ 8d. = £3\ 6s.$$

8d., the required cost price, per cwt.

EXERCISES.

19. Having sold 12 yards of cloth at 20*s*. per yard, and lost 10 per cent., what was the prime cost? *Ans.* 22*s*. 2⅘*d*.

20. Having sold 12 yards of cloth at 20*s*. per yard, and gained 10 per cent., what was the prime cost? *Ans.* 18*s*. 2²⁄₁₁*d*.

21. Having sold 12 yards of cloth for £5 14*s*., and gained 8 per cent., what was the prime cost per yard? *Ans.* 8*s*. 9½*d*.

22. For what did I buy 3 cwt. of sugar, which I sold for £6 3*s*., and lost 4 per cent.? *Ans.* For £6 8*s*. 1½*d*.

23. For what did I buy 53 yards of cloth, which I sold for £25, and gained £5 10*s*. per cent.? *Ans.* For £23 13*s*. 11¼*d*.

QUESTIONS.

1. What is the object of the rule? [17].

2. Given the prime cost and selling price, how is the profit or loss found? [17].

3. How do we find the profit or loss per cent? [18].

4. Given the prime cost and gain, how is the selling price found? [19].

5. Given the gain or loss per cent. and selling price, how do we find the cost price? [20].

FELLOWSHIP.

21. This rule enables us, when two or more persons are joined in partnership, to estimate the amount of profit or loss which belongs to the share of each.

Fellowship is either *single* (simple) or *double* (compound). It is single, or simple fellowship, when the different stocks have been in trade for the *same* time. It is double, or compound fellowship, when the different stocks have been employed for *different* times.

This rule also enables us to estimate how much of a bankrupt's stock is to be given to each creditor.

22. *Single Fellowship.*—RULE.—Say, as the whole stock is to the whole gain or loss, so is each person's contribution to the gain or loss which belongs to him.

EXAMPLE.—A put £720 into trade, B £340, and C £960; and they gained £47 by the traffic. What is B's share of it?

$$£$$
$$720$$
$$340$$
$$960$$

$$2020 : £47 :: £340 : \frac{£47 \times 340}{2020} = £7\ 18s.\ 2\tfrac{1}{2}d.$$

Each person's gain or loss must evidently be proportional to his contribution.

EXERCISES.

1. B and C buy certain merchandizes, amounting to £80, of which B pays £30, and C £50; and they gain £20. How is it to divided? *Ans.* B £7 10s., and C £12 10s.

2. B and C gain by trade £182; B put in £300, and C £400. What is the gain of each? *Ans.* B £78, and C £104.

3. 2 persons are to share £100 in the proportions of 2 to B and 1 to C. What is the share of each? *Ans.* B £66⅔, C £33⅓.

4. A merchant failing, owes to B £500, and to C £900; but has only £1100 to meet these demands. How much should each creditor receive? *Ans.* B £392⁶⁄₇, and C £707⅐.

5. Three merchants load a ship with butter; B gives 200 casks, C 300, and D 400; but when they are at sea it is found necessary to throw 180 casks overboard. How much of this loss should fall to the share of each merchant? *Ans.* B should lose 40 casks, C 60, and D 80.

6. Three persons are to pay a tax of £100 according to their estates. B's yearly property is £800, C's £600, and D's £400. How much is each person's share? *Ans.* B's is £44⁴⁄₉, C's £33⅓, and D's £22²⁄₉.

7. Divide 120 into three such parts as shall be to each other as 1, 2, and 3? *Ans.* 20, 40, and 60.

8. A ship worth £900 is entirely lost; $\frac{1}{8}$ of it belonged to B, $\frac{1}{4}$ to C, and the rest to D. What should be the loss of each, £540 being received as insurance? *Ans.* B £45, C £90, and D £225.

9. Three persons have gained £1320; if B were to take £6, C ought to take £4, and D £2. What is each person's share? *Ans.* B's £660, C's £440, and D's £220.

10. B and C have gained £600; of this B is to have 10 per cent. more than C. How much will each receive? *Ans.* B £314$\frac{2}{7}$, and C £285$\frac{5}{7}$.

11. Three merchants form a company; B puts in £150, and C £260; D's share of £62, which they gained, comes to £16. How much of the gain belongs to B, and how much to C; and what is D's share of the stock? *Ans.* B's profit is £16 16s. 7$\frac{7}{41}$d., C's £29 3s. 4$\frac{40}{41}$d.; and D put in £142 12s. 2$\frac{2}{23}$d.

12. Three persons join; B and C put in a certain stock, and D puts in £1090; they gain £110, of which B takes £35, and C £29. How much did B and C put in; and what is D's share of the gain? *Ans.* B put in £829 6s. 11$\frac{11}{23}$d., C £687 3s. 5$\frac{17}{23}$d.; and D's part of the profit is £46.

13. Three farmers hold a farm in common; one pays £97 for his portion, another £79, and the third £100. The county cess on the farm amounts to £34; what is each person's share of it? *Ans.* £11 18s. 11$\frac{19}{23}$d.; £9 14s. 7$\frac{15}{23}$d.; and £12 6s. 4$\frac{12}{23}$d.

23. *Compound Fellowship.*—RULE.—Multiply each person's stock by the time during which it has been in trade; and say, as the sum of the products is to the whole gain or loss, so is each person's product to his share of the gain or loss.

EXAMPLE.—A contributes £30 for 6 months, B £84 for 11 months, and C £96 for 8 months; and they lose £14. What is C's share of this loss?

30 × 6 = 180 for one month. ⎫
84 × 11 = 924 for one month. ⎬ = £1872 for one month.
96 × 8 = 768 for one month. ⎭

$$1872 : £14 :: £768 : \frac{£14 \times 768}{1872} = £6 \ 1s. \ 4\frac{1}{4}d., \text{ C's share.}$$

24. REASON OF THE RULE.—It is clear that £30 contributed for 6 months are, as far as the gain or the loss to be derived from it is concerned, the same as 6 times £30—or £180 contributed for 1 month. Hence A's contribution may be taken as £180 for 1 month; and, for the same reason, B's as £924 for the same time; and C's as £768 also for the same time. This reduces the question to one in simple fellowship [22].

EXERCISES.

14. Three merchants enter into partnership; B puts in £89 5s. for 5 months, C £92 15s. for 7 months, and D £38 10s. for 11 months; and they gain £86 16s. What should be each person's share of it? *Ans.* B's £25 10s., C's £37 2s., and D's £24 4s.

15. B, C, and D pay £40 as the year's rent of a farm. B puts 40 cows on it for 6 months, C 30 for 5 months, and D 50 for the rest of the time. How much of the rent should each person pay? *Ans.* B £21$\frac{9}{11}$, C £13$\frac{7}{11}$, and D £4$\frac{6}{11}$.

16. Three dealers, A, B, and C, enter into partnership, and in a certain time make £291 13s. 4d. A's stock, £150, was in trade 6 months; B's, £200, 3 months; and C's, £125, 16 months. What is each person's share of the gain? *Ans.* A's is £75, B's £50, and C's £166 13s. 4d.

17. Three persons have received £665 interest; B had put in £4000 for 12 months, C £3000 for 15 months, and D £5000 for 8 months; how much is each person's part of the interest? *Ans.* B's £240, C's £225, and D's £200.

18. X, Y, and Z form a company. X's stock is in trade 3 months, and he claims $\frac{1}{12}$ of the gain; Y's stock is 9 months in trade; and Z advanced £756 for 4 months, and claims half the profit. How much did X and Y contribute? *Ans.* X £168, and Y £280.

It follows that Y's gain was $\frac{5}{12}$. Then $\frac{1}{2} : \frac{1}{12} :: £756 \times 4 :$ 504 = X's product, which, being divided by his number of months, will give £168, as his contribution. Y's share of the stock may be found in the same way.

19. Three troops of horse rent a field, for which they pay £80; the first sent into it 56 horses for 12 days, the

second 64 for 15 days, and the third 80 for 18 days. What must each pay? *Ans.* The first must pay £17 10s., the second £25, and the third £37 10s.

20. Three merchants are concerned in a steam vessel; the first, A, puts in £240 for 6 months; the second, B, a sum unknown for 12 months; and the third, C, £160, for a time not known when the accounts were settled. A received £300 for his stock and profit, B £600 for his, and C £260 for his; what was B's stock, and C's time? *Ans.* B's stock was £400; and C's time was 15 months.

If £300 arise from £240 in 6 months, £600 (B's stock and profit) will be found to arise from £400 (B's stock) in 12 months.

Then £400 : £160 :: £200 (the profit on £400 in 12 months):£80 (the profit on £160 in 12 months). And £160+ 80 (£160 with its profit for 12 months) : £260 (£160 with its profit for some other time) :: 12 (the number of months in the one case) : $\frac{260 \times 12}{160+80}$ (the number of months in the other case)=13, the number of months required to produce the difference between £160, C's stock, and the £260, which he received.

21. In the foregoing question A's gain was £60 during 6 months, B's £200 during 12 months, and C's £100 during 13 months; and the sum of the products of their stocks and times is 8320. What were their stocks? *Ans.* A's was £240, B's £400, and C's £160.

22. In the same question the sum of the stocks is £800; A's stock was in trade 6 months, B's 12 months, and C's 15 months; and at the settling of accounts, A is paid £60 of the gain, B £200, and C £100. What was each person's stock? *Ans.* A's was £240, B's 400, and C's £160.

QUESTIONS.

1. What is fellowship? [21].
2. What is the difference between single and double fellowship; and are these ever called by any other names? [21].
3. What are the rules for single, and double fellowship? [22 and 23].

BARTER.

25. Barter enables the merchant to exchange one commodity for another, without either loss or gain—

RULE.—Find the price of the given quantity of one kind of merchandise to be bartered; and then ascertain how much of the other kind this price ought to purchase.

EXAMPLE 1.—How much tea, at 8s. per ℔, ought to be given for 3 cwt. of tallow, at £1 16s. 8d. per cwt.?

$$\begin{array}{ccc} £ & s. & d. \\ 1 & 16 & 8 \\ & & 3 \\ \hline 5 & 10 & 0 \end{array}$$ is the price of 3 cwt. of tallow.

And £5 10s. ÷ 8s. = 13⅔, is the number of pounds of tea which £5 10s., the price of the tallow, would purchase.

There must be so many pounds of tea, as will be equal to the number of times that 8s. is contained in the price of the tallow.

EXAMPLE 2.—I desire to barter 96 ℔ of sugar, which cost me 8d. per ℔, but which I sell at 13d., giving 9 months' credit, for calico which another merchant sells for 17d. per yard, giving 6 months' credit. How much calico ought I to receive?

I first find at what price I could sell my sugar, were I to give the same credit as he does—

If 9 months give me 5d. profit, what ought 6 months to give?

$$9 : 6 :: 5 : \frac{6 \times 5}{9} = \frac{30}{9} = 3\tfrac{1}{3}d.$$

Hence, were I to give 6 months' credit, I should charge 11⅓d. per ℔. Next—

As my selling price is to my buying price, so ought his selling to be to his buying price, both giving the same credit.

$$11\tfrac{1}{3} : 8 :: 17 : \frac{8 \times 17}{11\tfrac{1}{3}} = 12d.$$

The price of my sugar, therefore, is 96 × 8d., or 768d.; and of his calico, 12d. per yard.

Hence $\frac{768}{12}$ = 64, is the required number of yards.

EXERCISES.

1. A merchant has 1200 stones of tallow, at 2*s.* 3¼*d.* the stone ; B has 110 tanned hides, weight 3994 ℔, at 5¾*d.* the ℔ ; and they barter at these rates. How much money is A to receive of B, along with the hides ? *Ans.* £40 11*s.* 2½*d.*

2. A has silk at 14*s.* per ℔ ; B has cloth at 12*s.* 6*d.* which cost only 10*s.* the yard. How much must A charge for his silk, to make his profit equal to that of B ? *Ans.* 17*s.* 6*d.*

3. A has coffee which he barters at 10*d.* the ℔ more than it cost him, against tea which stands B in 10*s.*, but which he rates at 12*s.* 6*d.* per ℔. How much did the coffee cost at first ? *Ans.* 3*s.* 4*d.*

4. K and L barter. K has cloth worth 8*s.* the yard, which he barters at 9*s.* 3*d.* with L, for linen cloth at 3*s.* per yard, which is worth only 2*s.* 7*d.* Who has the advantage ; and how much linen does L give to K, for 70 yards of his cloth ? *Ans.* L gives K 215⁴⁄₆ yards ; and L has the advantage.

5. B has five tons of butter, at £25 10*s.* per ton, and 10½ tons of tallow, at £33 15*s.* per ton, which he barters with C ; agreeing to receive £150 1*s.* 6*d.* in ready money, and the rest in beef, at 21*s.* per barrel. How many barrels is he to receive ? *Ans.* 316.

6. I have cloth at 8*d.* the yard, and in barter charge for it at 13*d.*, and give 9 months' time for payment ; another merchant has goods which cost him 12*d.* per ℔, and with which he gives 6 months' time for payment. How high must he charge his goods to make an equal barter ? *Ans.* At 17*d.*

7. I barter goods which cost 8*d.* per ℔, but for which I charge 13*d.*, giving 9 months' time, for goods which are charged at 17*d.*, and with which 6 months' time are given. Required the cost of what I receive ? *Ans.* 12*d.*

8. Two persons barter ; A has sugar at 8*d.* per ℔, but charges it at 13*d.*, and gives 9 months time ; B has cocoa at 12*d.* per ℔, and charges it at 17*d.* per ℔. How much time must B give, to make the barter equal ? *Ans.* 6 months.

1. What is barter ? [25].
2. What is the rule for barter ? [25].

ALLIGATION.

26. This rule enables us to find what mixture will be produced by the union of certain ingredients—and then it is called alligation *medial ;* or what ingredients will be required to produce a certain mixture—when it is termed alligation *alternate ;* further division of the subject is unnecessary :—it is evident that any change in the amount of one ingredient of a given mixture must produce a proportional change in the amounts of the others, and of the entire quantity.

27. *Alligation Medial.*—Given the rates or kinds and quantities of certain ingredients, to find the mixture they will produce—

RULE.—Multiply the rate or kind of each ingredient by its amount ; divide the sum of the products by the number of the lowest denomination contained in the whole quantity, and the quotient will be the rate or kind of that denomination of the mixture. From this may be found the rate or kind of any other denomination.

EXAMPLE 1.—What ought to be the price per ℔, of a mixture containing 98 ℔ of sugar at 9*d.* per ℔, 87 ℔ at 5*d.*, and 34 ℔ at 6*d.*

$$
\begin{array}{rcl}
d. & & d. \\
9 \times 98 & = & 882 \\
5 \times 87 & = & 435 \\
6 \times 34 & = & 204 \\
\hline
219 & & 219)1521
\end{array}
$$

Ans. 7*d.* per ℔, nearly.

The price of each sugar, is the number of pence per pound multiplied by the number of pounds ; and the price of the whole is the sum of the prices. But if 219 ℔ of sugar have cost 1521*d.*, one ℔, or the 219th part of this, must cost the 219th part of 1521*d.*, or $\frac{1521}{219}d. = 7d.$, nearly.

EXAMPLE 2.—What will be the price per ℔ of a mixture containing 9 ℔ 6 oz. of tea at 5s. 6d. per ℔, 18 ℔ at 6s. per ℔, and 46 ℔ 3 oz. at 9s. 4½d. per ℔ ?

℔	oz.		s.	d.			£	s.	d.
9	6	at 5	6	per ℔ =	2	11	6¾
18	0	6	0	per ℔ =	5	8	0
46	3	9	4½	per ℔ =	21	13	0

73 9   1177)29 12 6¾

16   *Ans.* 6d. per oz. nearly

1177 ounces.

And 6d.×16=8s, is the price per pound.

In this case, the lowest denomination being ounces, we reduce the whole to ounces; and having found the price of an ounce, we multiply it by 16, to find that of a pound.

EXAMPLE 3.—A goldsmith has 3 ℔ of gold 22 carats fine, and 2 ℔ 21 carats fine. What will be the fineness of the mixture ?

In this case the value of each kind of ingredient is represented by a number of *carats*—

$$\begin{array}{l} \text{℔s} \\ 3\times22 = 66 \\ 2\times21 = 42 \\ \hline 5 \qquad 5)\overline{108} \end{array}$$

The mixture is 21⅗ carats fine.

EXERCISES.

1. A vintner mixed 2 gallons of wine, at 14s. per gallon, with 1 gallon at 12s., 2 gallons at 9s., and 4 gallons at 8s. What is one gallon of the mixture worth? *Ans.* 10s.

2. 17 gallons of ale, at 9d. per gallon, 14 at 7½d , 5 at 9½d., and 21 at 4½d., are mixed together. How much per gallon is the mixture worth? *Ans.* 7$\frac{1}{57}$d.

3. Having melted together 7 oz. of gold 22 carats fine, 12½ oz. 21 carats fine, and 17 oz. 19 carats fine, I wish to know the fineness of each ounce of the mixture? *Ans.* 20$\frac{19}{73}$ carats.

28. *Alligation Alternate.*—Given the nature of the mixture, and of the ingredients, to find the relative amounts of the latter—

RULE.—Put down the quantities *greater* than the given mean (each of them connected with the difference between it and the mean, by the sign —) in one column; put the differences between the remaining quantities and the mean (connected with the quantities to which they belong, by the sign +) in a column to the right hand of the former. Unite, by a line, each *plus* with some *minus* difference; and then each difference will express how much of the quantity, with whose difference it is connected, should be taken to form the required mixture.

If any difference is connected with *more than one* other difference, it is to be considered as *repeated* for each of the differences with which it is connected; and the sum of the differences with which it is connected is to be taken as the required amount of the quantity whose difference it is.

EXAMPLE 1.—How many pounds of tea, at 5*s.* and 8*s.* per ℔, would form a mixture worth 7*s.* per ℔?

Price. Differences. Price.

s. *s.*   *s.* *s.*
The mean $=8-1\text{———}2+5=$ the mean.

1 is connected with 2*s.*, the difference between the mean and 5*s.*; hence there must be 1 ℔ at 5*s.* 2 is connected with 1, the difference between 8*s.* and the mean; hence there must be 2 ℔ at 8*s.* Then 1 ℔ of tea at 5*s.* and 2 ℔ at 8*s.* per ℔, will form a mixture worth 7*s.* per ℔—as may be proved by the last rule.

It is evident that any equimultiples of these quantities would answer equally well; hence a great number of answers may be given to such a question.

EXAMPLE 2.—How much sugar at 9*d.*, 7*d.*, 5*d.*, and 10*d.*, will produce sugar at 8*d.* per ℔?

Prices. Differences. Prices.

d. *d.*   *d.* *d.*
The mean $= \left\{ \begin{matrix} 9--1\text{———}1+7 \\ 10-2\text{———}3+5 \end{matrix} \right\} =$ the mean.

1 is connected with 1, the difference between 7*d.* and the mean; hence there is to be 1 ℔ of sugar at 7*d.* per ℔. 2 is connected with 3, the difference between 5*d.* and the mean; hence there is to be 2 ℔ at 5*d.* 1 is connected with 1, the difference between 9*d.* and the mean; hence there is to be 1 ℔ at 9*d.* And 3 is connected with 2, the difference between 10*d.* and the mean; hence there are to be 3 ℔ at 10*d.* per ℔.

Consequently we are to take 1 ℔ at 7d., and 2 ℔ at 5d., 1 ℔ at 9d., and 3 ℔ at 10d. If we examine what mixture these will give [27], we shall find it to be the given mean.

EXAMPLE 3.—What quantities of tea at 4s., 6s., 8s., and 9s. per ℔, will produce a mixture worth 5s. ?

Prices. Differences. Prices.

$$\text{The mean} = \begin{cases} 8-3 \\ 6-1 \\ 9-4 \end{cases} 1+4 = \text{the mean.}$$

3, 1, and 4 are connected with 1s., the difference between 4s. and the mean ; therefore we are to take 3 ℔ + 1 ℔ + 4 ℔ of tea, at 4s. per ℔. 1 is connected with 3s., 1s., and 4s., the differences between 8s., 6s., and 9s., and the mean ; therefore we are to take 1 ℔ of tea at 8s., 1 ℔ of tea at 6s., and 1 ℔ of tea at 9s. per ℔.

We find in this example that 8s., 6s., and 9s. are all connected with the same 1 ; this shows that 1 ℔ of each will be required. 4s. is connected with 3. 1, and 4 ; there must be, therefore, 3 + 1 + 4 ℔ of tea at 4s.

EXAMPLE 4.—How much of anything, at 3s., 4s., 5s., 7s., 8s , 9s., 11s., and 12s. per ℔, would form a mixture worth 6s. per ℔ ?

Prices. Differences. Prices.

$$\begin{aligned} 7-1 &\quad 3+3 \\ 8-2 &\quad 2+4 \\ 9-3 &\quad 1+5 \\ 11-5 & \\ 12-6 & \end{aligned}$$

1 ℔ at 3s., 2 ℔ at 4s., 3 ℔ at 7s., 2 ℔ at 8s., 3+5+6 (14) ℔ at 5s., 1 ℔ at 9s., 1 ℔ at 11s., and 1 ℔ at 12s. per ℔, will form the required mixture.

29. REASON OF THE RULE.—The excess of one ingredient above the mean is made to counterbalance what the other wants of being equal to the mean. Thus in example 1, 1 ℔ at 5s. per ℔ gives a *deficiency* of 2s. : but this is corrected by 2s. *excess* in the 2 ℔ at 8s. per ℔.

In example 2, 1 ℔ at 7d. gives a *deficiency* of 1d., 1 ℔ at 9d. gives an *excess* of 1d. ; but the excess of 1d. and the deficiency of 1d. exactly neutralize each other.

Again, it is evident that 2 ℔ at 5d. and 3 ℔ at 10d. are worth just as much as 5 ℔ at 8d.—that is, 8d. will be the average price if we mix 2 ℔ at 5d. with 3 ℔ at 10d.

4　How much wine at 8*s*. 6*d*. and 9*s*. per gallon will make a mixture worth 8*s*. 10*d*. per gallon? *Ans*. 2 gallons at 8*s*. 6*d*., and 4 gallons at 9*s*. per gallon.

5. How much tea at 6*s*. and at 3*s*. 8*d*. per ℔, will make a mixture worth 4*s*. 4*d*. per ℔? *Ans*. 8 ℔ at 6*s*., and 20 ℔ at 3*s*. 8*d*. per ℔.

6. A merchant has sugar at 5*d*., 10*d*., and 12*d*. per ℔. How much of each kind, mixed together, will be worth 8*d*. per ℔? *Ans*. 6 ℔ at 5*d*., 3 ℔ at 10*d*., and 3 ℔ at 12*d*.

7. A merchant has sugar at 5*d*., 10*d*., 12*d*., and 16*d*. per ℔. How many ℔ of each will form a mixture worth 11*d*. per ℔? *Ans*. 5 ℔ at 5*d*., 1 ℔ at 10*d*., 1 ℔ at 12*d*., and 6 ℔ at 16*d*.

8. A grocer has sugar at 5*d*., 7*d*., 12*d*., and 13*d*. per ℔. How much of each kind will form a mixture worth 10*d*. per ℔? *Ans*. 3 ℔ at 5*d*., 2 ℔ at 7*d*., 3 ℔ at 12*d*., and 5 ℔ at 13*d*.

30. When a *given* amount of the mixture is required, to find the corresponding amounts of the ingredients—

RULE.—Find the amount of each ingredient by the last rule. Then add the amounts together, and say, as their sum is to the amount of any one of them, so is the required quantity of the mixture to the corresponding amount of that one.

EXAMPLE 1.—What must be the amount of tea at 4*s*. per ℔, in 736 ℔ of a mixture worth 5*s*. per ℔, and containing tea at 6*s*., 8*s*., and 9*s*. per ℔?

To produce a mixture worth 5*s*. per ℔, we require 8 ℔ at 4*s*., 1 at 8*s*., 1 at 6*s*., and 1 at 9*s*. per ℔. [28]. But all of these, added together, will make 11 ℔, in which there are 8 ℔ at 4*s*. Therefore

$$\overset{\text{℔}}{11} : 8 :: 736 : \overset{\text{℔}}{\frac{8 \times 736}{11}} = 526 \ \overset{\text{oz.}}{4\tfrac{4}{11}}, \text{ the required quantity}$$

of tea at 4*s*.

That is, in 736 ℔ of the mixture there will be 536 ℔ 4$\frac{4}{11}$ oz. at 4*s*. per ℔. The amount of each of the other ingredients may be found in the same way.

EXAMPLE 2.—Hiero, king of Syracuse, gave a certain quantity of gold to form a crown; but when he received it, suspecting that the goldsmith had taken some of the gold, and supplied its place by a baser metal, he commissioned Archimedes, the celebrated mathematician of Syracuse, to ascertain if his suspicion was well founded, and to what extent. Archimedes was for some time unsuccessful in his researches, until one day, going into a bath, he remarked that he displaced a quantity of water equal to his own bulk. Seeing at once that the same weight of different bodies would, if immersed in water, displace very different quantities of the fluid, he exclaimed with delight that he had found the desired solution of the problem. Taking a mass of gold equal in weight to what was given to the goldsmith, he found that it displaced less water than the crown; which, therefore, was made of a lighter, because a more bulky metal—and, consequently, was an *alloy* of gold.

Now supposing copper to have been the substance with which the crown was adulterated, to find its amount—

Let the gold given by Hiero have weighed 1 ℔, this would diplace about ·052 ℔ of water; 1 ℔ of copper would displace about ·1124 ℔ of water; but let the crown have displaced only ·072 ℔. Then

> Gold differs from ·072, the mean, by − ·020.
> Copper differs from it by . . + ·0404.

Copper.   Differences.   Gold.

Hence, the mean = ·1124 − ·0404——·020 + ·052 = the mean.

Therefore ·020 ℔ of copper and ·0404 ℔ of gold would produce the alloy in the crown.

But the crown was supposed to weigh 1 ℔; therefore

$$·0604 \text{ ℔ } (·020 + ·0404) : ·0404 \text{ ℔ } :: 1 \text{ ℔ } : \frac{·0404 \times 1 \text{ ℔}}{·060}$$

= ·669 ℔. the quantity of gold. And 1 − ·669 = ·331 ℔ is the quantity of copper.

EXERCISES.

9. A druggist is desirous of producing, from medicine at 5*s*., 6*s*., 8*s*., and 9*s*. per ℔, 1½ cwt. of a mixture worth 7*s*. per ℔. How much of each kind must he use for the purpose? *Ans.* 28 ℔ at 5*s*., 56 ℔ at 6*s*., 56 ℔ at 8*s*., and 28 ℔ at 9*s*. per ℔.

10. 27 ℔ of a mixture worth 4*s*. 4*d*. per ℔ are required. It is to contain tea at 5*s*. and at 3*s*. 6*d*. per

℔. How much of each must be used? *Ans.* 15 ℔ at 5*s.*, and 12 ℔ at 3*s.* 6*d.*

11. How much sugar, at 4*d.*, 6*d.*, and 8*d.* per ℔, must there be in 1 cwt. of a mixture worth 7*d.* per ℔ ? *Ans.* 18⅔ ℔ at 4*d.*, 18⅔ ℔ at 6*d.*, and 74⅔ ℔ at 8*d.* per ℔.

12. How much brandy at 12*s.*, 13*s.*, 14*s.*, and 14*s.* 6*d.* per gallon, must there be in one hogshead of a mixture worth 13*s.* 6*d.* per gallon ? *Ans.* 18 gals. at 12*s.*, 9 gals. at 13*s.*, 9 gals. at 14*s.*, and 27 gals. at 14*s.* 6*d.* per gallon.

31. When the amount of one ingredient is given, to find that of any other—

RULE.—Say, as the amount of one ingredient (found by the rule) is to the *given* amount of the same ingredient, so is the amount of any other ingredient (found by the rule) to the *required* quantity of that other.

EXAMPLE 1.—29 ℔ of tea at 4*s.* per ℔ is to be mixed with teas at 6*s.*, 8*s.*, and 9*s.* per ℔, so as to produce what will be worth 5*s.* per ℔. What quantities must be used?

8 ℔ of tea at 4*s.*, and 1 ℔ at 6*s.*, 1 ℔ at 8*s.*, and 1 ℔ at 9*s.*, will make a mixture worth 5*s.* per ℔ [27]. Therefore

8 ℔ (the quantity of tea at 4*s.* per ℔, as found by the rule) : 29 ℔ (the given quantity of the same tea) :: 1 ℔ (the quantity of tea at 6*s.* per ℔, as found by the rule) : $\dfrac{1 \times 29 \text{ ℔}}{8}$ (the quantity of tea at 6*s.*, which corresponds with 29 ℔ at 4*s* per ℔)=3⅝ ℔.

We may in the same manner find what quantities of tea at 8*s.* and 9*s.* per ℔ correspond with 29 ℔—or the *given* amount of tea at 4*s.* per ℔.

EXAMPLE 2.—A refiner has 10 oz. of gold 20 carats fine, and melts it with 16 oz. 18 carats fine. What must be added to make the mixture 22 carats fine?

10 oz. of 20 carats fine=10×20 = 200 carats.
16 oz. of 18 carats fine=16×18 = 288

$\overline{26}$: 1 :: $\overline{488}$: 18$\frac{10}{13}$ carats, the fineness of the mixture.

24−22=2 carats baser metal, in a mixture 22 carats fine.
24−18$\frac{10}{13}$=5$\frac{3}{13}$ carats baser metal, in a mixture 18$\frac{10}{13}$ carats fine.

Then 2 carats : 22 carats :: 5$\frac{3}{13}$: 57$\frac{7}{13}$ carats of pure gold—required to change $5\frac{3}{13}$ carats baser metal, into a mixture 22 carats fine. But there are already in the mixture $18\frac{10}{13}$ carats gold; therefore $57\frac{7}{13} - 18\frac{10}{13} = 38\frac{10}{13}$ carats gold are to be added to every ounce. There are 26 oz.; therefore $26 \times 38\frac{10}{13} = 1008$ carats of gold are wanting. There are 24 carats (page 5) in every oz.; therefore $\frac{1008}{24}$ carats $= 42$ oz. of gold must be added. There will then be a mixture containing

oz.	car.		car.
10	×20	=	200
16	×18	=	288
42	×24	=	1008

68 : 1 oz. :: 1496 : 22 carats, the required fineness.

EXERCISES.

13 How much tea at 6s. per ℔ must be mixed with 12 ℔ at 3s. 8d. per ℔, so that the mixture may be worth 4s. 4d. per ℔? *Ans.* $4\frac{4}{3}$ ℔.

14. How much brass, at 14d. per ℔, and pewter, at $10\frac{1}{2}d$. per ℔, must I melt with 50 ℔ of copper, at 16d. per ℔, so as to make the mixture worth 1s. per ℔? *Ans.* 50 ℔ of brass, and 200 ℔ of pewter.

15. How much gold of 21 and 23 carats fine must be mixed with 30 oz. of 20 carats fine, so that the mixture may be 22 carats fine? *Ans.* 30 of 21, and 90 of 23.

16. How much wine at 7s. 5d., at 5s. 2d., and at 4s. 2d. per gallon, must be mixed with 20 gallons at 6s. 8d. per gallon, to make the mixture worth 6s. per gallon? *Ans.* 44 gallons at 7s. 5d., 16 gallons at 5s. 2d., and 34 gallons at 4s. 2d.

QUESTIONS.

1. What is alligation medial? [26].
2. What is the rule for alligation medial? [27].
3. What is alligation alternate? [26].
4. What is the rule for alligation alternate? [28].
5. What is the rule, when a certain amount of the mixture is required? [30].
6. What is the rule, when the amount of one or more of the ingredients is given? [31].

SECTION IX.

INVOLUTION AND EVOLUTION, &c.

1. INVOLUTION.—A quantity which is the product of two or more factors, each of them the same number, is termed a *power* of that number ; and the number, multiplied by itself, is said to be *involved*. Thus $5 \times 5 \times 5$ ($=125$) is a "power of 5 ;" and 125, is 5 "involved." A power obtains its denomination from the number of times the *root* (or quantity involved) is taken as a factor. Thus 25 ($=5 \times 5$) is the *second* power of 5.—The second power of any number is also called its *square ;* because a square surface, one of whose sides is expressed by the given number, will have its area indicated by the second power of that number ; thus a square, 5 inches every way, will contain 25 (the square of 5) square inches ; a square 5 feet every way, will contain 25 square feet, &c. 216 ($6 \times 6 \times 6$) is the *third* power of 6.—The third power of any number is also termed its *cube ;* because a cube, the length of one of whose sides is expressed by the given number, will have its solid contents indicated by the third power of that number. Thus a cube 5 inches every way, will contain 125 (the cube of 5) cubic, or solid inches ; a cube 5 feet every way, will contain 125 cubic feet, &c.

2. In place of setting down all the factors, we put down only one of them, and mark how often they are *supposed* to be set down by a small figure, which, since it *points out* the number of the factors, is called the *index*, or *exponent*. Thus 5^2 is the abbreviation for 5×5 :—and 2 is the index. 5^5 means $5 \times 5 \times 5 \times 5 \times 5$, or 5 in the fifth power. 3^4 means $3 \times 3 \times 3 \times 3$, or 3 in the fourth power. 8^7 means $8 \times 8 \times 8 \times 8 \times 8 \times 8 \times 8$, or 8 in the seventh power, &c.

3. Sometimes the vinculum [Sec. II. 5] is used in conjunction with the index ; thus $\overline{5+8}^2$ means that the sum of 5 and 8 is to be raised to the second power—this

is very different from 5^2+8^2, which means the sum of the squares of 5 and 8 : $\overline{5+8}^2$ being 169; while 5^2+8^2 is only 89.

4. In multiplication the multiplier may be considered as a species of index. Thus in 187×5, 5 points out how often 187 should be set down as an *addend;* and 187×5 is merely an abbreviation for $187+187+187+187+187$ [Sec. II. 41]. In 187^5, 5 points out how often 187 should be set down as a *factor;* and 187^5 is an abbreviation for $187\times187\times187\times187\times187$:—that is, the "multiplier" tells the number of the *addends*, and the "index" or "exponent," the number of the *factors.*

5. To raise a number to any power—

RULE.—Find the product of so many factors as the index of the proposed power contains units—each of the factors being the number which is to be involved.

EXAMPLE 1.—What is the 5th power of 7 ?
$$7^5=7\times7\times7\times7\times7=16807.$$

EXAMPLE 2.—What is the amount of £1 at compound interest, for 6 years, allowing 6 per cent. per annum ?

The amount of £1 for 6 years, at 6 per cent. is—
$1\cdot06\times1\cdot06\times1\cdot06\times1\cdot06\times1\cdot06\times1\cdot06$ [Sec. VII. 20], or $\overline{1\cdot06}^6=1\cdot41852.$

We, as already mentioned [Sec. VII. 23], may abridge the process, by using one or more of the products, already obtained, as factors.

EXERCISES.

1. $3^5=243.$
2. $20^{10}=10240000000000.$
3. $3^7=2187.$
4. $105^6=1340095640625.$
5. $1\cdot05^6=1\cdot340095640625.$

6. To raise a fraction to any power—

RULE.—Raise both numerator and denominator to that power.

EXAMPLE.—$(\frac{3}{4})^3=\frac{3}{4}\times\frac{3}{4}\times\frac{3}{4}=\frac{27}{64}.$

To involve a fraction is to multiply it by itself. But to multiply it by itself any number of times, we must multiply its numerator by itself, and also its denominator by itself, that number of times [Sec. IV. 39].

6. $(\frac{3}{8})^4 = \frac{81}{8\,1}.$
7. $(\frac{3}{5})^7 = \frac{2187}{78125}.$
8. $(\frac{3}{4})^7 = \frac{2187}{16384}.$
9. $(\frac{5}{9})^5 = \frac{3125}{59049}.$

7. To raise a mixed number to any power—

RULE.—Reduce it to an improper fraction [Sec. IV. 24] ; and then proceed as directed by the last rule.

EXAMPLE.—$(2\frac{1}{2})^4 = (\frac{5}{2})^4 = \frac{625}{16}.$

10. $(11\frac{2}{5})^3 = \frac{185193}{125}.$
11. $(3\frac{3}{7})^5 = \frac{6436343}{16807}.$
12. $(5\frac{5}{8})^6 = \frac{22164361129}{531441}.$
13. $(4\frac{4}{7})^7 = \frac{4261844297\,7}{823543}.$

8. *Evolution* is a process exactly opposite to involution; since, by means of it, we find what number, raised to a given power, would produce a given quantity—the number so found is termed a *root*. Thus we " evolve" 25 when we take, for instance, its square root ; that is, when we find what number, multiplied by itself, will produce 25. Roots, also, are expressed by *exponents*—but as these exponents are fractions, the roots are called "*fractional powers*." Thus $4^{\frac{1}{2}}$ means the square root of 4 ; $4^{\frac{1}{3}}$ the cube root of 4 ; and $4^{\frac{5}{7}}$ the seventh root of the fifth power of 4. Roots are also expressed by $\sqrt{}$, called the *radical sign*. When used alone, it means the *square* root—thus $\sqrt{3}$, is the square root of 3; but other roots are indicated by a small figure placed within it—thus $\sqrt[3]{5}$; which means the cube root of 5. $\sqrt[3]{7^2}$ $(7^{\frac{2}{3}})$, is the cube root of the square of 7.

9. The fractional exponent, and radical sign are sometimes used in conjunction with the vinculum. Thus $\overline{4-3}^{\frac{1}{2}}$, is the square root of the difference between 4 and 3 ; $\sqrt[3]{5+7}$, or $\overline{5+7}^{\frac{1}{3}}$, is the cube root of the sum of 5 and 7.

10. To find the square root of any number—

RULE—I. Point off the digits in pairs, by dots ; putting one dot over the units' *place*, and then another dot over every second digit *both* to the right and left of the units' place—if there are digits at *both* sides of the decimal point.

II. Find the highest number the square of which will not exceed the amount of the highest period, or that which is at the extreme left—this number will be the first digit in the required square root. Subtract its square from the highest period, and to the remainder, considered as hundreds, add the next period.

III. Find the highest digit, which being multiplied into twice the part of the root already found (considered as so many tens), and into itself, the *sum* of the products will not exceed the *sum* of the last remainder and the period added to it. Put this digit in the root after the one last found, and subtract the former *sum* from the latter.

IV. To the remainder, last obtained, bring down another period, and proceed as before. Continue this process until the exact square root, or a sufficiently near approximation to it is obtained.

11. EXAMPLE.—What is the square root of 22420225 ?

$$22\overset{.}{4}2\overset{.}{0}2\overset{.}{2}5(4735, \text{ is the required root.}$$
$$16$$
$$87)\overline{642}$$
$$609$$
$$943)\overline{3302}$$
$$2829$$
$$9465)\overline{47325}$$
$$47325$$

22 is the highest period; and 4^2 is the highest square which does not exceed it—we put 4 in the root, and subtract 4^2, or 16 from 22. This leaves 6, which, along with 42, the next period, makes 642.

We subtract 87 (twice 4 tens + 7, the highest digit which we can now put in the root) × 7 from 642. This leaves 33, which, along with 02, the next period, makes 3302.

We subtract 943 (twice 47 tens + 3, the next digit of the root) × 3 from 3302. This leaves 473, which, along with 25, the only remaining period, makes 47325.

We subtract 9465 (twice 473 tens + 5, the next digit of the root) × 5. This leaves *no* remainder.

The given number, therefore, is exactly a square; and its square root is 4735.

12. REASON OF I.—We point off the digits of the given square in pairs, and consider the number of dots as indicating

the number of digits in the root, since neither one nor two
digits in the square can give more or less than one in the root;
neither three nor four digits in the square can give more or
less than two in the root, &c.—which the pupil may easily
ascertain by experiment. Thus 1, the smallest single digit,
will give one digit as its square root; and 99, the largest pair
of digits, can give only one—since 81, or the square of 9, is
the greatest square which does not exceed 99.

Pointing off the digits in pairs shows how many should be
brought down successively, to obtain the successive digits of
the root—since it will be necessary to bring down one period
for each new digit; but more than one will not be required.

REASON OF II.—We subtract from the highest period of the
given number the highest square which does not exceed it,
and consider the root of this square as the first or highest
digit of the required root; because, if we separate any number
into the parts indicated by its digits (563, for instance, into
500, 60, and 3), its square will be found to contain the square
of each of its parts.

REASON OF III.—We divide twice the quantity already in
the root (considered as expressing *tens* of the next denomination) into what is left after the preceding subtraction, &c., to
obtain a new digit of the root; because the square of any
quantity contains (besides the square of each of its parts)
twice the product of each part multiplied by each of the other
parts. Thus if 14 is divided into 1 ten and 4 units, its square
will contain not only 10^2 and 4^2, but also twice the product
of 10 and 4.—We subtract the square of the digit last put in
the root, at the *same* time that we subtract twice the product
obtained on multiplying it by the part of the root which precedes it. Thus in the example which illustrates the rule,
when we subtract 87×7, we really subtract $2 \times 40 \times 7 + 7^2$.

It will be easy to show, that the square of any quantity
contains the squares of the parts, along with twice the product of every two parts. Thus

$$22420225 = \overline{4735}^2 = \overline{4000 + 700 + 30 + 5}^2.$$
$$\overline{4000}^2 = 16000000$$
$$\overline{6420225}$$
$$2 \times 4000 \times 700 + \overline{700}^2 = 6090000$$
$$\overline{330225}$$
$$2 \times 4000 \times 30 + 2 \times 700 \times 30 + 30^2 = 282900$$
$$\overline{47325}$$
$$2 \times 4000 \times 5 + 2 \times 700 \times 5 + 2 \times 30 \times 5 + 5^2 = 47325$$

REASON OF IV.—Dividing twice the quantity already in
the root (considered as expressing tens of the next denomination) into the remainder of the given number, &c., gives
the next digit; because the square contains the sum of
twice the products (or, what is the same thing, the product

of twice the sum) of the parts cf the root already found, multiplied by the new digit. Thus 22420225, the square of 4735, contains $4000^2+700^2+30^2+5^2$; and *also* twice 4000×700 + twice 4000×30 + twice 4000×5; plus twice 700×30 + twice 700×5; plus twice 30×5:—that is, the square of each of its parts, with the sum of twice the product of every two of them (which is the same as each of them multiplied by twice the sum of all the rest). This would, on examination, be found the case with the square of any other number.

If we examine the example given, we shall find that it will not be necessary to bring down more than one period at a time, nor to add cyphers to the quantities subtracted.

13. When the given square contains decimals—

If any of the periods consist of decimals, the digits in the root obtained on bringing down *these* periods to the remainders will also be decimals. Thus, taking the example just given, but altering the decimal point, we shall have $\sqrt{224202 \cdot 25} = 473 \cdot 5$; $\sqrt{2242 \cdot 0225} = 47 \cdot 35$; $\sqrt{22 \cdot 420225} = 4 \cdot 735$; $\sqrt{\cdot 22420225} = \cdot 4735$; and $\sqrt{\cdot 0022420225} = \cdot 04735$, &c.: this is obvious. If there is an *odd* number of decimal places in the power, it must be made *even* by the addition of a cypher. Using the same figures, $\sqrt{2242022 \cdot 5} = 1497 \cdot 338$, &c.

$$
\begin{array}{l}
\overset{\displaystyle \cdot\ \cdot\ \cdot\ \cdot\ \cdot}{2242022 \cdot 50}\ (1497 \cdot 338,\ \&c. \\
1 \\
24)\overline{124} \\
\quad 96 \\
\cdot 289)\overline{2820} \\
\quad\ 2601 \\
2987)\overline{21922} \\
\quad\ 20909 \\
29943)\overline{101350} \\
\quad\ 89829 \\
299463)\overline{1152100} \\
\quad\ 898389 \\
2994668)\overline{25371100} \\
\quad\quad 23957344 \\
\hline
\quad\quad 1413756 /
\end{array}
$$

In this case the highest period consists but of a single digit; and the given number is not a perfect square.

There must be an *even* number of decimal places; since no number of decimals in the root will produce an odd number in the square [Sec. II. 48]—as may be proved by experiment.

14. $\sqrt{195364}=442$	20. $\sqrt{5}=2 \cdot 23607$
15. $\sqrt{328329}=573$	21. $\sqrt{\cdot 5}=\cdot 707106$
16. $\sqrt{\cdot 0676}=\cdot 26$	22. $\sqrt{91 \cdot 9681}=9 \cdot 59$
17. $\sqrt{87 \cdot 65}=9 \cdot 3622$	23. $\sqrt{238144}=488$
18. $\sqrt{861}=29 \cdot 3428$	24. $\sqrt{32 \cdot 3761}=5 \cdot 69$
19. $\sqrt{984064}=992$	25. $\sqrt{\cdot 331776}=\cdot 576$

14. To extract the square root of a fraction—

RULE.—Having reduced the fraction to its lowest terms, make the square root of its numerator the numerator, and the square root of its denominator the denominator of the required root.

EXAMPLE.—$\sqrt{\frac{4}{9}}=\frac{2}{3}$.

15. REASON OF THE RULE.—The square root of any quantity must be such a number as, multiplied by itself, will produce that quantity. Therefore $\frac{2}{3}$ is the square root of $\frac{4}{9}$; for $\frac{2}{3} \times \frac{2}{3}=\frac{4}{9}$. The same might be shown by any other example.

Besides, to square a fraction, we must multiply its numerator by itself, and its denominator by itself [6]; therefore, to take its square root—that is, to bring back both numerator and denominator to what they were before—we must take the square root of each.

16. Or, when the numerator and denominator are not squares—

RULE.—Multiply the numerator and denominator together; then make the square root of the product the numerator of the required root, and the given denominator its denominator; or make the square root of the product the denominator of the required root, and the given numerator its numerator.

EXAMPLE.—What is the square root of $\frac{4}{5}$? $\left(\frac{4}{5}\right)^{\frac{1}{2}}=\frac{\sqrt{4 \times 5}}{5}$ or $\sqrt{\frac{4}{5 \times 4}}=4 \cdot 472136 \div 5=\cdot 894427$.

17. We, in this case, only multiply the numerator and denominator by the same number, and then extract the square root of each product. For $\frac{4}{5}=\frac{4 \times 5}{5 \times 5}$, or $\frac{4 \times 4}{5 \times 4}$. Therefore $\left(\frac{4}{5}\right)^{\frac{1}{2}}$ $=\left(\frac{4 \times 5}{5 \times 5}\right)^{\frac{1}{2}}=\frac{\sqrt{4 \times 5}}{5}$, or $\left(\frac{4 \times 4}{5 \times 4}\right)^{\frac{1}{2}}=\frac{4}{\sqrt{5 \times 4}}$.

18. Or, lastly—
RULE.—Reduce the given fraction to a decimal
Sec. IV. 63], and extract its square root [13].

EXERCISES.

26. $\left(\dfrac{22}{37}\right)^{\frac{1}{2}} = \dfrac{28\cdot5306852}{37}$ 29. $\left(\dfrac{5}{9}\right)^{\frac{1}{2}} = \cdot745356$

27. $\left(\dfrac{14}{16}\right)^{\frac{1}{2}} = \dfrac{14}{14\cdot9666295}$ 30. $\left(\dfrac{9}{12}\right)^{\frac{1}{2}} = \cdot8660254$

28. $\left(\dfrac{3}{13}\right)^{\frac{1}{2}} = \dfrac{6\cdot244998}{13}$ 31. $\left(\dfrac{5}{7}\right)^{\frac{1}{2}} = \cdot8451542$

19. To extract the square root of a mixed number—
RULE.—Reduce it to an improper fraction, and then
proceed as already directed [14, &c.]

EXAMPLE.—$\sqrt{2\frac{1}{4}} = \sqrt{\frac{9}{4}} = \frac{3}{2} = 1\frac{1}{2}$.

EXERCISES.

32. $\sqrt{51\frac{9}{25}} = 7\frac{1}{5}$ 35. $\sqrt{17\frac{3}{8}} = 4\cdot1683$

33. $\sqrt{27\frac{9}{16}} = 5\frac{1}{4}$ 36. $\sqrt{6\frac{2}{5}} = 2\cdot5298$

34. $\sqrt{1\frac{3}{80}} = 1\cdot01858$ 37. $\sqrt{13\frac{1}{5}} = 3\cdot6332$

20. To find the cube root of any number—
RULE—I. Point off the digits in threes, by dots—
putting the first dot over the units' *place*, and then
proceeding *both* to the right and left hand, if there are
digits at *both* sides of the decimal point.

II. Find the highest digit whose cube will not exceed the highest period, or that which is to the left hand
side—this will be the highest digit of the required root;
subtract its cube, and bring down the next period to
the remainder.

III. Find the highest *digit*, which, being multiplied
by 300 times the square of that part of the root,
already found—being squared and then multiplied by
30 times the part of the root already found—and being
multiplied by its own square—the *sum* of all the products will not exceed the *sum* of the last remainder and
the period brought down to it.—Put this *digit* in the
root after what is already there, and subtract the former
sum from the latter.

IV. To what now remains, bring down the next

period, and proceed as before. Continue this process until the exact cube root, or a sufficiently near approximation to it, is obtained.

EXAMPLE.—What is the cube root of 179597069288 ?

$$\overset{\centerdot\quad\centerdot\quad\centerdot\quad\centerdot}{179597069288}(5642, \text{ the required root.}$$
$$125$$

$$
\left.\begin{array}{l}
300\times5^2\times6 \\
30\times5\times6^2 \\
6^2\times6
\end{array}\right\} = \begin{array}{l} \overline{54597} \\ 50616 \end{array}
$$

$$
\left.\begin{array}{l}
300\times56^2\times4 \\
30\times56\times4^2 \\
4^2\times4
\end{array}\right\} = \begin{array}{l} \overline{3981069} \\ 3790144 \end{array}
$$

$$
\left.\begin{array}{l}
300\times564^2\times2 \\
30\times564\times2^2 \\
2^2\times2
\end{array}\right\} = \begin{array}{l} \overline{190925288} \\ 190925288 \end{array}
$$

We find (by trial) that 5 is the first, 6 the second, 4 the third, and 2 the last digit of the root. And the given number is exactly a cube.

21. REASON OF I.—We point off the digits in threes, for a reason similar to that which caused us to point them off in twos, when extracting the square root [12].

REASON OF II.—Each cube will be found to contain the cube of each part of its cube root.

REASON OF III.—The cube of a number divided into any two parts, will be found to contain, besides the sum of the cubes of its parts, the sum of 3 times the product of each part by the other part, and 3 times the product of each part by the square of the other part. This will appear from the following :—

$$
\begin{array}{r}
179597069288 \\
5000^3 = 125000000000 \\
\hline
54597069288
\end{array}
$$

$$3\times5000^2\times600+3\times5000\times600^2+600^3 = 50616000000$$

$$
\begin{array}{r}
\hline
3981069288
\end{array}
$$

$$3\times5600^2\times40+3\times5600\times40^2+40^3 = 3790144000$$

$$
\begin{array}{r}
\hline
190925288
\end{array}
$$

$$3\times5640^2\times2+3\times5640\times2^2+2^3 = 190925288$$

Hence, to find the second digit of the root, we must find *by trial* some number which—being multiplied by 3 times the square of the part of the root already found—its square being multiplied by 3 times the part of the root already found—and being multiplied by the square of itself—the sum of the products will not exceed what remains of the given number.

Instead of considering the part of the root already found as so many *tens* [12] of the denomination next following (as it really is), which would add one cypher to it, and two cyphers to its square, we consider it as so many units, and multiply it, not by 3, but by 30, and its square, not by 3, but by 300. For $300 \times 5^2 \times 6 + 30 \times 5 \times 6^2 + 6^2 \times 6$ is the same thing as $3 \times 50^2 \times 6 + 3 \times 50 \times 6^2 + 6^2 \times 6$; since we only change the *position* of the factors 100 and 10, which does not alter the product [Sect. II. 35].

It is evidently unnecessary to bring down more than one period at a time; or to add cyphers to the subtrahends.

REASON OF IV.—The portion of the root already found may be treated as if it were a single digit. Since into whatever two parts we divide any number, its cube root will contain the cube of each part, with 3 times the square of each multiplied into the other.

22. When there are decimals in the given cube—

If any of the periods consist of decimals, it is evident that the digits found on bringing down these periods must be decimals. Thus $\sqrt[3]{179597 \cdot 069288} = 56 \cdot 42$, &c.

When the decimals do not form complete periods, the periods are to be completed by the addition of cyphers.

EXAMPLE.—What is the cube root of ·3 ?

$$\overset{\bullet}{0} \cdot \overset{\bullet}{3}00(\cdot 669, \&c.$$
$$216$$

$$\left. \begin{array}{l} 300 \times 6^2 \times 6 \\ 30 \times 6 \times 6^2 \\ 6 \times 6^2 \end{array} \right\} = \begin{array}{l} 84000 \\ 71496 \end{array}$$

$$\left. \begin{array}{l} 300 \times 66^2 \times 9 \\ 30 \times 66 \times 9^2 \\ 9 \times 9^2 \end{array} \right\} = \begin{array}{l} 12504000 \\ 11922309 \\ \hline 581691, \&c. \end{array}$$

$\sqrt[3]{\cdot 3} = \cdot 669$, &c. And ·3 is not exactly a cube.

It is necessary, in this case, to add cyphers; since one decimal in the root will give 3 decimal places in the cube; two decimal places in the root will give six in the cube, &c. [Sec. II. 48.]

EXERCISES.

38. $\sqrt[3]{33} = 3 \cdot 207534$
39. $\sqrt[3]{39} = 3 \cdot 391211$
40. $\sqrt[3]{212} = 5 \cdot 962731$
41. $\sqrt[3]{123505992} = 498$
42. $\sqrt[3]{190109375} = 575$

43. $\sqrt[3]{458314011} = 771$
44. $\sqrt[3]{483 \cdot 736625} = 7 \cdot 85$
45. $\sqrt[3]{\cdot 636056} = \cdot 86$
46. $\sqrt[3]{999} = 9 \cdot 996666$
47. $\sqrt[3]{\cdot 979146657} = \cdot 993$

23. To extract the cube root of a fraction—

RULE.—Having reduced the given fraction to its lowest terms, make the cube root of its numerator the numerator of the required fraction, and the cube root of its denominator, the denominator.

EXAMPLE.—$\sqrt[3]{\frac{8}{125}} = \frac{\sqrt[3]{8}}{\sqrt[3]{125}} = \frac{2}{5}$.

24. REASON OF THE RULE.—The cube root of any number must be such as that, taken three times as a factor, it will produce that number. Therefore $\frac{2}{5}$ is the cube root of $\frac{8}{125}$; for $\frac{2}{5} \times \frac{2}{5} \times \frac{2}{5} = \frac{8}{125}$.—The same thing might be shown by any other example.

Besides, to cube a fraction, we must cube both numerator and denominator; therefore, to take its cube root—that is to reduce it to what it was before—we must take the cube root of both.

25. Or, when the numerator and denominator are not cubes—

RULE.—Multiply the numerator by the square of the denominator; and then divide the cube root of the product by the given denominator; or divide the given numerator by the cube root of the product of the given denominator multiplied by the square of the given numerator.

EXAMPLE.—What is the cube root of $\frac{3}{7}$?

$$\left(\frac{3}{7}\right)^{\frac{1}{3}} = \sqrt[3]{\frac{3 \times 7^2}{7}} \text{ or } \frac{3}{\sqrt[3]{7 \times 3^2}} = 5\cdot277632 \div 7 = \cdot753947.$$

This rule depends on a principle already explained [16].

26. Or, lastly—

RULE.—Reduce the given fraction to a decimal [Sec. IV. 63], and extract its cube root [22].

EXERCISES.

48. $\left(\dfrac{8}{9}\right)^{\frac{1}{3}} = \dfrac{8\cdot653497}{9}$

49. $\left(\dfrac{4}{11}\right)^{\frac{1}{3}} = \dfrac{4}{5\cdot604079}$

50. $\left(\dfrac{7}{8}\right)^{\frac{1}{3}} = \dfrac{7\cdot651725}{8}$

51. $\left(\dfrac{5}{6}\right)^{\frac{1}{3}} = \cdot941036$

52. $\left(\dfrac{3}{17}\right)^{\frac{1}{3}} = \cdot560907$

53. $\left(\dfrac{2}{19}\right)^{\frac{1}{3}} = \cdot472163$

27. To find the cube root of a mixed number—

RULE.—Reduce it to an improper fraction; and then proceed as already directed [23, &c.]

EXAMPLE.—$\sqrt[3]{3\frac{64}{91}} = \sqrt[3]{\frac{340}{91}} = 1\cdot54$

54. $(28\frac{2}{3})^{\frac{1}{3}}=3\cdot0635$ | 57. $(71\frac{3}{4})^{\frac{1}{3}}=4\cdot1553$
55. $(7\frac{1}{4})^{\frac{1}{3}}=1\cdot93098$ | 58. $(32\frac{8}{11})^{\frac{1}{3}}=3\cdot1987$
56. $(9\frac{1}{3})^{\frac{1}{3}}=2\cdot0928$ | 59. $(5\frac{4}{9})^{\frac{1}{3}}=1\cdot7592$

28. To extract any root whatever—

RULE.—When the index of the root is some power of 2, extract the square root, when it is some power of 3, extract the cube root of the given number so many times, successively, as that power of 2, or 3 contains unity.

EXAMPLE 1.—The 8th root of $65536=\sqrt{\sqrt{\sqrt{65536}}}=4$.
Since 8 is the *third* power of 2, we are to extract the square root *three* times, successively.

EXAMPLE 2.—$134217728^{\frac{1}{9}}=\sqrt[3]{\sqrt[3]{134217728}}=8$.
Since 9 is the *second* power of 3, we are to extract the cube root *twice*, successively.

29. In other cases we may use the following (Hutton Mathemat. Dict. vol. i. p. 135).

RULE.—Find, by trial, some number which, raised to the power indicated by the index of the given root, will not be far from the given number. Then say, as one less than the index of the root, multiplied by the given number—plus one more than the index of the root, multiplied by the assumed number raised to the power expressed by the index of the root : one more than the index of the root, multiplied by the given number— plus one less than the index of the root, multiplied by the assumed number raised to the power indicated by the index of the root, :: the assumed root : a still nearer approximation. Treat the fourth proportional thus obtained in the same way as the assumed number was treated, and a still nearer approximation will be found. Proceed thus until an approximation as near as desirable is discovered.

EXAMPLE.—What is the 13th root of 923 ?
Let 2 be the assumed root, and the proportion will be
$12\times923+14\times2^{13}$: $14\times923+12\times2^{13}$ :: 2 : a nearer approximation. Substituting this nearer approximation for 2, in the above proportion, we get another approximation, which we may treat in the same way.

P

60. $(96698)^{\frac{1}{6}} = 6 \cdot 7749$

61. $(66457)^{\frac{1}{11}} = 2 \cdot 7442$

62. $(2365)^{\frac{4}{9}} = 31 \cdot 585$

63. $(87426)^{\frac{3}{4}} = 5084 \cdot 29$

64. $(8 \cdot 965)^{\frac{1}{7}} = 1 \cdot 368$

65. $(\cdot 075426)^{\frac{13}{14}} = \cdot 046988$

30. To find the squares and cubes, the square and cube roots of numbers, by means of the table at the end of the treatise—

This table contains the squares and cubes, the square and cube roots of all numbers which do not exceed 1000; but it will be found of considerable utility even when very high numbers are concerned—provided the pupil bears in mind that [12] the square of any number is equal to the sum of the squares of its parts (which may be found by the table) plus twice the product of each part by the sum of all the others; and that [21] the cube of a number divided into any two parts is equal to the sum of the cubes of its parts (which may be found by the table) plus three times the product of each part multiplied by the square (found by means of the table) of the other. One or two illustrations will render this sufficiently clear.

EXAMPLE 1.—Find the square of 873456.

873456 may be divided into two parts, 873 (thousand) and 456 (units). But we find by the table that $\overline{873}^2 = 762129$ and $\overline{456}^2 = 207936$.

Therefore $762129000000 = \overline{873000}^2$

$$796176000 = 873000 \times \text{twice } 456$$

$$207936 = \overline{456}^2$$

And $762925383936 = \overline{873456}^2$

EXAMPLE 2.—Find the cube of 864379. Dividing this into 864 (thousand) and 379 (units), we find $\overline{864}^3 = 644972544$. $\overline{864}^2 = 746496$, $\overline{379}^3 = 54439939$, and $\overline{379}^2 = 143641$.

Therefore $644972544000000000 = \overline{864000}^3$

$$848765952000000 = 3 \times \overline{864000}^2 \times 379$$

$$372317472000 = 3 \times 864000 \times \overline{379}^2$$

$$54439939 = \overline{379}^3$$

And $645821682323911939 = \overline{864379}^3$

31. In finding the square and cube roots of large numbers, we obtain their *three* highest digits at once, if we look in the table for the highest cube or square, the highest period of which (the required cyphers being added) does not exceed the highest period of the given number. The remainder of the process, also, may often be greatly abbreviated by means of the table.

QUESTIONS.

1. What are involution and evolution? [1].
2. What are a power, index, and exponent? [1 & 2].
3. What is the meaning of square and cube, of the square and cube roots? [1 and 8].
4. What is the difference between an integral and a fractional index? [2 and 8).
5. How is a number raised to any power? [5].
6. What is the rule for finding the square root? [10].
7. What is the rule for finding the cube root? [20].
8. How is the square or cube root of a fraction or of a mixed number found? [14, &c., 19, 23, &c., 27].
9. How is *any* root found? [28 and 29].
10. How are the squares and cubes, the square roots and cube roots, of numbers found, by the table? [30].

LOGARITHMS.

32. Logarithms are a set of *artificial* numbers, which represent the ordinary or *natural* numbers. Taken along with what is called the *base* of the system to which they belong, they are the *equals* of the corresponding natural numbers, but without it, they are merely their *representatives*. Since the base is unchangeable, it is not written along with the logarithm. The logarithm of any number is that power of the base which is equal to it. Thus 10^2 is *equal* to 100; 10 is the *base*, 2 (the index) is the *logarithm*, and 100 is the corresponding natural number.—Logarithms, therefore, are merely the *indices* which designate certain powers of some base.

33. Logarithms afford peculiar facilities for calculation. For, as we shall see presently, the multiplication of numbers is performed by the addition of their

logarithms ; one number is divided by another if we subtract the logarithm of the divisor from that of the dividend ; numbers are involved if we multiply their logarithms by the index of the proposed power ; and evolved if we divide their logarithms by the index of the proposed root.—But it is evident that addition and subtraction are much easier than multiplication and division ; and that multiplication and division (particularly when the multipliers and divisors are very small) are much easier than involution and evolution.

34. To use the properties of logarithms, they must be exponents of the same base—that is, the quantities raised to those powers which they indicate must be the *same*. Thus $10^4 \times 12^3$ is neither 10^7 nor 12^7, the former being too small, the latter too great. If, therefore, we desire to multiply 10^4 and 12^3 by means of *indices*, we must find some power of 10 which will be equal to 12^3, or some power of 12 which will be equal to 10^4, or finally, two powers of some other number which will be equal respectively to 10^4 and 12^3, and then, adding these powers of the *same* number, we shall have that power of it which will represent the product of 10^4 and 12^3. This explains the necessity for a *table* of logarithms— we are obliged to find the powers of some *one* base which will be either equal to all possible numbers, or so nearly equal that the inaccuracy is not deserving of notice. The base of the ordinary system is 10 ; but it is clear that there may be as many different systems of logarithms as there are different bases, that is, as there are different numbers.

35. In the ordinary system—which has been calculated with great care, and with enormous labour, 1 is the logarithm of 10 ; 2 that of 100 ; 3 that of 1000, &c. And, since to divide numbers by means of these logarithms (as we shall find presently), we are to subtract the logarithm of the divisor from that of the dividend, 0 is the logarithm of 1, for $1 = \frac{10}{10} = 10^{1-1} = 10^0$; -1 is the logarithm of $\cdot 1$, for $\cdot 1 = \frac{1}{10} = \frac{10^0}{10^1} = 10^{0-1} = 10^{-1}$; and for the same reason, -2 is the logarithm of $\cdot 01$; -3 that of $\cdot 001$, &c.

36. The logarithms of numbers *between* 1 and 10, must be more than 0 and less than 1 ; that is, must be some decimal. The logarithms of numbers between 10 and 100 must be more than 1, and less than 2; that is, unity with some decimal, &c. ; and the logarithms of numbers between ·1 and ·01 must be −1 and some decimal; between ·01 and ·001, −2 and some decimal, &c.· The decimal part of a logarithm is *always* positive.

37. As the integral part or *characteristic* of a positive logarithm is so easily found—being [35] one *less* than the number of *integers in* its corresponding number, and of a negative logarithm one *more* than the number of *cyphers prefixed* in its natural number, it is not set down in the tables. Thus the logarithm corresponding to the digits 9872 (that is, its decimal part) is 994405 ; hence, the logarithm of 9872 is 3 ·994405 ; that of 987·2 is 2·994405 ; that of 9·872 is 0·994405 ; that of ·9872 is − 1·994405 (since there is no integer, nor prefixed cypher) ; of ·009872 − 3·994405, &c. :—The same digits, whatever may be their value, have the same *decimals* in their logarithms; since it is the integral part, only, which changes. Thus the logarithm of 57864000 is 7·762408 ; that of 57864, is 4·762408 ; and that of ·0000057864, is − 6·762408.

38. To find the logarithm of a given number, by the table—

The integral part, or characteristic, of the logarithm may be found at once, from what has been just said [37]—

When the number is not greater than 100, it will be found in the column at the top of which is N, and the decimal part of its logarithm immediately opposite to it in the next column to the right hand.

If the number is greater than 100, and less than 1000, it will also be found in the column marked N, and the decimal part of its logarithm opposite to it, in the column at the top of which is 0.

If the number contains 4 digits, the first three of them will be found in the column under N, and the fourth at the top of the page ; and then its logarithm in the same horizontal line as the three first digits of the given number, and in the same column as its fourth

If the number contains more than 4 digits, find the logarithm of its first four, and also the difference between that and the logarithm of the next higher number, in the table; multiply this difference by the remaining digits, and cutting off from the product so many digits as were in the multiplier (but at the same time adding unity if the highest cut off is not less than 5), add it to the logarithm corresponding to the four first digits.

EXAMPLE 1.—The logarithm of 59 is 1·770852 (the characteristic being positive, and *one less* than the number of *integers*).

EXAMPLE 2.—The logarithm of 338 is 2·528917.

EXAMPLE 3.—The logarithm of ·0004587 is −4·661529 (the characteristic being negative, and *one more* than the number of *prefixed cyphers*).

EXAMPLE 4.—The logarithm of 28434 is 4·453838.

For, the difference between 453777 the logarithm of 2843, the four first digits of the given number, and 453930 the logarithm of 2844, the next number, is 153; which, multiplied by 4, the remaining digit of the given number, produces 612; then cutting off one digit from this (since we have multiplied by only *one* digit) it becomes 61, which being added to 453777 (the logarithm of 2844) makes 453838, and, with the characteristic, 4·453838, the required logarithm.

EXAMPLE 5.—The logarithm of 873457 is 5·941242.

For, the difference between the logarithms of 8734 and 8735 is 50, which, being multiplied by 57, the remaining digits of the given number, makes 2850; from this we cut off *two* digits to the right (since we have multiplied by *two* digits), when it becomes 28; but as the highest digit cut off is 5, we add unity, which makes 29. Then 5·941213 (the logarithm of 8734) + 29 = 5·941242, is the required logarithm.

39. Except when the logarithms increase very rapidly—that is, at the commencement of the table—the differences may be taken from the right hand column (and opposite the three first digits of the given number) where the *mean* differences will be found.

Instead of multiplying the mean difference by the remaining digits (the fifth, &c., to the right) of the given number, and cutting off so many places from the product as are equal to the number of digits in the multiplier, to obtain the *proportional part*—or what is to be added to the logarithm of the first four digits, we may take the proportional part corresponding to each of the remaining digits from that part of the column at the *left* hand side of the page, which is in the same horizontal division as that in which the first three digits of the given number have been found.

EXAMPLE.—What is the logarithm of 839785 ?

The (decimal part of the) logarithm of 839700 is 924124. Opposite to 8, in the same horizontal division of the page, we find 42, or rather, (since it is 80) 420, and opposite to 5, 26. Hence the required logarithm is 924124 + 420 + 26 = 924570 ; and, with the characteristic, 5·924570.

40. The method given for finding the *proportional part*—or what is to be added to the next lower logarithm, in the table— arises from the difference of numbers being proportional to the difference of their logarithms. Hence, using the last example,

100 : 85 :: 52 (924176, the logarithm of 839800 − 924124,

the logarithm of 839700) : $\frac{52 \times 85}{100}$, or the difference (the *mean* difference may generally be used) × by the *remaining* digits of the given number ÷ 100 (the division being performed by cutting off *two* digits to the right). It is evident that the number of digits to be cut off depends on the number of digits in the multiplier. The logarithm found is not *exactly* correct, because numbers are not *exactly* proportional to the differences of their logarithms.

The *proportional parts* set down in the left hand column, have been calculated by making the necessary multiplications and divisions.

41. To find the logarithm of a fraction—

RULE.—Find the logarithms of both **numerator** and denominator, and then subtract the former from the latter; this will give the logarithm of the quotient.

EXAMPLE.—Log. $\frac{4.7}{5.6}$ is 1·672098 − 1·748188 = −1·923910. We find that 2 is to be subtracted from 1 (the characteristic of the numerator) ; but 2 from 1 leaves 1 still to be subtracted, or [Sect. II. 15] − 1, the characteristic of the quotient.

We shall find presently that to divide one quantity by another, we have merely to subtract the logarithm of the latter from that of the former.

42. To find the logarithm of a mixed number—

RULE.—Reduce it to an improper fraction, and proceed as directed by the last rule.

43. To find the number which corresponds to a given logarithm—

If the logarithm itself is found in the table—

RULE.—Take from the table the number which corresponds to it, and place the decimal point so that there may be the requisite number of integral, or decimal places—according to the characteristic [37].

EXAMPLE.—What number corresponds to the logarithm 4·214314?

We find 21 opposite the natural number 163; and looking along the horizontal line, we find the rest of the logarithm under the figure 8 at the top of the page; therefore the digits of the required number are 1638. But as the characteristic is 4, there must in it be 5 places of integers. Hence the required number is 16380.

44. If the given logarithm is not found in the table—

RULE.—Find that logarithm in the table which is next lower than the given one, and its digits will be the highest digits of the required number; find the difference between this logarithm and the given one, annex to it a cypher, and then divide it by that difference in the table, which corresponds to the four highest digits of the required number—the quotient will be the next digit; add another cypher, divide again by the tabular difference, and the quotient will be the next digit. Continue this process as long as necessary.

EXAMPLE.—What number corresponds to the logarithm 5·654329?

654273, which corresponds with the natural number 4511, is the logarithm next less than the given one; therefore the first *four* digits of the required number are 4511. Adding a cypher to 56, the difference between 654273 and the given logarithm, it becomes 560, which, being divided by 96, the *tabular difference* corresponding with 4511, gives 5 as quotient, and 80 as remainder. Therefore, the first *five* digits of the required number are 45115. Adding a cypher to 80, it becomes 800; and, dividing this by 96, we obtain 8 as the next digit of the required number, and 32 as remainder. The *integers* of the required number (one more than 5, the characteristic) are, therefore, 451158. We may obtain the decimals, by continuing the addition of cyphers to the remainders, and the division by 96.

45. We arrive at the same result, by subtracting
- from the *difference* between the given logarithm and
the next less in the table, the highest (which does not
exceed it) of those proportional parts found at the right
hand side of the page and in the same horizontal *division* with the first *three* digits of the given number—
continuing the process by the addition of cyphers, until
nothing, or almost nothing, remains.

EXAMPLE.—Using the last, 4511 is the natural number
corresponding to the logarithm 654273, which differs from
the given logarithm by 56. The proportional parts, in the
same horizontal division as 4511, are 10, 19, 29, 38, 48, 58,
67, 77, and 86. The highest of these, contained in 56, is
48, which we find opposite to, and therefore corresponding
with, the natural number 5: hence 5 is the next of the
required digits. 48 subtracted from 56, leaves 8; this, when
a cypher is added, becomes 80, which contains 77 (corresponding to the natural number 8); therefore 8 is the next
of the required digits. 77, subtracted from 80, leaves 3;
this, when a cypher is added, becomes 30, &c. The integers, therefore, of the required number, are found to be
451158, the same as those obtained by the other method.

The rules for finding the numbers corresponding to
given logarithms are merely the converse of those used
for finding the logarithms of given numbers.

Use of Logarithms in Arithmetic.

46. To multiply numbers, by means of their logarithms—

RULE.—Add the logarithms of the factors; and the
natural number corresponding to the result will be the
required product.

EXAMPLE.—$87 \times 24 = 1\cdot939519$ (the log. of 87) $+ 1\cdot380211$
(the log. of 24) $= 3\cdot319730$; which is found to correspond
with the natural number, 2088. Therefore $87 \times 24 = 2088$.

REASON OF THE RULE.—This mode of multiplication arises
from the very nature of indices. Thus $5^4 \times 5^8 = 5 \times 5 \times 5 \times 5$
multiplied $5 \times 5 \times 5 \times 5 \times 5 \times 5 \times 5 \times 5$; and the abbreviation for
this [2] is 5^{12}. But 12 is equal to the *sum* of the indices
(logarithms). The rule might, in the same way, be proved
correct by any other example.

47. When the characteristics of the logarithms to be added are both positive, it is evident that their sum will be positive. When they are both negative, their sum (diminished by what is to be carried from the sum of the positive [36] decimal parts) will be negative. When one is negative, and the other positive, subtract the less from the greater, and prefix to the difference the sign belonging to the greater—bearing in mind what has been already said [Sec. II. 15] with reference to the subtraction of a greater from a less quantity.

48. To divide numbers, by means of their logarithms—

RULE.—Subtract the logarithm of the divisor from that of the dividend; and the natural number, corresponding to the result, will be the required quotient.

EXAMPLE.—$1134 \div 42 = 3\cdot054613$ (the log. of 1134) − $1\cdot623249$ (the log. of 42) = $1\cdot431364$, which is found to correspond with the natural number, 27. Therefore $1134 \div 42 = 27$.

REASON OF THE RULE.—This mode of division arises from the nature of indices. Thus $4^5 \div 4^3 = [2]$ $4 \times 4 \times 4 \times 4 \times 4 \div 4 \times 4 \times 4 = \dfrac{4 \times 4 \times 4 \times 4 \times 4}{4 \times 4 \times 4} = 4 \times 4 \times \dfrac{4 \times 4 \times 4}{4 \times 4 \times 4} = 4 \times 4$, the abbreviation for which is 4^2. But 2 is equal to the index (logarithm) of the dividend minus that of the divisor. The rule might, in the same way, be proved correct by any other example.

49. In subtracting the logarithm of the divisor, if it is negative, change the sign of its characteristic or integral part, and then proceed as if this were to be added to the characteristic of the dividend; but before making the characteristic of the divisor positive, subtract what was borrowed (if any thing), in subtracting its decimal part. For, since the decimal part of a logarithm is positive, what is *borrowed*, in order to make it possible to subtract the decimal part of the logarithm of the divisor from that of the dividend, must be so much taken away from what is positive, or added to what is negative in the remainder.

We change the sign of the negative characteristic, and then *add* it; for, adding a positive, is the same as taking away a negative quantity.

50. To raise a quantity to any power, by means of ts logarithm—

RULE.—Multiply the logarithm of the quantity by the index of the power ; and the natural number corresponding to the result will be the required power.

EXAMPLE.—Raise 5 to the 5th power.

The logarithm of 5 is 0·69897, which, multiplied by 5, gives 3·49485, the logarithm of 3125. Therefore, the 5th power of 5^2 is 3125.

REASON OF THE RULE.—This rule also follows from the nature of indices. 5^2 raised to the 5th power is 5×5 multiplied by 5×5 multiplied by 5×5 multiplied by 5×5 multiplied by 5×5, or $5 \times 5 \times 5 \times 5 \times 5 \times 5 \times 5 \times 5 \times 5 \times 5$, the abbreviation for which is $[2] \, 5^{10}$. But 10 is equal to 2, the index (logarithm) of the quantity, multiplied by 5, that of the power. The rule might, in the same way, be proved correct by any other example.

51. It follows from what has been said [47] that when a negative characteristic is to be multiplied, the product is negative ; and that what is to be carried from the multiplication of the decimal part (always positive) is to be *subtracted* from this negative result.

52. To evolve any quantity, by means of its logarithm—

RULE.—Divide the logarithm of the given quantity by that number which expresses the root to be taken ; and the natural number corresponding to the result will be the required root.

EXAMPLE.—What is the 4th root of 2401.

The logarithm of 2401 is 3·380392, which, divided by 4, the number expressing the root, gives ·845098, the logarithm of 7. Therefore, the fourth root of 2401 is 7.

REASON OF THE RULE.—This rule follows, likewise, from the nature of indices. Thus the 5th root of 16^{10} is such a number as, raised to the 5th power—that is, taken 5 times as a factor—would produce 16^{10}. But $16^{\frac{10}{5}}$, taken 5 times as a factor, would produce 16^{10}. The rule might be proved correct, equally well, by any other example.

53. When a negative characteristic is to be divided—

RULE I.—If the characteristic is *exactly* divisible by the divisor, divide in the ordinary way, but make the characteristic of the quotient negative.

II.—If the negative characteristic is *not exactly* divisible, add what will make it so, both to it and to the decimal part of the logarithm. Then proceed with the division.

EXAMPLE.—Divide the logarithm -4.837564 by 5.

4 wants 1 of being divisible by 5; then $-4.837564 \div 5 =$ $\overline{-5} + 1.837564 \div 5 = 1.367513$, the required logarithm.

REASON OF I.—The quotient multiplied by the divisor must give the dividend; but [51]-a negative quotient multiplied by a positive divisor will give a negative dividend.

REASON OF II.—In example 2, we have merely added $+1$ and -1 to the same quantity—which, of course, does not alter it.

QUESTIONS.

1. What are logarithms ? [32].
2. How do they facilitate calculation ? [33].
3. Why is a table of logarithms necessary ? [34].
4. What is the characteristic of a logarithm ; and how is it found ? [37].
5. How is the logarithm of a number found, by the table ? [38].
6. How are the " differences," given in the table used ? [39].
7. What is the use of " proportional parts ?" [39].
8. How is the logarithm of a fraction found ? [41].
9. How do we find the logarithm of a mixed number ? [42].
10. How is the number corresponding to a given logarithm found ? [43].
11. How is a number found when its corresponding logarithm is not in the table ? [44].
12. How are multiplication, division, involution and evolution effected, by means of logarithms ? [46, 48, 50, and 52].
13. When negative characteristics are added, what is the sign of their sum ? [47].
14. What is the process for division, when the characteristic of the divisor is negative ? [49].
15. How is a negative characteristic multiplied? [51].
16. How is a negative characteristic divided ? [53].

SECTION X.

PROGRESSION, &c.

1. A progression consists of a number of quantities increasing, or decreasing by a certain law, and forming what are called *continued proportionals*. When the terms of the series constantly increase, it is said to be an *ascending*, but when they decrease (increase to the *left*), a *descending series*.

2. In an *equidifferent* or *arithmetical* progression, the quantities increase, or decrease by a *common difference*. Thus 5, 7, 9, 11, &c., is an ascending, and 15, 12, 9, 6, &c., is a descending *arithmetical* series or progression. The common difference in the former is 2, and in the latter 3. A continued proportion may be formed out of such a series. Thus—

5 : 7 :: 7 : 9 :: 9 : 11, &c.; and 15 : 12 :: 12 : 9 :: 9 : 6, &c. Or we may say 5 : 7 :: 9 : 11 :: &c.; and 15 : 12 :: 9 : 6 :: &c.

3. In a *geometrical* or *equirational* progression, the quantities increase by a *common ratio* or multiplier. Thus 5, 10, 20, 40, &c.; and 10000, 1000, 100, 10, &c., are geometrical series. The common ratio in the former case is 2, and the quantities increase to the *right;* in the latter it is 10, and the quantities increase to the *left*. A continued proportion may be formed out of such a series. Thus—

5 : 10 :: 10 : 20 :: 20 : 40, &c.; and 10000 : 1000 :: 1000 : 100 :: 100 : 10, &c. Or we may say 5 : 10 :: 20 : 40 :: &c.; and 10000 : 1000 :: 100 : 10 :: &c.

4. The first and last terms of a progression are called its *extremes*, and all the intermediate terms its *means*.

5. *Arithmetical Progression.*—To find the sum of a series of terms in arithmetical progression—

RULE.—Multiply the sum of the extremes by half the number of terms.

EXAMPLE.—What is the sum of a series of 10 terms, the first being 2, and last 20 ? *Ans.* $2 + 20 \times \frac{10}{2} = 110$.

6. REASON OF THE RULE.—This rule can be easily proved. For this purpose, set down the progression twice over—but in such a way as that the last term of one shall be under the first term of the other series.

Then, $24 + 21 + 18 + 15 + 12 + 9 =$ the sum.
$9 + 12 + 15 + 18 + 21 + 24 =$ the sum. And,

adding the equals, $33 + 33 + 33 + 33 + 33 + 33 =$ twice the sum.

That is, *twice* the sum of the series will be equal to the sum of as many quantities as there are terms in the series—each of the quantities being equal to the sum of the extremes. And the sum of the series itself will be equal to half as much, or to the sum of the extremes taken *half* as many times as there are terms in the series. The rule might be proved correct by any other example, and, therefore, is general.

EXERCISES.

1. One extreme is 3, the other 15, and the number of terms is 7. What is the sum of the series ? *Ans.* 63.

2. One extreme is 5, the other 93, and the number of terms is 49. What is the sum ? *Ans.* 2401.

3. One extreme is 147, the other $\frac{3}{4}$, and the number of terms is 97. What is the sum ? *Ans.* 7165·875.

4. One extreme is $4\frac{3}{8}$, the other 143, and the number of terms is 42. What is the sum ? *Ans.* 3094·875.

7. Given the extremes, and number of terms—to find the common difference—

RULE.—Find the difference between the given extremes, and divide it by one less than the number of terms. The quotient will be the common difference.

EXAMPLE.—In an arithmetical series, the extremes are 21 and 3, and the number of terms is 7. What is the common difference ?

$$\overline{21 - 3} \div \overline{7 - 1} = 18 \div 6 = 3, \text{ the required number.}$$

8. REASON OF THE RULE.—The difference between the greater and lesser extreme arises from the common difference being added to the lesser extreme once for every term, except the lowest; that is, the greater contains the lesser extreme plus the common difference taken once less than the number of terms. Therefore, if we subtract the lesser from the greater extreme, the difference obtained will be equal to the *common* difference multiplied by one less than the number of terms. And if we divide the difference by one less than the number of terms, we will have the *common* difference.

EXERCISES.

5. The extremes of an arithmetical series are 21 and 497, and the number of terms is 41. What is the common difference? *Ans.* 11·9.

6. The extremes of an arithmetical series are $127\frac{25}{28}$ and $9\frac{1}{7}$, and the number of terms is 26. What is the common difference? *Ans.* $4\frac{3}{4}$.

7. The extremes of an arithmetical series are $77\frac{23}{28}$ and $\frac{3}{4}$, and the number of terms is 84. What is the common difference? *Ans.* $1\frac{5}{14}$.

9. To find *any number* of arithmetical means between two given numbers—

RULE.—Find the common difference [7]; and, according as it is an ascending or a descending series, add it to, or subtract it from the first, to form the second term; add it to, or subtract it from the second, to form the third. Proceed in the same way with the remaining terms.

We must remember that one *less* than the number of terms is one *more* than the number of means.

EXAMPLE 1.—Find **4** arithmetical means between 6 and 21. $21-6=15$. $\frac{15}{4+1}=3$, the common difference. And the series is—

$6 . 6+3 . 6+2\times3 . 6+3\times3 . 6+4\times3 . 6+5\times3.$
Or 6 . 9 . 12 . 15 . 18 . 21

EXAMPLE 2.—Find 4 arithmetical means between 30 and 10. $30-10=20$. $\frac{20}{4+1}=4$, the common difference. And the series is—

30 . 26 . 22 . 18 . 14 . 10
This rule is evident.

EXERCISES.

8. Find 11 arithmetical means between 2 and 26. *Ans.* 4, 6, 8, 10, 12, 14, 16, 18, 20, 22, and 24.

9. Find 7 arithmetical means between 8 and 32. *Ans.* 11, 14, 17, 20, 23, 26, 29.

10. Find 5 arithmetical means between $4\frac{1}{2}$, and $13\frac{1}{2}$. *Ans.* 6, $7\frac{1}{2}$, 9, $10\frac{1}{2}$, 12.

10. Given the extremes, and the number of terms—
to find *any term* of an arithmetical progression—

RULE.—Find the common difference by the last rule,
and if it is an ascending series, the required term will
be the lesser extreme *plus*—if a descending series, the
greater extreme *minus* the common difference multiplied
by one less than the number of the term.

EXAMPLE 1.—What is the 5th term of a series containing
9 terms, the first being 4, and the last 28 ?

$\frac{28-4}{8}=3$, is the common difference.   And $4+3\times\overline{5-1}=$
16, is the required term.

EXAMPLE 2.—What is the 7th term of a series of 10 terms,
the extremes being 20 and 2 ?

$\frac{20-2}{9}=2$, is the common difference. $20-2\times\overline{7-1}=8$,
is the required term.

11. REASON OF THE RULE.—In an ascending series the
required term is greater than the given lesser extreme to the
amount of all the differences found in it.   But the number of
differences it contains is equal only to the number of terms
which *precede* it—since the common difference is not found in
the *first* term.

In a descending series the required term is less than the
given greater extreme, to the amount of the differences subtracted from the greater extreme—but one has been subtracted
from it, for each of the terms which *precede* the required term.

EXERCISES.

11. In an arithmetical progression the extremes are
14 and 86, and the number of terms is 19.   What is
the 11th term ?   *Ans.* 54.

12. In an arithmetical series the extremes are 22 and
4, and the number of terms is 7.   What is the 4th
term ?   *Ans.* 13.

13. In an arithmetical series 49 and $\frac{3}{4}$ are the extremes, and 106 is the number of terms.   What is the
94th term ?   *Ans.* 6·2643.

12. Given the extremes, and common difference—to
find the number of terms—

RULE.—Divide the difference between the given extremes by the *common* difference, and the quotient plus
unity will be the number of terms.

EXAMPLE.—How many terms in an arithmetical series of which the extremes are 5 and 26, and the common difference 3 ?

$$\frac{26-5}{3}=7. \text{ And } 7+1=8, \text{ is the number of terms.}$$

13. REASON OF THE RULE.—The greater differs from the lesser extreme to the amount of the differences found in all the terms. But the common difference is found in all the terms except the lesser extreme. Therefore the difference between the extremes contains the common difference once less than will be expressed by the number of terms.

EXERCISES.

14. In an arithmetical series, the extremes are 96 and 12, and the common difference is 6. What is the number of terms ? *Ans.* 15.

15. In an arithmetical series, the extremes are 14 and 32, and the common difference is 3. What is the number of terms ? *Ans.* 7.

16. In an arithmetical series, the common difference is $\frac{5}{9}$, and the extremes are $14\frac{8}{9}$ and 11. What is the number of terms ? *Ans.* 8.

14. Given the sum of the series, the number of terms, and one extreme—to find the other—

RULE.—Divide twice the sum by the number of terms, and take the given extreme from the quotient. The difference will be the required extreme.

EXAMPLE.—One extreme of an arithmetical series is 10, the number of terms is 6, and the sum of the series is 42. What is the other extreme ?

$$\frac{2\times42}{6}-10=4, \text{ is the required extreme.}$$

15. REASON OF THE RULE.—We have seen [5] that $2\times$ the sum=sum of the extremes$\times$ the number of terms. But if we divide each of these equal quantities by the number of terms, we shall have

$$\frac{2\times\text{the sum}}{\text{the number of terms}}=\frac{\text{sum of extremes}\times\text{the number of terms}}{\text{the number of terms}}$$

Or $\dfrac{2\times\text{the sum}}{\text{the number of terms}}=$sum of the extremes. And subtracting the same extreme from each of these equals, we shall have

$$\frac{2 \times \text{the sum}}{\text{the number of terms}} - \text{one extreme} = \text{the sum of the extremes}$$
— the same extreme.

Or $\dfrac{\text{twice the sum}}{\text{the number of terms}}$ minus one extreme = the other extreme.

EXERCISES.

17. One extreme is 4, the number of terms is 17, and the sum of the series is 884. What is the other extreme? *Ans.* 100.

18. One extreme is 3, the number of terms is 63, and the sum of the series is 252. What is the other extreme? *Ans.* 5.

19. One extreme is 27, the number of terms is 26, and the sum of the series is 1924. What is the other extreme? *Ans.* 121.

16. *Geometrical Progression.*—Given the extremes and common ratio—to find the sum of the series—

RULE.—Subtract the lesser extreme from the product of the greater and the common ratio; and divide the difference by one less than the common ratio.

EXAMPLE.—In a geometrical progression, 4 and 312 are the extremes, and the common ratio is 2. What is the sum of the series.

$$\frac{312 \times 2 - 4}{2 - 1} = 620, \text{ the required number.}$$

17. REASON OF THE RULE.—The rule may be proved by setting down the series, and placing over it (but in a reverse order) the product of each of the terms and the common ratio. Then

Sum × common ratio =   8+16+32, &c. . +312+624
Sum = ... 4+8+16+32, &c. . +312 .

And, subtracting the lower from the upper line, we shall have

$$\overline{\text{Sum} \times \text{common ratio}} - \text{Sum} = 624 - 4. \quad \text{Or}$$
$$\overline{\text{Common ratio} - 1} \times \text{Sum} = 624 - 4.$$

And, dividing each of the equal quantities by the common ratio minus 1

$$\text{Sum} = \frac{642 \text{ (last term} \times \text{common ratio)} - 4 \text{ (the first term)}}{\text{common ratio} - 1}$$

Which is the rule.

20. The extremes of a geometrical series are 512 and 2, and the common ratio is 4. What is the sum? *Ans.* 682.

21. The extremes of a geometrical series are 12 and 175692, and the common ratio is 11. What is the sum? *Ans.* 193260.

22. The extremes of an infinite geometrical series are $\frac{1}{10}$ and 0, and $\frac{1}{10}$ is the common ratio. What is the sum? *Ans.* $\frac{1}{9}$. [Sec. IV. 74.]

Since the series is infinite, the lesser extreme $=0$.

23. The extremes of a geometrical series are ·3 and 937·5, and the common ratio is 5. What is the sum? *Ans.* 1171·875.

18. Given the extremes, and number of terms in a geometrical series—to find the common ratio—

RULE.—Divide the greater of the given extremes by the lesser; and take that root of the quotient which is indicated by the number of terms minus 1. This will be the required number.

EXAMPLE.—5 and 80 are the extremes of a geometrical progression, in which there are 5 terms. What is the common ratio?

$\frac{80}{5}=16$. And $\sqrt[3]{16}=2$, the required common ratio.

19. REASON OF THE RULE.—The greater extreme is equal to the lesser multiplied by a product which has for its factors the common ratio taken once less than the number of terms— since the common ratio is not found in the *first* term. That is, the greater extreme contains the common ratio raised to a power indicated by 1 less than the number of terms, and multiplied by the lesser extreme. Consequently if, after dividing by the lesser extreme, we take that root of the quotient, which is indicated by one less than the number of terms, we shall obtain the common ratio itself.

24. The extremes of a geometrical series are 49152 and 3, and the number of terms is 8. What is the common ratio? *Ans.* 4.

25. The extremes of a geometrical series are 1 and

15625, and the number of terms is 7. What is the common ratio ? *Ans.* 5.

26. The extremes of a geometrical series are 201768035 and 5, and the number of terms is 10. What is the common ratio ? *Ans.* 7.

20. To find *any number* of geometrical means between two quantities—

RULE.—Find the common ratio (by the last rule), and—according as the series is ascending, or descending—multiply or divide it into the first term to obtain the second; multiply or divide it into the second to obtain the third; and so on with the remaining terms.

We must remember that one *less* than the number of terms is one *more* than the number of means.

EXAMPLE 1.—Find 3 geometrical means between 1 and 81.

$\sqrt[4]{\dfrac{81}{1}}=3$, the common ratio. And 3, 9, 27, are the required means.

EXAMPLE 2.—Find 3 geometrical means between 1250 and 2.

$\sqrt[4]{\dfrac{1250}{2}}=5$. And $\dfrac{1250}{5}$ $\dfrac{1250}{5\times5}$ $\dfrac{1250}{5\times5\times5}$, or 250, 50, 10, are the required means.

This rule requires no explanation.

EXERCISES.

27. Find 7 geometrical means between 3 and 19683 ? *Ans.* 9, 27, 81, 243, 729, 2187, 6561.

28. Find 8 geometrical means between 4096 and 8 ? *Ans.* 2048, 1024, 512, 256, 128, 64, 32, and 16.

29. Find 7 geometrical means between 14 and 23514624 ? *Ans.* 84, 504, 3024, 18144, 108864, 653184, and 3919104.

21. Given the first and last term, and the number of terms—to find *any* term of a geometrical series—

RULE.—If it be an ascending series, multiply, if a descending series, divide the first term by that power of the common ratio which is indicated by the number of the term minus 1.

EXAMPLE 1.—Find the 3rd term of a geometrical series, of which the first term is 6, the last 1458, and the number of terms 6.

The common ratio is $\sqrt[5]{\dfrac{1458}{6}} = 3$. Therefore the required term is $6 \times 3^2 = 54$.

EXAMPLE 2.—Find the 5th term of a series, of which the extremes are 524288 and 2, and the number of terms is 10.

The common ratio $\sqrt[9]{\dfrac{524288}{2}} = 4$. And $\dfrac{524288}{4^4} = 2048$, is the required term.

22. REASON OF THE RULE.—In an ascending series, any term is the product of the first and the common ratio taken as a factor so many times as there are *preceding* terms—since it is not found in the *first* term.

In a descending series, any term is equal to the first term. divided by a product containing the common ratio as a factor so many times as there are preceding terms—since every term but that which is required adds it once to the factors which constitute the divisor.

EXERCISES.

30. What is the 6th term of a series having 3 and 5859375 as extremes, and containing 10 terms? *Ans.* 9375.

31. Given 39366 and 2 as the extremes of a series having 10 terms. What is the 8th term? *Ans.* 18.

32. Given 1959552 and 7 as the extremes of a series having 8 terms. What is the 6th term? *Ans.* 252.

23. Given the extremes and common ratio—to find the number of terms—

RULE.—Divide the greater by the lesser extreme, and one more than the number expressing what power ,of common ratio is equal to the quotient, will be the required quantity.

EXAMPLE.—How many terms in a series of which the extremes are 2 and 256, and the common ratio is 2?

$\dfrac{256}{2} = 128$. But $2^7 = 128$. There are, therefore, 8 terms

The common ratio is found as a factor (in the quotient of the greater divided by the lesser extreme) once less than the number of terms.

33. How many terms in a series of which the first is 78732 and the last 12, and the common ratio is 9? *Ans.* 5.

34. How many terms in a series of which the extremes and common ratio are 4, 470596, and 7? *Ans.* 7.

35. How many terms in a series of which the extremes and common ratio, are 196608, 6, and 8? *Ans.* 6.

24. Given the common ratio, number of terms, and one extreme—to find the other—

RULE.—If the lesser extreme is given, multiply, if the greater, divide it by the common ratio raised to a power indicated by one less than the number of terms.

EXAMPLE 1.—In a geometrical series, the lesser extreme is 8, the number of terms is 5, and the common ratio is 6; what is the other extreme? *Ans.* $8 \times 6^{5-1} = 10368$.

EXAMPLE 2.—In a geometrical series, the greater extreme is 6561, the number of terms is 7, and the common ratio is 3; what is the other extreme? *Ans.* $6561 \div 3^{7-1} = 9$.

This rule does not require any explanation.

36. The common ratio is 3, the number of terms is 7, and one extreme is 9; what is the other? *Ans.* 6561.

37. The common ratio is 4, the number of terms is 6, and one extreme is 1000; what is the other? *Ans.* 1024000.

38. The common ratio is 8, the number of terms is 10, and one extreme is 402653184; what is the other? *Ans.* 3.

In progression, as in many other rules, the application of algebra to the reasoning would greatly simplify it.

MISCELLANEOUS EXERCISES IN PROGRESSION.

1. The clocks in Venice, and some other places strike the 24 hours, not beginning again, as ours do, after 12. How many strokes do they give in a day? *Ans.* 300.

2. A butcher bought 100 sheep; for the first he gave 1*s.*, and for the last £9 19*s.* What did he pay for

all, supposing their prices to form an arithmetical series? *Ans.* £500.

3. A person bought 17 yards of cloth; for the first yard he gave 2s., and for the last 10s. What was the price of all? *Ans.* £5 2s.

4. A person travelling into the country went 3 miles the first day, 8 miles the second, 13 the third, and so on, until he went 58 miles in one day. How many days did he travel? *Ans.* 12.

5. A man being asked how many sons he had, said that the youngest was 4 years old, and the eldest 32, and that he had added one to his family every fourth year. How many had he? *Ans.* 8.

6. Find the sum of an infinite series, $\frac{1}{3}$, $\frac{1}{9}$, $\frac{1}{27}$, &c. *Ans.* $\frac{1}{2}$

7. Of what value is the decimal ˙463˙? *Ans.* $\frac{463}{999}$.

8. What debt can be discharged in a year by monthly payments in geometrical progression, the first term being £1, and the last £2048; and what will be the common ratio? *Ans.* The debt will be £4095; and the ratio 2.

9. What will be the price of a horse sold for 1 farthing for the first nail in his shoes, 2 farthings for the second, 4 for the third, &c., allowing 8 nails in each shoe? *Ans.* £4473924 5s. $3\frac{3}{4}d.$

10. A nobleman dying left 11 sons, to whom he bequeathed his property as follows; to the youngest he gave £1024; to the next, as much and a half; to the next, $1\frac{1}{2}$ of the preceding son's share; and so on. What was the eldest son's fortune; and what was the amount of the nobleman's property? *Ans.* The eldest son received £59049, and the father was worth £175099.

QUESTIONS.

1. What is meant by ascending and descending series? [1].

2. What is meant by an arithmetical and a geometrical progression; and are they designated by any other names? [2 and 3].

3. What are the common difference and common ratio? [2 and 3].

4. Show that a continued proportion may be formed from a series of either kind ? [2 and 3].

5. What are means and extremes ? [4].

6. How is the sum of an arithmetical or a geometrical series found ? [5 and 16].

7. How is the common difference or common ratio found ? [7 and 18].

8. How is any number of arithmetical or geometrical means found ? [9 and 20].

9. How is any particular arithmetical or geometrical mean found ? [10 and 21].

10. How is the number of terms in an arithmetical or geometrical series found ? [12 and 23].

11. How is one extreme of an arithmetical or geometrical series found ? [14 and 24].

ANNUITIES.

25. An annuity is an income to be paid at stated times, yearly, half-yearly, &c. It is either *in possession*, that is, entered upon already, or to be entered upon immediately ; or it is *in reversion*, that is, not to commence until after some period, or after something has occurred. An annuity is *certain* when its commencement and termination are assigned to definite periods, contingent when its beginning, or end, or both are uncertain ; is *in arrears* when one, or more payments are retained after they have become due. The *amount* of an annuity is the sum of the payments forborne (in arrears), and the interest due upon them.

When an annuity is paid off at once, the price given for it is called its *present worth*, or *value*—which ought to be such as would—if left at compound interest until the annuity ceases—produce a sum equal to what would be due from the annuity left unpaid until that time. This value is said to be so many *years' purchase ;* that is, so many annual payments of the income as would be just equivalent to it.

26. To find the amount of a certain number of payments in arrears, and the interest due on them—

RULE.—Find the interest due on each payment; then the sum of the payments and interest due on them, will be the required amount.

EXAMPLE 1.—What will be the amount of £1 per annum, unpaid for 6 years, 5 per cent. simple interest being allowed?

The last, and preceding payments, with the interest due on them, form the *arithmetical* series £1 + £·05 × 5. £1 + £·05 × 4 .. £1 + £·05. £1. And its sum is $\overline{£1 + £1 + £·05 \times 5} \times \frac{6}{2} = £2 + £·25 \times 3 = £6·75 = £6$ 15s., the required amount.

EXAMPLE 2.—If the rent of a farm worth £60 per annum is unpaid for 19 years, how much does it amount to, at 5 per cent. per an. compound interest?

In this case the series is *geometrical*; and the last payment with its interest is the *amount* of £1 for 18 (19 — 1) years multiplied by the given annuity, the preceding payment with its interest is the amount of £1 for 17 years multiplied by the given annuity, &c.

The amount of £1 (as we find by the table at the end of the treatise) for 18 years is £2·40662. Then the sum of the series is—

$$\frac{£2·40662 \times 1·05 \times 60 - 60}{1·05 - 1} \quad [16] = 1832·4, \text{the required amount.}$$

The amount of £1 for 18 years multiplied by 1·05 is the same as the amount of £1 for 19, or the *given* number of years, which is found to be £2·527. And 1·05 — 1, the divisor, is equal to the amount of £1 for one payment minus £1; that is, to the interest of £1 for one payment. Hence the required sum will be $\dfrac{£2·527 \times 60 - 60}{·05} = £1832·4.$

It would evidently be the same thing to consider the annuity as £1, and then multiply the result by 60. Thus $\dfrac{2·527 - 1}{·05} \times 60 = £1832·4.$ For an annuity of £60 ought to be 60 times as productive as one of only £1.

Hence, briefly, to find the amount of any number of payments in arrears, and the *compound* interest due on them—

Subtract £1 from the amount of £1 for the given number of payments, and divide the difference by the interest of £1 for one payment; then multiply the quotient by the given sum.

Q

27. Reason of the Rule.—Each payment, with its interest, evidently constitute a *separate* amount; and the sum due must be the sum of these amounts—which form a *decreasing* series, because of the decreasing interest, arising from the decreasing number of times of payment.

When simple interest is allowed, it is evident that what is due will be the sum of an *arithmetical* series, one extreme of which is the first payment plus the interest due upon it at the time of the last, the other the last payment; and its common difference the interest on one payment due at the next.

But when *compound* interest is allowed, what is due will be the sum of a *geometrical* series, one extreme of which is the first payment plus the interest due on it at the last, the other the last payment; and its common ratio £1 plus its interest for the interval between two payments. And in each case the interest due on the first payment at the time of the last will be the interest due for *one less* than the number of payments, since interest is not due on the first until the time of the second payment.

EXERCISES.

1. What is the amount of £37 per annum unpaid for 11 years, at 5 per cent. per an. simple interest? *Ans.* £508 15s.

2. What is the amount of an annuity of £100, to continue 5 years at 6 per cent. per an. compound interest? *Ans.* £563 14s. 2¼d.

3. What is the amount of an annuity of £356, to continue 9 years, at 6 per cent. per an. simple interest? *Ans.* £3972 19s. 2½d.

4. What is the amount of £49 per annum unpaid for 7 years, 6 per cent. compound interest being allowed? *Ans.* £411 5s. 11½d.

28. To find the present value of an annuity—

Rule.—Find (by the last rule) the amount of the given annuity if not paid up to the time it will cease. Then ascertain how often this sum contains the amount of £1 up to the same time, at the interest allowed.

Example.—What is the present worth of an annuity of £12 per annum, to be paid for 18 years, 5 per cent. compound interest being allowed?

An annuity of £12 unpaid for 18 years would amount to £28·13238 × 12 = £337·58856

But £1 put to interest for 18 years at the same rate would amount to £2·40662. Therefore

$$\frac{£337·58856}{2·40662} = £140 \; 5s. \; 6d. \text{ is the required value.}$$

The sum to be paid for the annuity should evidently be such as would produce the same as the annuity itself, in the same time.

EXERCISES.

5. What is the present worth of an annuity of £27, to be paid for 13 years, 5 per cent. compound interest being allowed? *Ans.* £253 12s. 6¼d.

6. What is the present worth of an annuity of £324, to be paid for 12 years, 5 per cent. compound interest being allowed? *Ans.* £2871 13s. 10¼d.

7. What is the present worth of an annuity of £22, to be paid for 21 years, 4 per cent. compound interest being allowed? *Ans.* £308 12s. 10d.

29. To find the present value, when the annuity is in perpetuity—

RULE.—Divide the interest which £1 would produce in perpetuity into £1, and the quotient will be the sum required to produce an annuity of £1 per annum in perpetuity. Multiply the quotient by the number of pounds in the given annuity, and the product will be the required present worth.

EXAMPLE.—What is the value of an income of £17 for ever?

Let us suppose that £100 would produce £5 per cent. per an. for ever:—then £1 would produce £·05. Therefore, to produce £1, we require as many pounds as will be equal to the number of times £·05 is contained in £1. But $\frac{£1}{·05} =$ £20, therefore £20 would produce an annuity of £1 for ever. And 17 times as much, or £20×17=340, which would produce an annuity of £17 for ever, is the required present value.

EXERCISES.

8. A small estate brings £25 per annum; what is its present worth, allowing 4 per cent. per annum interest? *Ans.* £625.

9. What is the present worth of an income of £347

in perpetuity, allowing 6 per cent. interest? *Ans.*
£5783 6s. 8d.

10. What is the value of a perpetual annuity of £46,
allowing 5 per cent. interest? *Ans.* £920.

30. To find the present value of an annuity in reversion—

RULE.—Find the amount of the annuity as if it were
forborne until it should cease. Then find what sum,
put to interest now, would at that time produce the
same amount.

EXAMPLE.—What is the value of an annuity of £10 per
annum, to continue for 6, but not to commence for 12 years,
5 per cent. compound interest being allowed?

An annuity of £10 for 6 years if left unpaid, would be
worth £68·0191; and £1 would, in 18 years, be worth
£11·68959. Therefore

$$\frac{£68·0191}{11·68959} = £28 \ 5s. \ 3d.,$$ is the required present worth.

EXERCISES.

11. What is the present worth of £75 per annum,
which is not to commence for 10 years, but will continue 7 years after, at 6 per cent. compound interest?
Ans. £155 9s. 7¾d.

12. The reversion of an annuity of £175 per annum,
to continue 11 years, and commence 9 years hence, is to
be sold; what is its present worth, allowing 6 per cent.
per annum compound interest? *Ans.* £430 7s. 1d.

13. What is the present worth of a rent of £45 per
annum, to commence in 8, and last for 12 years, 6 per
cent. compound interest, payable half-yearly, being
allowed? *Ans.* 117 2s. 8½d.

31. When the annuity is contingent, its value depends
on the probability of the contingent circumstance, or
circumstances.

A life annuity is equal to its amount multiplied by
the value of an annuity of £1 (found by tables) for the
given age. The tables used for the purpose are calculated on principles derived from the doctrine of chances,
observations on the duration of life in different circumstances, the rates of compound interest, &c.

QUESTIONS.

1. What is an annuity ? [25],
2. What is an annuity in possession—in reversion—certain—contingent—or in arrears ? [25].
3. What is meant by the present worth of an annuity? [25].
4. How is the amount of any number of payments in arrears found, the interest allowed being simple or compound? [26].
5. How is the present value of an annuity in possession found ? [28].
6. How is the present value of an annuity in perpetuity found ? [29].
7. How is the present value of an annuity in reversion found ? [30].

POSITION.

32. Position, called also the " rule of false," is a rule which, by the use of one or more assumed, but *false* numbers, enables us to find the true one. By means of it we can obtain the answers to certain questions, which we could not resolve by the ordinary direct rules.

When the results are really proportional to the supposition—as, for instance, when the number sought is to be *multiplied* or *divided* by some proposed number; or is to be increased or diminished by *itself,* or by some given *multiple* or *part* of itself—and when the question contains only one *proposition,* we use what is called *single* position, assuming only *one* number ; and the quantity found is *exactly* that which is required. Otherwise—as, for instance, when the number sought is to be increased or diminished by some *absolute* number, which is not a known multiple, or part of it—or when *two* propositions, neither of which can be banished, are contained in the problem, we use *double* position, assuming *two* numbers. If the number sought is, during the process indicated by the question, to be involved or evolved, we obtain only an *approximation* to the quantity required.

33. *Single Position.*—RULE. Assume a number, and perform with it the operations described in the question; then say, as the result obtained is to the number *used*, so is the true or *given* result to the number *required*.

EXAMPLE.—What number is that which, being multiplied by 5, by 7, and by 9, the sum of the results shall be 231 ?

Let us assume 4 as the quantity sought. $4 \times 5 + 4 \times 7 + 4 \times 9 = 84$. And $84 : 4 :: 231 : \dfrac{4 \times 231}{84} = 11$, the required number.

34. REASON OF THE RULE.—It is evident that two numbers, *multiplied*, or *divided* by the same, should produce proportionate results.—It is otherwise, however, when the same quantity is *added* to, or *subtracted* from them. Thus let the given question be changed into the following. What number is that which being multiplied by 5, by 7, and by 9, the sum of the products, plus 8, shall be equal to 239 ?

Assuming 4, the result will be 92. Then we cannot say

$$92 \ (84+8) : 4 :: 239 \ (231+8) : 11.$$

For though $84 : 4 :: 231 : 11$, it does not follow that $84+8 : 4 :: 231+8 : 11$. Since, while [Sec. V. 29] we may *multiply* or *divide* the first and third terms of a geometrical proportion by the same number, we cannot, without destroying the proportion, *add* the same number to, or *subtract* it from them. The question in this latter form belongs to the rule of *double* position.

EXERCISES.

1. A teacher being asked how many pupils he had, replied, if you add $\frac{1}{3}$, $\frac{1}{4}$, and $\frac{1}{6}$ of the number together, the sum will be 18 ; what was their number ? *Ans.* 24.

2. What number is it, which, being increased by $\frac{1}{2}$, $\frac{1}{3}$, and $\frac{1}{4}$ of itself, shall be 125 ? *Ans.* 60.

3. A gentleman distributed 78 pence among a number of poor persons, consisting of men, women, and children; to each man he gave 6*d.*, to each woman, 4*d.*, and to each child, 2*d.* ; there were twice as many women as men, and three times as many children as women. How many were there of each? *Ans.* 3 men, 6 women, and 18 children.

4. A person bought a chaise, horse, and harness, for £60 ; the horse came to twice the price of the harness, and the chaise to twice the price of the horse and har-

ness. What did he give for each? *Ans.* He gave for the harness, £6 13*s.* 4*d.*; for the horse, £13 6*s.* 8*d.*; and for the chaise, £40.

5. A's age is. double that of B's; B's is treble that of C's; and the sum of all their ages is 140. What is the age of each? *Ans.* A's is 84, B's 42, and C's 14.

6. After paying away ¼ of my money, and then ⅓ of the remainder, I had 72 guineas left. What had I at first? *Ans.* 120 guineas.

7. A can do a piece of work in 7 days; B can do the same in 5 days; and C in 6 days. In what time will all of them execute it? *Ans.* in $1\frac{103}{107}$ days.

8. A and B can do a piece of work in 10 days; A by himself can do it in 15 days. In what time will B do it? *Ans.* In 30 days.

9. A cistern has three cocks; when the first is opened all the water runs out in one hour; when the second is opened, it runs out in two hours; and when the third is opened, in three hours. In what time will it run out, if all the cocks are kept open together? *Ans.* In $\frac{6}{11}$ hours.

10. What is that number whose ⅓, ⅙, and ¼ parts, taken together, make 27? *Ans.* 42.

11. There are 5 mills; the first grinds 7 bushels of corn in 1 hour, the second 5 in the same time, the third 4, the fourth 3, and the fifth 1. In what time will the five grind 500 bushels, if they work together? *Ans.* In 25 hours.

12. There is a cistern which can be filled by a cock in 12 hours; it has another cock in the bottom, by which it can be emptied in 18 hours. In what time will it be filled, if both are left open? *Ans.* In 36 hours.

35. *Double Position.*—RULE I. Assume two convenient numbers, and perform upon them the processes supposed by the question, marking the error derived from each with + or −, according as it is an error of *excess*, or of *defect*. Multiply each assumed number into the error which belongs to the other; and, if the errors are *both* plus, or *both* minus, divide the *difference* of the products by the *difference* of the errors. But, if *one* is a plus, and the *other* is a minus error, divide the *sum* of

the products by the *sum* of the errors. In either case, the result will be the number sought, or an approximation to it.

EXAMPLE 1.—If to 4 times the price of my horse £10 is added, the sum will be £100. What did it cost?

Assuming numbers which give two errors of *excess*—

First, let 28 be one of them.

Multiply by 4

112

Add 10

From 122, the result obtained, subtract 100, the result required,

and the remainder, +22, is an error of *excess*.

Multiply by 31, the other assumed number,

and 682 will be the product.

Next, let the assumed number be 31

Multiply by 4

124

Add 10

From 134, the result obtained, subtract 100, the result required,

and the remainder, +34, is an error of *excess*.

Multiply by 28, the other assumed num.

and 952 will be the product.

From this subtract 682, the product found above,

divide by 12)270

and the required quantity is 22·5 = £22 10s.

Difference of errors = 34 − 22 = 12, the number by which we have divided.

36. REASON OF THE RULE.—When in example 1, we multiply 28 and 31 by 4, we multiply the error belonging to each by 4. Hence 122 and 134 are, respectively, equal to the true result, plus 4 times one of the errors. Subtracting 100, the true result, from each of them, we obtain 22 (4 times the error in 28) and 34 (4 times the error in 31).

But, as numbers are proportional to their *equi*multiples, the error in 28 : the error in 31 :: 22 (a multiple of the former) : 34 (an equimultiple of the latter).

And from the nature of proportion [Sec. V. 21]—

The error in $28 \times 34 =$ the error in 31×22.

But $682 = \overline{\text{the error in } 31 + \text{the required number}} \times 22$.

And $952 = \overline{\text{the error in } 28 + \text{the required number}} \times 34$.

Or, since to multiply quantities under the vinculum [Sec. II. 34], we are to multiply each of them—

$682 = 22$ times the error in $31 + 22$ times the required number.

$952 = 34$ times the error in $28 + 34$ times the required number.

Subtracting the upper from the lower line, we shall have $952 - 682 = 34$ times the error in $28 - 22$ times the error in $31 + 34$ times the required number $- 22$ times the required number.

But, as we have seen above, 34 times the error in $28 = 22$ times the error in 31. Therefore, 34 times the error in $28 - 22$ times the error in $31 = 0$; that is, the two quantities cancel each other, and may be omitted. We shall then have $952 - 682 = 34$ times the required number $- 22$ times the required number; or $270 = 34 - 22$ $(= 12)$ times the required number. And, [Sec. V. 6] dividing both the equal quantities by 12,

$$\frac{270}{12} \ (22 \cdot 5) = \frac{34 - 22}{12} \text{ times (once) the required number.}$$

37. EXAMPLE 2.—Using the same example, and assuming numbers which give two errors of *defect*.

Let them be 14, and 16—

14	16
4	4
56	64
10	10
66, the result obtained,	74, the result obtained,
100, the result required,	100, the result required,
-34, an error of *defect*.	-26, an error of *defect*.
16	14
544	364
364	Difference of errors $= 34 - 26 = 8$.

$8) 180$

$22 \cdot 5 = £22$ $10s.$, is the required quantity.

In this example $34 =$ four times the error (of defect) in 14; and $26 =$ four times the error (of defect) in 16. And, since numbers are proportional to their equimultiples,

The error in 14 : the error in 16 :: 34 : 26. Therefore

The error in $14 \times 26 =$ the error in 16×34.

But $544 = \overline{\text{the required number} - \text{the error in } 16} \times 34$

And $364 = \overline{\text{the required number} - \text{the error in } 14} \times 26$

If we subtract the lower from the upper line, we shall have
544−364=(removing the vinculum, and changing the sign
[Sec. II. 16]) 34 times the required number−26 times the
required number−34 times the error in 16+26 times the error
in 14.

But we found above that 34 times the error in 16=26 times
the error in 14. Therefore−34 times the error in 16, and+26
times the error in 14=0, and may be omitted. We will then
have 544−364=34 times the required number−26 times the
required number; or 180=8 times the required number; and,
dividing both these equal quantities by 8,

$$\frac{180}{8}\ (22\cdot5)=\frac{8}{8}\ \text{times (once) the required number.}$$

38. EXAMPLE 3.—Using still the same example, and assuming numbers which will give an error of *excess*, and an
error of *defect*.

Let them be 15, and 23—

15			23	
4			4	

60   92
10   10

70, the result obtained.   102, the result obtained.
100, the result required.   100, the result required.

−30, an error of *defect*.   +2, an error of *excess*.
23   15

690   30
30   Sum of errors=30+2=32.

32)720

.22·5=£22 10*s*., the required quantity.

In this example 30 is 4 times the error (of defect) in 15;
and 2, 4 times the error (of excess) in 23. And, since numbers
are proportioned to the equimultiples,

The error in 23 : the error in 15 :: 2 : 30.   Therefore
The error in 23×30=the error in 15×2.

But 690=the required number+the error in 23×30.
And 30=the required number−the error in 15×2.

If we add these two lines together, we shall have 690+30=
(removing the vinculum) 30 times the required number+
twice the required number + 30 times the error in 23 − twice
the error in 15.

But we found above that 30×the error in 23=2×the error
in 15. Therefore 30×the error in 23−2×the error in 15=0,

and may be omitted. We shall then have 690+30=the required number × 30 + the required number × 2; or 720 = 32 times the required number. And dividing each of these equal quantities by 32.

$$\frac{720}{32}(22\ 5) = \frac{32}{32} \text{ times (once) the required number.}$$

The given questions might be changed into one belonging to *single* position, thus—

Four times the price of my horse is equal to £100 − £10; or four times the price of my horse is equal to £90. What did it cost? This change, however, supposes an effort of the mind not required when the question is solved by double position.

39. EXAMPLE 4.—What is that number which is equal to 4 times its square root +21?

Assume 64 and 81—

$\sqrt{64}=8$	$\sqrt{81}=\ 9$
4	4
32	36
21	21
53, result obtained.	57, result obtained.
64, result required.	81, result required.
−11	−24
81	64
891	1536
	891
	13)645

The first approximation is 49·6154

It is evident that 11 and 24 are not the errors in the assumed numbers multiplied or divided by the same quantity, and therefore, as the reason upon which the rule is founded, does not apply, we obtain only an approximation. Substituting this, however, for one of the assumed numbers, we obtain a still nearer approximation.

40. RULE—II. Find the errors by the last rule; then divide their difference (if they are both of the same kind), or their sum (if they are of different kinds), into the product of the difference of the numbers and one of the errors. The quotient will be the correction of that error which has been used as multiplier.

EXAMPLE.—Taking the same as in the last rule, and assuming 19 and 25 as the required number.

19	25
4	4
76	100
10	10
86 the result obtained.	110 the result obtained.
100 the result required.	100 the result required.
−14, is error of *defect*.	+10, is error of *excess*.

The errors are of *different* kinds; and their *sum* is 14+10=24; and the difference of the assumed numbers is 25−19=6. Therefore

14 one of the errors,

is multiplied by 6, by the difference of the numbers. Then divide by   24)84

and 3·5 is the correction for 19, the number

which gave an error of 14.

19+(the error being one of *defect*, the correction is to be *added*) 3·5=22·5=£22 10s. is the required quantity.

41. REASON OF THE RULE.—The difference of the results arising from the use of the different assumed numbers (the difference of the errors) : the difference between the result obtained by using one of the assumed numbers and that obtained by using the true number (one of the errors) :: the difference between the numbers in the former case (the difference between the assumed numbers) : the difference between the numbers in the latter case (the difference between the true number, and that assumed number which produced the error placed in the third term—that is the correction required by that assumed number).

It is clear that the difference between the numbers used produces a proportional difference in the results. For the results are different, only because the difference between the assumed numbers has been multiplied, or divided, or both—in accordance with the conditions of the question. Thus, in the present instance, 25 produces a greater result than 19, because 6, the difference between 19 and 25, has been multiplied by 4. For 25×4=19×4+6×4. And it is this 6× which makes up 24, the *real* difference of the errors.—The difference between a negative and positive result being the sum of the differences between each of them and no result. Thus, if I gain 10s., I am richer to the amount of 24s. than if I lose 14s.

EXERCISES.

13. What number is it which, being multiplied by 3, the product being increased by 4, and the sum divided by 8, the quotient will be 32 ? *Ans.* 84.

14. A son asked his father how old he was, and received the following answer. Your age is now $\frac{1}{4}$ of mine, but 5 years ago it was only $\frac{1}{5}$. What are their ages ? *Ans.* 80 and 20.

15. A workman was hired for 30 days at 2s. 6d. for every day he worked, but with this condition, that for every day he did not work, he should forfeit a shilling. At the end of the time he received £2 14s., how many days did he work ? *Ans.* 24.

16. Required what number it is from which, if 34 be taken, 3 times the remainder will exceed it by $\frac{1}{4}$ o itself ? *Ans.* $58\frac{2}{7}$.

17. A and B go out of a town by the same road. A goes 8 miles each day ; B goes 1 mile the first day, 2 the second, 3 the third, &c. When will B overtake A ?

	A.	B.			A.	B.
Suppose	5	1		Suppose	7	1
	8	2			8	2
	—	3			—	3
	40	4			56	4
	15	5			28	5
	—				—	6
	5)25	15			7)28	7
	−5				−4	—
	—				5	28
	7				—	
	—				20	
	35					
	20		5 − 4 = 1			
	1)15					

We divide the entire error by the number of days in each case, which gives the error in one day.

18. A gentleman hires two labourers ; to the one he gives 9d. each day ; to the other, on the first day, 2d., on the second day, 4d., on the third day, 6d., &c. In how many days will they earn an equal sum ? *Ans.* In 8.

19. What are those numbers which, when added,

make 25; but when one is halved and the other doubled, give equal results? *Ans.* 20 and 5.

20. Two contractors, A and B, are each to build a wall of equal dimensions; A employs as many men as finish $22\frac{1}{2}$ perches in a day; B employs the first day as many as finish 6 perches, the second as many as finish 9, the third as many as finish 12, &c. In what time will they have built an equal number of perches? *Ans.* In 12 days.

21. What is that number whose $\frac{1}{2}$, $\frac{1}{4}$, and $\frac{3}{8}$, multiplied together, make 24?

Suppose 12	Suppose 4
$\frac{1}{2}=6$	$\frac{1}{2}=2$
$\frac{1}{4}=3$	$\frac{1}{4}=1$
Product $= 18$	Product $=2$
$\frac{3}{8}=4\frac{1}{2}$	$\frac{3}{8}=1\frac{1}{2}$
81 result obtained.	3 result obtained.
24 result required.	24 result required.
$+57$	-21
64, the cube of 4.	1728, the cube of 12.
3648, product.	36288 To this product
	3648 is added.
$57+21=78$	
$57-21=78.$	78)39936 is the sum.
	And 512 the quotient.

$\sqrt[3]{512}=8$, is the required number.

We multiply the alternate error by the *cube* of the supposed number, because the errors belong to the $\frac{3}{64}$th part of the *cube* of the assumed numbers, and not to the numbers themselves; for, in reality, it is the cube of some number that is required —since, 8 being assumed, according to the question we have $\frac{8}{2}\times\frac{8}{4}\times\frac{3\times8}{8}=24$; or $\frac{3}{64}\times 8^3=24.$

22. What number is it whose $\frac{1}{2}$, $\frac{1}{4}$, $\frac{1}{5}$, and $\frac{1}{6}$, multiplied together, will produce $6998\frac{2}{5}$? *Ans.* 36.

23. A said to B, give me one of your shillings, and I shall have twice as many as you will have left. B answered, if you give me 1s., I shall have as many as you. How many had each? *Ans.* A 7, and B 5.

24. There are two numbers which, when added together, make 30; but the $\frac{1}{2}$, $\frac{1}{3}$, and $\frac{1}{6}$, of the greater are equal to $\frac{1}{2}$, $\frac{3}{4}$, and $\frac{1}{4}$, of the lesser. What are they? *Ans.* 12 and 18.

25. A gentleman has 2 horses and a saddle worth £50. The saddle, if set on the back of the first horse, will make his value double that of the second; but if set on the back of the second horse, it will make his value treble that of the first. What is the value of each horse? *Ans.* £30 and £40.

26. A gentleman finding several beggars at his door, gave to each 4*d.* and had 6*d.* left, but if he had given 6*d.* to each, he would have had 12*d.* too little. How many beggars were there? *Ans.* 9.

It is so likely that those who are desirous of studying this subject further will be acquainted with the method of treating algebraic equations—which in many cases affords a so much simpler and easier mode of solving questions belonging to position—that we do not deem it necessary to enter further into it.

QUESTIONS.

1. What is the difference between single and double position? [32].

2. In what cases may we expect an exact answer by these rules? [32].

3. What is the rule for single position? [33].

4. What are the rules for double position? [35 and 40].

MISCELLANEOUS EXERCISES.

1. A father being asked by his son how old he was; replied, your age is now $\frac{1}{5}$ of mine; but 4 years ago it was only $\frac{1}{7}$ of what mine is now; what is the age of each? *Ans.* 70 and 14.

2. Find two numbers, the difference of which is 30, and the relation between them as $7\frac{1}{4}$ is to $3\frac{1}{2}$? *Ans.* 58 and 28.

3. Find two numbers whose sum and product are equal, neither of them being 2? *Ans.* 10 and $1\frac{1}{9}$.

4. A person being asked the hour of the day, answered, It is between 5 and 6, and both the hour and minute hands are together. Required what it was? *Ans.* $27\frac{3}{11}$ minutes past 5.

5. What is the sum of the series $\frac{1}{2}$, $\frac{1}{4}$, $\frac{1}{8}$, &c.? *Ans.* 1.

6. What is the sum of the series $\frac{2}{5}$, $\frac{4}{15}$, $\frac{8}{45}$, $\frac{16}{135}$, &c.? *Ans.* $1\frac{1}{5}$.

7. A person had a salary of £75 a year, and let it remain unpaid for 17 years. How much had he to receive at the end of that time, allowing 6 per cent. per annum compound interest, payable half-yearly? *Ans.* £204 17 s. $10\frac{1}{4}d$.

8. Divide 20 into two such parts as that, when the greater is divided by the less, and the less by the greater, and the greater quotient is multiplied by 4, and the less by 64, the products shall be equal? *Ans.* 4 and 16.

9. Divide 21 into two such parts, as that when the less is divided by the greater, and the greater by the less, and the greater quotient is multiplied by 5, and the less by 125, the products shall be equal? *Ans.* $3\frac{1}{2}$ and $17\frac{1}{2}$.

10. A, B, and C, can finish a piece of work in 10 days; B and C will do it in 16 days. In what time will A do it by himself? *Ans.* $26\frac{2}{3}$ days.

11. A can trench a garden in 10 days, B in 12, and C in 14. In what time will it be done by the three if they work together? *Ans.* In $3\frac{99}{107}$ days.

12. What number is it which, divided by 16, will leave 3; but which, divided by 9, will leave 4? *Ans.* 67.

13. What number is it which, divided by 7, will leave 4; but divided by 4, will leave 2? *Ans.* 18.

14. If £100, put to interest at a certain rate, will, at the end of 3 years, be augmented to £115·7625 (compound interest being allowed), what principal and interest will be due at the end of the first year? *Ans.* £105.

15. An elderly person in trade, desirous of a little respite, proposes to admit a sober, and industrious young person to a share in the business; and to encourage him, he offers, that if his circumstances allow him to

advance £100, his salary shall be £40 a year; that if
he is able to advance £200, he shall have £55; but
that if he can advance £300, he shall receive £70
annually. In this proposal, what was allowed for his
attendance simply? *Ans.* £25 a year.

16. If 6 apples and 7 pears cost 33 pence, and 10
apples and 8 pears 44 pence, what is the price of one
apple and one pear? *Ans.* 2*d.* is the price of an apple,
and 3*d.* of a pear.

17. Find three such numbers as that the first and $\frac{1}{2}$
the sum of the other two, the second and $\frac{1}{3}$ the sum of
the other two, the third and $\frac{1}{4}$ the sum of the other
two will make 34? *Ans.* 10, 22, 26.

18. Find a number, to which, if you add 1, the sum
will be divisible by 3; but if you add 3, the sum will
be divisible by 4? *Ans.* 17.

19. A market woman bought a certain number of
eggs, at two a penny, and as many more at 3 a penny;
and having sold them all at the rate of five for 2*d.*, she
found she had lost fourpence. How many eggs did she
buy? *Ans.* 240.

20. A person was desirous of giving 3*d.* a piece to
some beggars, but found he had 8*d.* too little; he therefore gave each of them 2*d.*, and had then 3*d.* remaining. Required the number of beggars? *Ans.* 11.

21. A servant agreed to live with his master for £8
a year, and a suit of clothes. But being turned out
at the end of 7 months, he received only £2 13*s.* 4*d.*
and the suit of clothes; what was its value? *Ans.*
£4 16*s.*

22. There is a number, consisting of two places of
figures, which is equal to four times the sum of its
digits, and if 18 be added to it, its digits will be inverted. What is the number? *Ans.* 24.

23. Divide the number 10 into three such parts, that
if the first is multiplied by 2, the second by 3, and the
third by 4, the three products will be equal? *Ans.*
$4\frac{8}{13}$, $3\frac{1}{13}$, $2\frac{4}{13}$.

24. Divide the number 90 into four such parts that,
if the first is increased by 2, the second diminished by
2, the third multiplied by 2, and the fourth divided by

2, the sum, difference, product, and quotient will be equal ? *Ans.* 18, 22, 10, 40.

25. What fraction is that, to the numerator of which, if 1 is added, its value will be $\frac{1}{3}$; but if 1 be added to the denominator, its value will be $\frac{1}{4}$? *Ans.* $\frac{4}{15}$.

26. 21 gallons were drawn out of a cask of wine, which had · leaked away a third part, and the cask being then guaged, was found to be half full. How much did it hold ? *Ans.* 126 gallons.

27. There is a number, $\frac{1}{2}$ of which, being divided by 6, $\frac{1}{3}$ of it by 4, and $\frac{1}{4}$ of it by 3, each quotient will be 9 ? *Ans.* 108.

28. Having counted my books, I found that when I multiplied together $\frac{1}{2}$, $\frac{1}{4}$, and $\frac{3}{4}$ of their number, the product was 162000. How many had I ? *Ans.* 120.

29. Find the sum of the series $1+\frac{1}{2}+\frac{1}{4}+\frac{1}{8}$, &c. ? *Ans.* 2.

30. A can build a wall in 12 days, by getting 2 days' assistance from B ; and B can build it in 8 days, by getting 4 days' assistance from A. In what time will both together build it ? *Ans.* In $6\frac{6}{7}$ days.

31. A and B can perform a piece of work in 8 days, when the days are 12 hours long ; A, by himself, can do it in 12 days, of 16 hours each. In how many days of 14 hours long will B do it ? *Ans.* $13\frac{5}{7}$.

32. In a mixture of spirits and water, $\frac{1}{2}$ of the whole plus 25 gallons was spirits, but $\frac{1}{3}$ of the whole minus 5 gallons was water. How many gallons were there of each ? *Ans.* 85 of spirits, and 35 of water.

33. A person passed $\frac{1}{6}$ of his age in childhood, $\frac{1}{12}$ of it in youth, $\frac{1}{7}$ of it +5 years in matrimony ; he ha'd then a son whom he survived 4 years, and who reached only $\frac{1}{2}$ the age of his father. At what age did this person die ? *Ans.* At the age of 84.

34. What number is that whose $\frac{1}{5}$ exceeds its $\frac{1}{8}$ by 72 ? *Ans.* 540.

35. A vintner has a vessel of wine containing 500 gallons ; drawing 50 gallons, he then fills up the cask with water. After doing this five times, how much wine and how much water are in the cask ? *Ans.* $295\frac{49}{200}$ gallons of wine, and $204\frac{151}{200}$ gallons of water.

36. A mother and two daughters working together can spin 3 ℔ of flax in one day; the mother, by herself, can do it in $2\frac{1}{2}$ days; and the eldest daughter in $2\frac{1}{4}$ days. In what time can the youngest do it? *Ans.* In $6\frac{3}{7}$ days.

37. A merchant loads two vessels, A and B; into A he puts 150 hogsheads of wine, and into B 240 hogsheads. The ships, having to pay toll, A gives 1 hogshead, and receives 12*s.*; B gives 1 hogshead and 36*s.* besides. At how much was each hogshead valued? *Ans.* £4 12*s.*

38. Three merchants traffic in company, and their stock is £400; the money of A continued in trade 5 months, that of B six months, and that of C nine months; and they gained £375, which they divided equally. What stock did each put in? *Ans.* A £$167\frac{19}{43}$, B £$139\frac{23}{43}$, and C £$93\frac{1}{43}$.

39. A fountain has 4 'cocks, A, B, C, and D, and under it stands a cistern, which can be filled by A in 6, by B in 8, by C in 10, and by D in 12 hours; the cistern has 4 cocks, E, F, G, and H; and can be emptied by E in 6, by F in 5, by G in 4, and by H in 3 hours. Suppose the cistern is full of water, and that the 8 cocks are all open, in what time will it be emptied? *Ans.* In $2\frac{2}{19}$ hours.

40. What is the value of ·2'97'? *Ans.* $\frac{1}{3}\frac{1}{3}$.

41. What is the value of ·5416'? *Ans.* $\frac{13}{24}$.

42. What is the value of ·0'76923'? *Ans.* $\frac{1}{13}$.

43. There are three fishermen, A, B, and C, who have each caught a certain number of fish; when A's fish and B's are put together, they make 110; when B's and C's are put together, they make 130; and when A's and C's are put together, they make 120. If the fish is divided equally among them, what will be each man's share; and how many fish did each of them catch? *Ans.* Each man had 60 for his share; A caught 50, B 60, and C 70.

44. There is a golden cup valued at 70 crowns, and two heaps of crowns. The cup and first heap, are worth 4 times the value of the second heap; but the cup and second heap, are worth double the value of the first

heap. How many crowns are there in each heap? *Ans.* 50 in one, and 30 in another.

45. A certain number of horse and foot soldiers are to be ferried over a river; and they agree to pay $2\frac{1}{2}d.$ for two horse, and $3\frac{1}{2}d.$ for seven foot soldiers; seven foot always followed two horse soldiers; and when they were all over, the ferryman received £25. How many horse and foot soldiers were there? *Ans.* 2000 horse, and 7000 foot.

46. The hour and minute hands of a watch are together at 12; when will they be together again? *Ans.* at $5\frac{5}{11}$ minutes past 1 o'clock.

47. A and B are at opposite sides of a wood 135 fathoms in compass. They begin to go round it, in the same direction, and at the same time; A goes at the rate of 11 fathoms in 2 minutes, and B at that of 17 in 3 minutes. How many rounds will each make, before one overtakes the other? *Ans.* A will go 17, and B $16\frac{1}{2}.$

48. A, B, and C, start at the same time, from the same point, and in the same direction, round an island 73 miles in circumference; A goes at the rate of 6, B at the rate of 10, and C at the rate of 16 miles per day. In what time will they be all together again? *Ans.* In $36\frac{1}{2}$ days.

MATHEMATICAL TABLES.

LOGARITHMS OF NUMBERS FROM 1 TO 10,000, WITH
DIFFERENCES AND PROPORTIONAL PARTS.

Numbers from 1 to 100.									
No.	Log.	No.	Log.	No.	Log.	No.	Log.	No.	Log.
1	0·000000	21	1·322219	41	1·612784	61	1·785330	81	1·908485
2	0·301030	22	1·342423	42	1·623249	62	1·792392	82	1·913814
3	0·477121	23	1·361728	43	1·633468	63	1·799341	83	1·919078
4	0·602060	24	1·380211	44	1·643453	64	1·806180	84	1·924279
5	0·698970	25	1·397940	45	1·653213	65	1·812913	85	1·929419
6	0·778151	26	1·414973	46	1·662758	66	1·819544	86	1·934498
7	0·845098	27	1·431364	47	1·672098	67	1·826075	87	1·939519
8	0·903090	28	1·447158	48	1·681241	68	1·832509	88	1·944483
9	0·954243	29	1·462398	49	1·690196	69	1·838849	89	1·949390
10	1·000000	30	1·477121	50	1·698970	70	1·845098	90	1·954243
11	1·041393	31	1·491362	51	1·707570	71	1·851258	91	1·959041
12	1·079181	32	1·505150	52	1·716003	72	1·857332	92	1·963788
13	1·113943	33	1·518514	53	1·724276	73	1·863323	93	1·968483
14	1·146128	34	1·531479	54	1·732394	74	1·869232	94	1·973128
15	1·176091	35	1·544068	55	1·740363	75	1·875061	95	1·977724
16	1·204120	36	1·556303	56	1·748188	76	1·880814	96	1·982271
17	1·230449	37	1·568202	57	1·755875	77	1·886491	97	1·986772
18	1·255273	38	1·579784	58	1·763428	78	1·892095	98	1·991226
19	1·278754	39	1·591065	59	1·770852	79	1·897627	99	1·995635
20	1·301030	40	1·602060	60	1·778151	80	1·903090	100	2·000000

P P	N.	0	1	2	3	4	5	6	7	8	9	D.
	100	000000	000434	000868	001301	001734	002166	002598	003029	003461	003891	432
41	1	4321	4751	5181	5609	6038	6466	6894	7321	7748	8174	428
83	2	8600	9026	9451	9876	010300	010724	011147	011570	011993	012415	424
124	3	012837	013259	013680	014100	4521	4940	5360	5779	6197	6616	420
166	4	7033	7451	7868	8284	8700	9116	9532	9947	020361	020775	416
207	5	021189	021603	022016	022428	022841	023252	023664	024075	4486	4896	412
248	6	5306	5715	6125	6533	6942	7350	7757	8164	8571	8978	408
290	7	9384	9789	030195	030600	031004	031408	031812	032216	032619	033021	404
331	8	033424	033826	4227	4628	5029	5430	5830	6230	6629	7028	400
373	9	7426	7825	8228	8620	9017	9414	9811	040207	040602	040998	397
	110	041393	041787	042182	042576	042969	043362	043755	044148	044540	044932	393
38	1	5323	5714	6105	6495	6885	7275	7664	8053	8442	8830	390
76	2	9218	9606	9993	050380	050766	051153	051538	051924	052309	052694	386
113	3	053078	053463	053846	4230	4613	4996	5378	5760	6142	6524	383
151	4	6905	7286	7666	8046	8426	8805	9185	9563	9942	060320	379
189	5	060698	061075	061452	061829	062206	062582	062958	063333	063709	4083	376
227	6	4458	4832	5206	5580	5953	6326	6699	7071	7443	7815	373
265	7	8186	8557	8928	9298	9668	070033	070407	070776	071145	071514	370
302	8	071882	072250	072617	072985	073352	3718	4085	4451	4816	5182	366
340	9	5547	5912	6276	6640	7004	7368	7731	8094	8457	8819	363
	120	079181	079543	079904	080266	080626	080987	081347	081707	082067	082426	360
35	1	082785	083144	083503	3861	4219	4576	4934	5291	5647	6004	357
70	2	6360	6716	7071	7426	7781	8136	8490	8845	9198	9552	355
104	3	9905	090258	090611	090963	091315	091667	092018	092370	092721	093071	352
139	4	093422	3772	4122	4471	4820	5169	5518	5866	6215	6562	349
174	5	6910	7257	7604	7951	8298	8644	8990	9335	9681	100026	346
209	6	100371	100715	101059	101403	101747	102091	102434	102777	103119	3462	343
244	7	3804	4146	4487	4828	5169	5510	5851	6191	6531	6871	341
278	8	7210	7549	7888	8227	8565	8903	9241	9579	9916	110253	338
313	9	110590	110926	111263	111599	111934	112270	112605	112940	113275	3609	335
	130	113943	114277	114611	114944	115278	115611	115943	116276	116608	116940	333
32	1	7271	7603	7934	8265	8595	8926	9256	9586	9915	120245	330
64	2	120574	120903	121231	121560	121888	122216	122544	122871	123198	3525	328
97	3	3852	4178	4504	4830	5156	5481	5806	6131	6456	6781	325
129	4	7105	7429	7753	8076	8399	8722	9045	9368	9690	130012	323
161	5	130334	130655	130977	131298	131619	131939	132260	132580	132900	3219	321
193	6	3539	3858	4177	4496	4814	5133	5451	5769	6085	6403	318
225	7	6721	7037	7354	7671	7987	8303	8618	8934	9249	9564	316
258	8	9879	140194	140508	140822	141136	141450	141763	142076	142389	142702	314
290	9	143015	3327	3639	3951	4263	4574	4885	5196	5507	5818	311
	140	146128	146438	146748	147058	147367	147676	147985	148294	148603	148911	309
30	1	9219	9527	9835	150142	150449	150756	151063	151370	151676	151982	307
60	2	152288	152594	152900	3205	3510	3815	4120	4424	4728	5032	305
90	3	5336	5640	5943	6246	6549	6852	7154	7457	7759	8061	303
120	4	8362	8664	8965	9266	9567	9868	160168	160469	160769	161068	301
150	5	161368	161667	161967	162266	162564	162863	3161	3460	3758	4055	299
180	6	4353	4650	4947	5244	5541	5838	6134	6430	6726	7022	297
210	7	7317	7613	7908	8203	8497	8792	9036	9380	9674	9968	295
240	8	170262	170555	170848	171141	171434	171726	172019	172311	172603	172895	293
270	9	3186	3478	3769	4060	4351	4641	4932	5222	5512	5802	291
	150	176091	176381	176670	176959	177248	177536	177825	178113	178401	178689	289
28	1	8977	9264	9552	9839	180126	180413	180699	180986	181272	181558	287
56	2	181844	182129	182415	182700	2985	3270	3555	3839	4123	4407	285
84	3	4691	4975	5259	5542	5825	6108	6391	6674	6956	7239	283
112	4	7521	7803	8084	8366	8647	8928	9209	9490	9771	190051	281
140	5	190332	190612	190892	191171	191451	191730	192010	192289	192567	2846	279
168	6	3125	3403	3681	3959	4237	4514	4792	5069	5346	5623	278
196	7	5900	6176	6453	6729	7005	7281	7556	7832	8107	8382	276
224	8	8657	8932	9206	9481	9755	200029	200303	200577	200850	201124	274
252	9	201397	201670	201943	202216	202488	2761	3033	3305	3577	3848	272

P P	N.	0	1	2	3	4	5	6	7	8	9	D.
	160	204120	204391	204663	204934	205204	205475	205746	206016	206286	206556	271
26	1	6826	7096	7365	7634	7904	8173	8441	8710	8979	9247	269
53	2	9515	9783	210051	210319	210586	210853	211121	211388	211654	211921	267
79	3	212188	212454	2720	2986	3252	3518	3783	4049	4314	4579	266
105	4	4844	5109	5373	5638	5902	6166	6430	6694	6957	7221	264
132	5	7484	7747	8010	8273	8536	8798	9060	9323	9585	9846	262
158	6	220108	220370	220631	220892	221153	221414	221675	221936	222196	222456	261
184	7	2716	2976	3236	3496	3755	4015	4274	4533	4792	5051	259
210	8	5309	5568	5826	6084	6342	6600	6858	7115	7372	7630	258
237	9	7887	8144	8400	8657	8913	9170	9426	9682	9938	230193	256
	170	230449	230704	230960	231215	231470	231724	231979	232234	232488	232742	254
25	1	2996	3250	3504	3757	4011	4264	4517	4770	5023	5276	253
50	2	5528	5781	6033	6285	6537	6789	7041	7292	7544	7795	252
74	3	8046	8297	8548	8799	9049	9299	9550	9800	240050	240300	250
99	4	240549	240799	241048	241297	241546	241795	242044	242293	2541	2790	249
124	5	3038	3286	3534	3782	4030	4277	4525	4772	5019	5266	248
149	6	5513	5759	6006	6252	6499	6745	6991	7237	7482	7728	246
174	7	7973	8219	8464	8709	8954	9198	9443	9687	9932	250176	245
198	8	250420	250664	250908	251151	251395	251638	251881	252125	252368	2610	243
223	9	2853	3096	3338	3580	3322	4064	4306	4548	4790	5031	242
	180	255273	255514	255755	255996	256237	256477	256718	256958	257198	257439	241
24	1	7679	7918	8158	8398	8637	8877	9116	9355	9594	9833	239
47	2	260071	260310	260548	260787	261025	261263	261501	261739	261976	262214	238
71	3	2451	2688	2925	3162	3399	3636	3373	4109	4346	4582	237
94	4	4818	5054	5290	5525	5761	5996	6232	6467	6702	6937	235
118	5	7172	7406	7641	7875	8110	8344	8578	8812	9046	9279	234
141	6	9513	9746	9980	270213	270446	270679	270912	271144	271377	271609	233
165	7	271842	272074	272306	2538	2770	3001	3233	3464	3696	3927	232
188	8	4158	4389	4620	4850	5081	5311	5542	5772	6002	6232	230
212	9	6462	6692	6921	7151	7380	7609	7838	8067	8296	8525	229
	190	278754	278982	279211	279439	279667	279895	280123	280351	280578	280806	228
22	1	281033	281261	281488	281715	281942	282169	2396	2622	2849	3075	227
45	2	3301	3527	3753	3979	4205	4431	4656	4882	5107	5332	226
67	3	5557	5782	6007	6232	6456	6681	6905	7130	7354	7578	225
89	4	7802	8026	8249	8473	8696	8920	9143	9366	9589	9812	223
112	5	290035	290257	290480	290702	290925	291147	291369	291591	291813	292034	222
134	6	2256	2478	2699	2920	3141	3363	3584	3804	4025	4246	221
156	7	4466	4687	4907	5127	5347	5567	5787	6007	6226	6446	220
178	8	6665	6884	7104	7323	7542	7761	7979	8198	8416	8635	219
201	9	8853	9071	9289	9507	9725	9943	300161	300378	300595	300813	218
	200	301030	301247	301464	301681	301898	302114	302331	302547	302764	302980	217
21	1	3196	3412	3628	3844	4059	4275	4491	4706	4921	5136	216
42	2	5351	5566	5781	5996	6211	6425	6639	6854	7068	7282	215
64	3	7496	7710	7924	8137	8351	8564	8778	8991	9204	9417	213
85	4	9630	9843	310056	310268	310481	310693	310906	311118	311330	311542	212
106	5	311754	311966	2177	2389	2600	2812	3023	3234	3445	3656	211
127	6	3867	4078	4289	4499	4710	4920	5130	5340	5551	5760	210
148	7	5970	6180	6390	6599	6809	7018	7227	7436	7646	7854	209
170	8	8063	8272	8481	8689	8898	9106	9314	9522	9730	9938	208
191	9	320146	320354	320562	320769	320977	321184	321391	321598	321805	322012	207
	210	322219	322426	322633	322839	323046	323252	323458	323665	323871	324077	206
20	1	4282	4488	4694	4899	5105	5310	5516	5721	5926	6131	205
40	2	6336	6541	6745	6950	7155	7359	7563	7767	7972	8176	204
61	3	8330	8583	8787	8991	9194	9398	9601	9805	330008	330211	203
81	4	330414	330617	330819	331022	331225	331427	331630	331832	2034	2236	202
101	5	2438	2640	2842	3044	3246	3447	3649	3850	4051	4253	202
121	6	4454	4655	4856	5057	5257	5458	5658	5859	6059	6260	201
141	7	6460	6660	6860	7060	7260	7459	7659	7858	8058	8257	200
162	8	8456	8656	8855	9054	9253	9451	9650	9849	340047	340246	199
182	9	340444	340642	340841	341039	341237	341435	341632	341830	2028	2225	198

PP	N.	0	1	2	3	4	5	6	7	8	9	D.
	220	342423	342620	342817	343014	343212	343409	343606	343802	343999	344196	197
19	1	4392	4589	4785	4981	5178	5374	5570	5766	5962	6157	196
39	2	6353	6549	6744	6939	7135	7330	7525	7720	7915	8110	195
58	3	8305	8500	8694	8889	9083	9278	9472	9666	9860	350054	194
77	4	350248	350442	350636	350829	351023	351216	351410	351603	351796	1989	193
97	5	2183	2375	2568	2761	2954	3147	3339	3532	3724	3916	193
116	6	4108	4301	4493	4685	4876	5068	5260	5452	5643	5834	192
135	7	6026	6217	6408	6599	6790	6981	7172	7363	7554	7744	191
154	8	7935	8125	8316	8506	8696	8886	9076	9266	9456	9646	190
174	9	9835	360025	360215	360404	360593	360783	360972	361161	361350	361539	189
	230	361728	361917	362105	362294	362482	362671	362859	363048	363236	363424	188
19	1	3612	3800	3988	4176	4363	4551	4739	4926	5113	5301	188
37	2	5488	5675	5862	6049	6236	6423	6610	6796	6983	7169	187
56	3	7356	7542	7729	7915	8101	8287	8473	8659	8845	9030	186
74	4	9216	9401	9587	9772	9958	370143	370328	370513	370698	370883	185
93	5	371068	371253	371437	371622	371806	1991	2175	2360	2544	2728	184
111	6	2912	3096	3280	3464	3647	3831	4015	4198	4382	4565	184
130	7	4748	4932	5115	5298	5481	5664	5846	6029	6212	6394	183
148	8	6577	6759	6942	7124	7306	7488	7670	7852	8034	8216	182
167	9	8398	8580	8761	8943	9124	9306	9487	9668	9849	380030	181
	240	380211	380392	380573	380754	380934	381115	381296	381476	381656	381837	181
18	1	2017	2197	2377	2557	2737	2917	3097	3277	3456	3636	180
35	2	3815	3995	4174	4353	4533	4712	4891	5070	5249	5428	179
53	3	5606	5785	5964	6142	6321	6499	6677	6856	7034	7212	178
71	4	7390	7568	7746	7923	8101	8279	8456	8634	8811	8989	178
89	5	9166	9343	9520	9698	9875	390051	390228	390405	390582	390759	177
106	6	390935	391112	391288	391464	391641	1817	1993	2169	2345	2521	176
124	7	2697	2873	3048	3224	3400	3575	3751	3926	4101	4277	176
142	8	4452	4627	4802	4977	5152	5326	5501	5676	5850	6025	175
159	9	6199	6374	6548	6722	6896	7071	7245	7419	7592	7766	174
	250	397940	398114	398287	398461	398634	398808	398981	399154	399328	399501	173
17	1	9674	9847	400020	400192	400365	400538	400711	400883	401056	401228	173
34	2	401401	401573	1745	1917	2089	2261	2433	2605	2777	2949	172
51	3	3121	3292	3464	3635	3807	3978	4149	4320	4492	4663	171
68	4	4834	5005	5176	5346	5517	5688	5858	6029	6199	6370	171
85	5	6540	6710	6881	7051	7221	7391	7561	7731	7901	8070	170
102	6	8240	8410	8579	8749	8918	9087	9257	9426	9595	9764	169
119	7	9933	410102	410271	410440	410609	410777	410946	411114	411283	411451	169
136	8	411620	1788	1956	2124	2293	2461	2629	2796	2964	3132	168
153	9	3300	3467	3635	3803	3970	4137	4305	4472	4639	4806	167
	260	414973	415140	415307	415474	415641	415808	415974	416141	416308	416474	167
16	1	6641	6807	6973	7139	7306	7472	7638	7804	7970	8135	166
33	2	8301	8467	8633	8798	8964	9129	9295	9460	9625	9791	165
49	3	9956	420121	420286	420451	420616	420781	420945	421110	421275	421439	165
66	4	421604	1768	1933	2097	2261	2426	2590	2754	2918	3082	164
82	5	3246	3410	3574	3737	3901	4065	4228	4392	4555	4718	164
98	6	4882	5045	5208	5371	5534	5697	5860	6023	6186	6349	163
115	7	6511	6674	6836	6999	7161	7324	7486	7648	7811	7973	162
131	8	8135	8297	8459	8621	8783	8944	9106	9268	9429	9591	162
148	9	9752	9914	430075	430236	430398	430559	430720	430881	431042	431203	161
	270	431364	431525	431685	431846	432007	432167	432328	432488	432649	432809	161
16	1	2969	3130	3290	3450	3610	3770	3930	4090	4249	4409	160
32	2	4569	4729	4888	5048	5207	5367	5526	5685	5844	6004	159
47	3	6163	6322	6481	6640	6799	6957	7116	7275	7433	7592	159
63	4	7751	7909	8067	8226	8384	8542	8701	8859	9017	9175	158
79	5	9333	9491	9648	9806	9964	440122	440279	440437	440594	440752	158
95	6	440909	441066	441224	441381	441538	1695	1852	2009	2166	2323	157
111	7	2480	2637	2793	2950	3106	3263	3419	3576	3732	3889	157
126	8	4045	4201	4357	4513	4669	4825	4981	5137	5293	5449	156
142	9	5604	5760	5915	6071	6226	6382	6537	6692	6848	7003	155

P P	N.	0	1	2	3	4	5	6	7	8	9	D.
	280	447158	447313	447468	447623	447778	447933	448088	448242	448397	448552	155
15	1	8706	8861	9015	9170	9324	9478	9633	9787	9941	450095	154
31	2	450249	450403	450557	450711	450865	451018	451172	451326	451479	1633	154
46	3	1786	1940	2093	2247	2400	2553	2706	2859	3012	3165	153
61	4	3318	3471	3624	3777	3930	4082	4235	4387	4540	4692	153
77	5	4845	4997	5150	5302	5454	5606	5758	5910	6062	6214	152
92	6	6366	6518	6670	6821	6973	7125	7276	7428	7579	7731	152
107	7	7882	8033	8184	8336	8487	8638	8789	8940	9091	9242	151
122	8	9392	9543	9694	9845	9995	460146	460296	460447	460597	460748	151
138	9	460898	461048	461198	461348	461499	1649	1799	1948	2098	2248	150
	290	462398	462548	462697	462847	462997	463146	463296	463445	463594	463744	150
15	1	3893	4042	4191	4340	4490	4639	4788	4936	5085	5234	149
29	2	5383	5532	5680	5829	5977	6126	6274	6423	6571	6719	149
44	3	6868	7016	7164	7312	7460	7608	7756	7904	8052	8200	148
59	4	8347	8495	8643	8790	8938	9085	9233	9380	9527	9675	148
74	5	9822	9969	470116	470263	470410	470557	470704	470851	470998	471145	147
88	6	471292	471438	1585	1732	1878	2025	2171	2318	2464	2610	146
103	7	2756	2902	3049	3195	3341	3487	3633	3779	3925	4071	146
118	8	4216	4362	4508	4653	4799	4944	5090	5235	5381	5526	146
132	9	5671	5816	5962	6107	6252	6397	6542	6687	6832	6976	145
	300	477121	477266	477411	477555	477700	477844	477989	478133	478278	478422	145
14	1	8566	8711	8855	8999	9143	9287	9431	9575	9719	9863	144
29	2	480007	480151	480294	480438	480582	480725	480869	481012	481156	481299	144
43	3	1443	1586	1729	1872	2016	2159	2302	2445	2588	2731	143
57	4	2874	3016	3159	3302	3445	3587	3730	3872	4015	4157	143
72	5	4300	4442	4585	4727	4869	5011	5153	5295	5437	5579	142
86	6	5721	5863	6005	6147	6289	6430	6572	6714	6855	6997	142
100	7	7138	7280	7421	7563	7704	7845	7986	8127	8269	8410	141
114	8	8551	8692	8833	8974	9114	9255	9396	9537	9677	9818	141
129	9	9958	490099	490239	490380	490520	490661	490801	490941	491081	491222	140
	310	491362	491502	491642	491782	491922	492062	492201	492341	492481	492621	140
14	1	2760	2900	3040	3179	3319	3458	3597	3737	3876	4015	139
28	2	4155	4294	4433	4572	4711	4850	4989	5128	5267	5406	139
41	3	5544	5683	5822	5960	6099	6238	6376	6515	6653	6791	139
55	4	6930	7068	7206	7344	7483	7621	7759	7897	8035	8173	138
69	5	8311	8448	8586	8724	8862	8999	9137	9275	9412	9550	138
83	6	9687	9824	9962	500099	500236	500374	500511	500648	500785	500922	137
97	7	501059	501196	501333	1470	1607	1744	1880	2017	2154	2291	137
110	8	2427	2564	2700	2837	2973	3109	3246	3382	3518	3655	136
124	9	3791	3927	4063	4199	4335	4471	4607	4743	4878	5014	136
	320	505150	505286	505421	505557	505693	505828	505964	506099	506234	506370	136
13	1	6505	6640	6776	6911	7046	7181	7316	7451	7586	7721	135
27	2	7856	7991	8126	8260	8395	8530	8664	8799	8934	9068	135
40	3	9203	9337	9471	9606	9740	9874	510009	510143	510277	510411	134
54	4	510545	510679	510813	510947	511081	511215	1349	1482	1616	1750	134
67	5	1883	2017	2151	2284	2418	2551	2684	2818	2951	3084	133
80	6	3218	3351	3484	3617	3750	3883	4016	4149	4282	4415	133
94	7	4548	4681	4813	4946	5079	5211	5344	5476	5609	5741	133
107	8	5874	6006	6139	6271	6403	6535	6668	6800	6932	7064	132
121	9	7196	7328	7460	7592	7724	7855	7987	8119	8251	8382	132
	330	518514	518646	518777	518909	519040	519171	519303	519434	519566	519697	131
13	1	9828	9959	520090	520221	520353	520484	520615	520745	520876	521007	131
26	2	521138	521269	1400	1530	1661	1792	1922	2053	2183	2314	131
39	3	2444	2575	2705	2835	2966	3096	3226	3356	3486	3616	130
52	4	3746	3876	4006	4136	4266	4396	4526	4656	4785	4915	130
65	5	5045	5174	5304	5434	5563	5693	5822	5951	6081	6210	129
78	6	6339	6469	6598	6727	6856	6985	7114	7243	7372	7501	129
91	7	7630	7759	7888	8016	8145	8274	8402	8531	8660	8788	129
104	8	8917	9045	9174	9302	9430	9559	9687	9815	9943	530072	128
117	9	530200	530328	530456	530584	530712	530840	530968	531096	531223	1351	128

k

P P	N.	0	1	2	3	4	5	6	7	8	9	D.
	340	531479	531607	531734	531862	531990	532117	532245	532372	532500	532627	128
13	1	2754	2882	3009	3136	3264	3391	3518	3645	3772	3899	127
25	2	4026	4153	4280	4407	4534	4661	4787	4914	5041	5167	127
38	3	5294	5421	5547	5674	5800	5927	6053	6180	6306	6432	126
50	4	6558	6685.	6811	6937	7063	7189	7315	7441	7567	7693	126
63	5	7819	7945	8071	8197	8322	8448	8574	8699	8825	8951	126
76	6	9076	9202	9327	9452	9578	9703	9829	9954	540079	540204	125
88	7	540329	540455	540580	540705	540830	540955	541080	541205	1330	1454	125
101	8	1579	1704	1829	1953	2078	2203	2327	2452	2576	2701	125
113	9	2825	2950	3074	3199	3323	3447	3571	3696	3820	3944	124
	350	544068	544192	544316	544440	544564	544688	544812	544936	545060	545183	124
12	1	5307	5431	5555	5678	5802	5925	6049	6172	6296	6419	124
24	2	6543	6666	6789	6913	7036	7159	7282	7405	7529	7652	123
37	3	7775	7898	8021	8144	8267	8389	8512	8635	8758	8881	123
49	4	9003	9126	9249	9371	9494	9616	9739	9861	9984	550106	123
61	5	550228	550351	550473	550595	550717	550840	550962	551084	551206	1328	122
73	6	1450	1572	1694	1816	1938	2060	2181	2303	2425	2547	122
85	7	2668	2790	2911	3033	3155	3276	3398	3519	3640	3762	121
98	8	3883	4004	4126	4247	4368	4489	4610	4731	4852	4973	121
110	9	5094	5215	5336	5457	5578	5699	5820	5940	6061	6182	121
	360	556303	556423	556544	556664	556785	556905	557026	557146	557267	557387	120
12	1	7507	7627	7748	7868	7988	8108	8228	8349	8469	8589	120
24	2	8709	8829	8948	9068	9188	9308	9428	9548	9667	9787	120
36	3	9907	560026	560146	560265	560385	560504	560624	560743	560863	560982	119
48	4	561101	1221	1340	1459	1578	1698	1817	1936	2055	2174	119
60	5	2293	2412	2531	2650	2769	2887	3006	3125	3244	3362	119
71	6	3481	3600	3718	3837	3955	4074	4192	4311	4429	4548	119
83	7	4666	4784	4903	5021	5139	5257	5376	5494	5612	5730	118
95	8	5848	5966	6084	6202	6320	6437	6555	6673	6791	6909	118
107	9	7026	7144	7262	7379	7497	7614	7732	7849	7967	8084	118
	370	568202	568319	568436	568554	568671	568788	568905	569023	569140	569257	117
12	1	9374	9491	9608	9725	9842	9959	570076	570193	570309	570426	117
23	2	570543	570660	570776	570893	571010	571126	1243	1359	1476	1592	117
35	3	1709	1825	1942	2058	2174	2291	2407	2523	2639	2755	116
46	4	2872	2988	3104	3220	3336	3452	3568	3684	3800	3915	116
58	5	4031	4147	4263	4379	4494	4610	4726	4841	4957	5072	116
70	6	5188	5303	5419	5534	5650	5765	5880	5996	6111	6226	115
81	7	6341	6457	6572	6687	6802	6917	7032	7147	7262	7377	115
93	8	7492	7607	7722	7836	7951	8066	8181	8295	8410	8525	115
104	9	8639	8754	8868	8983	9097	9212	9326	9441	9555	9669	114
	380	579784	579898	580012	580126	580241	580355	580469	580583	580697	580811	114
11	1	580925	581039	1153	1267	1381	1495	1608	1722	1836	1950	114
23	2	2063	2177	2291	2404	2518	2631	2745	2858	2972	3085	114
34	3	3199	3312	3426	3539	3652	3765	3879	3992	4105	4218	113
45	4	4331	4444	4557	4670	4783	4896	5009	5122	5235	5348	113
57	5	5461	5574	5686	5799	5912	6024	6137	6250	6362	6475	113
68	6	6587	6700	6812	6925	7037	7149	7262	7374	7486	7599	112
79	7	7711	7823	7935	8047	8160	8272	8384	8496	8608	8720	112
90	8	8832	8944	9056	9167	9279	9391	9503	9615	9726	9838	112
102	9	9950	590061	590173	590284	590396	590507	590619	590730	590842	590953	112
	390	591065	591176	591287	591399	591510	591621	591732	591843	591955	592066	111
11	1	2177	2288	2399	2510	2621	2732	2843	2954	3064	3175	111
22	2	3286	3397	3508	3618	3729	3840	3950	4061	4171	4282	111
33	3	4393	4503	4614	4724	4834	4945	5055	5165	5276	5386	110
44	4	5496	5606	5717	5827	5937	6047	6157	6267	6377	6487	110
55	5	6597	6707	6817	6927	7037	7146	7256	7366	7476	7586	110
66	6	7695	7805	7914	8024	8134	8243	8353	8462	8572	8681	110
77	7	8791	8900	9009	9119	9228	9337	9446	9556	9665	9774	109
88	8	9883	9992	600101	600210	600319	600428	600537	600646	600755	600864	109
99	9	600973	601082	1191	1299	1408	1517	1625	1734	1843	1951	109

P P	N.	0	1	2	3	4	5	6	7	8	9	D.
	400	602060	602169	602277	602386	602494	602603	602711	602819	602928	603036	108
11	1	3144	3253	3361	3469	3577	3686	3794	3902	4010	4118	108
21	2	4226	4334	4442	4550	4658	4766	4874	4982	5089	5197	108
32	3	5305	5413	5521	5628	5736	5844	5951	6059	6166	6274	108
43	4	6381	6489	6596	6704	6811	6919	7026	7133	7241	7348	107
54	5	7455	7562	7669	7777	7884	7991	8098	8205	8312	8419	107
64	6	8526	8633	8740	8847	8954	9061	9167	9274	9381	9488	107
75	7	9594	9701	9808	9914	610021	610128	610234	610341	610447	610554	107
86	8	610660	610767	610873	610979	1086	1192	1298	1405	1511	1617	106
96	9	1723	1829	1936	2042	2148	2254	2360	2466	2572	2678	106
	410	612784	612890	612996	613102	613207	613313	613419	613525	613630	613736	106
11	1	3842	3947	4053	4159	4264	4370	4475	4581	4686	4792	106
21	2	4897	5003	5108	5213	5319	5424	5529	5634	5740	5845	105
32	3	5950	6055	6160	6265	6370	6476	6581	6686	6790	6895	105
42	4	7000	7105	7210	7315	7420	7525	7629	7734	7839	7943	105
53	5	8048	8153	8257	8362	8466	8571	8676	8780	8884	8989	105
63	6	9093	9198	9302	9406	9511	9615	9719	9824	9928	620032	104
74	7	620136	620240	620344	620448	620552	620656	620760	620864	620968	1072	104
84	8	1176	1280	1384	1488	1592	1695	1799	1903	2007	2110	104
95	9	2214	2318	2421	2525	2628	2732	2835	2939	3043	3146	104
	420	623249	623353	623456	623559	623663	623766	623869	623973	624076	624179	103
10	1	4282	4385	4488	4591	4695	4798	4901	5004	5107	5210	103
20	2	5312	5415	5518	5621	5724	5827	5929	6032	6135	6238	103
31	3	6340	6443	6546	6648	6751	6853	6956	7058	7161	7263	103
41	4	7366	7468	7571	7673	7775	7878	7980	8082	8185	8287	102
51	5	8389	8491	8593	8695	8797	8900	9002	9104	9206	9308	102
61	6	9410	9512	9613	9715	9817	9919	630021	630123	630224	630326	102
71	7	630428	630530	630631	630733	630835	630936	1038	1139	1241	1342	102
82	8	1444	1545	1647	1748	1849	1951	2052	2153	2255	2356	101
92	9	2457	2559	2660	2761	2862	2963	3064	3165	3266	3367	101
	430	633468	633569	633670	633771	633872	633973	634074	634175	634276	634376	101
10	1	4477	4578	4679	4779	4880	4981	5081	5182	5283	5383	101
20	2	5484	5584	5685	5785	5886	5986	6087	6187	6287	6388	100
30	3	6488	6588	6688	6789	6889	6989	7089	7189	7290	7390	100
40	4	7490	7590	7690	7790	7890	7990	8090	8190	8290	8389	100
50	5	8489	8589	8689	8789	8888	8988	9088	9188	9287	9387	100
60	6	9486	9586	9686	9785	9885	9984	640083	640183	640283	640382	99
70	7	640481	640581	640680	640779	640879	640978	1077	1177	1276	1375	99
80	8	1474	1573	1672	1771	1871	1970	2069	2168	2267	2366	99
90	9	2465	2563	2662	2761	2860	2959	3058	3156	3255	3354	99
	440	643453	643551	643650	643749	643847	643946	644044	644143	644242	644340	98
10	1	4439	4537	4636	4734	4832	4931	5029	5127	5226	5324	98
20	2	5422	5521	5619	5717	5815	5913	6011	6110	6208	6306	98
29	3	6404	6502	6600	6698	6796	6894	6992	7089	7187	7285	98
39	4	7383	7481	7579	7676	7774	7872	7969	8067	8165	8262	98
49	5	8360	8458	8555	8653	8750	8848	8945	9043	9140	9237	97
59	6	9335	9432	9530	9627	9724	9821	9919	650016	650113	650210	97
69	7	650308	650405	650502	650599	650696	650793	650890	0987	1084	1181	97
78	8	1278	1375	1472	1569	1666	1762	1859	1956	2053	2150	97
88	9	2246	2343	2440	2536	2633	2730	2826	2923	3019	3116	97
	450	653213	653309	653405	653502	653598	653695	653791	653888	653984	654080	96
10	1	4177	4273	4369	4465	4562	4658	4754	4850	4946	5042	96
19	2	5138	5235	5331	5427	5523	5619	5715	5810	5906	6002	96
29	3	6098	6194	6290	6386	6482	6577	6673	6769	6864	6960	96
38	4	7056	7152	7247	7343	7438	7534	7629	7725	7820	7916	96
48	5	8011	8107	8202	8298	8393	8488	8584	8679	8774	8870	95
58	6	8965	9060	9155	9250	9346	9441	9536	9631	9726	9821	95
67	7	9916	660011	660106	660201	660296	660391	660486	660581	660676	660771	95
77	8	660865	0960	1055	1150	1245	1339	1434	1529	1623	1718	95
86	9	1813	1907	2002	2096	2191	2286	2380	2475	2569	2663	95

P P	N.	0	1	2	3	4	5	6	7	8	9	D.
	460	662758	662852	662947	663041	663135	663230	663324	663418	663512	663607	94
9	1	3701	3795	3889	3983	4078	4172	4266	4360	4454	4548	94
19	2	4642	4736	4830	4924	5018	5112	5206	5299	5393	5487	94
28	3	5581	5675	5769	5862	5956	6050	6143	6237	6331	6424	94
38	4	6518	6612	6705	6799	6892	6986	7079	7173	7266	7360	94
47	5	7453	7546	7640	7733	7826	7920	8013	8106	8199	8293	93
56	6	8386	8479	8572	8665	8759	8852	8945	9038	9131	9224	93
66	7	9317	9410	9503	9596	9689	9782	9875	9967	670060	670153	93
75	8	670246	670339	670431	670524	670617	670710	670802	670895	0988	1080	93
85	9	1173	1265	1358	1451	1543	1636	1728	1821	1913	2005	93
	470	672098	672190	672283	672375	672467	672560	672652	672744	672836	672929	92
9	1	3021	3113	3205	3297	3390	3482	3574	3666	3758	3850	92
18	2	3942	4034	4126	4218	4310	4402	4494	4586	4677	4769	92
28	3	4861	4953	5045	5137	5228	5320	5412	5503	5595	5687	92
37	4	5778	5870	5962	6053	6145	6236	6328	6419	6511	6602	92
46	5	6694	6785	6876	6968	7059	7151	7242	7333	7424	7516	91
55	6	7607	7698	7789	7881	7972	8063	8154	8245	8336	8427	91
64	7	8518	8609	8700	8791	8882	8973	9064	9155	9246	9337	91
74	8	9428	9519	9610	9700	9791	9882	9973	680063	680154	680245	91
83	9	680336	680426	680517	680607	680698	680789	680879	0970	1060	1151	91
	480	681241	681332	681422	681513	681603	681693	681784	681874	681964	682055	90
9	1	2145	2235	2326	2416	2506	2596	2686	2777	2867	2957	90
18	2	3047	3137	3227	3317	3407	3497	3587	3677	3767	3857	90
27	3	3947	4037	4127	4217	4307	4396	4486	4576	4666	4756	90
36	4	4845	4935	5025	5114	5204	5294	5383	5473	5563	5652	90
45	5	5742	5831	5921	6010	6100	6189	6279	6368	6458	6547	89
54	6	6636	6726	6815	6904	6994	7083	7172	7261	7351	7440	89
63	7	7529	7618	7707	7796	7886	7975	8064	8153	8242	8331	89
72	8	8420	8509	8598	8687	8776	8865	8953	9042	9131	9220	89
81	9	9309	9398	9486	9575	9664	9753	9841	9930	690019	690107	89
	490	690196	690285	690373	690462	690550	690639	690728	690816	690905	690993	89
9	1	1081	1170	1258	1347	1435	1524	1612	1700	1789	1877	88
18	2	1965	2053	2142	2230	2318	2406	2494	2583	2671	2759	88
26	3	2847	2935	3023	3111	3199	3287	3375	3463	3551	3639	88
35	4	3727	3815	3903	3991	4078	4166	4254	4342	4430	4517	88
44	5	4605	4693	4781	4868	4956	5044	5131	5219	5307	5394	88
53	6	5482	5569	5657	5744	5832	5919	6007	6094	6182	6269	87
62	7	6356	6444	6531	6618	6706	6793	6880	6968	7055	7142	87
70	8	7229	7317	7404	7491	7578	7665	7752	7839	7926	8014	87
79	9	8101	8188	8275	8362	8449	8535	8622	8709	8796	8883	87
	500	698970	699057	699144	699231	699317	699404	699491	699578	699664	699751	87
9	1	9838	9924	700011	700098	700184	700271	700358	700444	700531	700617	87
17	2	700704	700790	0877	0963	1050	1136	1222	1309	1395	1482	86
26	3	1568	1654	1741	1827	1913	1999	2086	2172	2258	2344	86
34	4	2431	2517	2603	2689	2775	2861	2947	3033	3119	3205	86
43	5	3291	3377	3463	3549	3635	3721	3807	3893	3979	4065	86
52	6	4151	4236	4322	4408	4494	4579	4665	4751	4837	4922	86
60	7	5008	5094	5179	5265	5350	5436	5522	5607	5693	5778	86
69	8	5864	5949	6035	6120	6206	6291	6376	6462	6547	6632	85
77	9	6718	6803	6888	6974	7059	7144	7229	7315	7400	7485	85
	510	707570	707655	707740	707826	707911	707996	708081	708166	708251	708336	85
8	1	8421	8506	8591	8676	8761	8846	8931	9015	9100	9185	85
17	2	9270	9355	9440	9524	9609	9694	9779	9863	9948	710033	85
25	3	710117	710202	710287	710371	710456	710540	710625	710710	710794	0879	85
34	4	0963	1048	1132	1217	1301	1385	1470	1554	1639	1723	84
42	5	1807	1892	1976	2060	2144	2229	2313	2397	2481	2566	84
50	6	2650	2734	2818	2902	2986	3070	3154	3238	3323	3407	84
59	7	3491	3575	3659	3742	3826	3910	3994	4078	4162	4246	84
67	8	4330	4414	4497	4581	4665	4749	4833	4916	5000	5084	84
76	9	5167	5251	5335	5418	5502	5586	5669	5753	5836	5920	84

PP	N.	0	1	2	3	4	5	6	7	8	9	D.
	520	716003	716087	716170	716254	716337	716421	716504	716588	716671	716754	83
8	1	6838	6921	7004	7088	7171	7254	7338	7421	7504	7587	83
17	2	7671	7754	7837	7920	8003	8086	8169	8253	8336	8419	83
25	3	8502	8585	8668	8751	8834	8917	9000	9083	9165	9248	83
33	4	9331	9414	9497	9580	9663	9745	9828	9911	9994	720077	83
41	5	720159	720242	720325	720407	720490	720573	720655	720738	720821	0903	83
50	6	0986	1068	1151	1233	1316	1398	1481	1563	1646	1728	82
58	7	1811	1893	1975	2058	2140	2222	2305	2387	2469	2552	82
66	8	2634	2716	2798	2881	2963	3045	3127	3209	3291	3374	82
75	9	3456	3538	3620	3702	3784	3866	3948	4030	4112	4194	82
	530	724276	724358	724440	724522	724604	724685	724767	724849	724931	725013	82
8	1	5095	5176	5258	5340	5422	5503	5585	5667	5748	5830	82
16	2	5912	5993	6075	6156	6238	6320	6401	6483	6564	6646	82
24	3	6727	6809	6890	6972	7053	7134	7216	7297	7379	7460	81
32	4	7541	7623	7704	7785	7866	7948	8029	8110	8191	8273	81
41	5	8354	8435	8516	8597	8678	8759	8841	8922	9003	9084	81
49	6	9165	9246	9327	9408	9489	9570	9651	9732	9813	9893	81
57	7	9974	730055	730136	730217	730298	730378	730459	730540	730621	730702	81
65	8	730782	0863	0944	1024	1105	1186	1266	1347	1428	1508	81
73	9	1589	1669	1750	1830	1911	1991	2072	2152	2233	2313	81
	540	732394	732474	732555	732635	732715	732796	732876	732956	733037	733117	80
8	1	3197	3278	3358	3438	3518	3598	3679	3759	3839	3919	80
16	2	3999	4079	4160	4240	4320	4400	4480	4560	4640	4720	80
24	3	4800	4880	4960	5040	5120	5200	5279	5359	5439	5519	80
32	4	5599	5679	5759	5838	5918	5998	6078	6157	6237	6317	80
40	5	6397	6476	6556	6635	6715	6795	6874	6954	7034	7113	80
48	6	7193	7272	7352	7431	7511	7590	7670	7749	7829	7908	79
56	7	7987	8067	8146	8225	8305	8384	8463	8543	8622	8701	79
64	8	8781	8860	8939	9018	9097	9177	9256	9335	9414	9493	79
72	9	9572	9651	9731	9810	9889	9968	740047	740126	740205	740284	79
	550	740363	740442	740521	740600	740678	740757	740836	740915	740994	741073	79
8	1	1152	1230	1309	1388	1467	1546	1624	1703	1782	1860	79
16	2	1939	2018	2096	2175	2254	2332	2411	2489	2568	2647	79
23	3	2725	2804	2882	2961	3039	3118	3196	3275	3353	3431	78
31	4	3510	3588	3667	3745	3823	3902	3980	4058	4136	4215	78
39	5	4293	4371	4449	4528	4606	4684	4762	4840	4919	4997	78
47	6	5075	5153	5231	5309	5387	5465	5543	5621	5699	5777	78
55	7	5855	5933	6011	6089	6167	6245	6323	6401	6479	6556	78
62	8	6634	6712	6790	6868	6945	7023	7101	7179	7256	7334	78
70	9	7412	7489	7567	7645	7722	7800	7878	7955	8033	8110	78
	560	748188	748266	748343	748421	748498	748576	748653	748731	748808	748885	77
8	1	8963	9040	9118	9195	9272	9350	9427	9504	9582	9659	77
15	2	9736	9814	9891	9968	750045	750123	750200	750277	750354	750431	77
23	3	750508	750586	750663	750740	0817	0894	0971	1048	1125	1202	77
31	4	1279	1356	1433	1510	1587	1664	1741	1818	1895	1972	77
39	5	2048	2125	2202	2279	2356	2433	2509	2586	2663	2740	77
46	6	2816	2893	2970	3047	3123	3200	3277	3353	3430	3506	77
54	7	3583	3660	3736	3813	3889	3966	4042	4119	4195	4272	77
62	8	4348	4425	4501	4578	4654	4730	4807	4883	4960	5036	76
69	9	5112	5189	5265	5341	5417	5494	5570	5646	5722	5799	76
	570	755875	755951	756027	756103	756180	756256	756332	756408	756484	756560	76
8	1	6636	6712	6788	6864	6940	7016	7092	7168	7244	7320	76
15	2	7396	7472	7548	7624	7700	7775	7851	7927	8003	8079	76
23	3	8155	8230	8306	8382	8458	8533	8609	8685	8761	8836	76
30	4	8912	8988	9063	9139	9214	9290	9366	9441	9517	9592	76
38	5	9668	9743	9819	9894	9970	760045	760121	760196	760272	760347	75
46	6	760422	760498	760573	760649	760724	0799	0875	0950	1025	1101	75
53	7	1176	1251	1326	1402	1477	1552	1627	1702	1778	1853	75
61	8	1928	2003	2078	2153	2228	2303	2378	2453	2529	2604	75
68	9	2679	2754	2829	2904	2978	3053	3128	3203	3278	3353	75

PP	N.	0	1	2	3	4.	5	6	7	8	9	D.
	580	763428	763503	763578	763653	763727	763802	763877	763952	764027	764101	75
7	1	4176	4251	4326	4400	4475	4550	4624	4699	4774	4848	75
15	2	4923	4998	5072	5147	5221	5296	5370	5445	5520	5594	75
22	3	5669	5743	5818	5892	5966	6041	6115	6190	6264	6338	74
30	4	6413	6487	6562	6636	6710	6785	6859	6933	7007	7082	74
37	5	7156	7230	7304	7379	7453	7527	7601	7675	7749	7823	74
44	6	7898	7972	8046	8120	8194	8268	8342	8416	8490	8564	74
52	7	8638	8712	8786	8860	8934	9008	9082	9156	9230	9303	74
59	8	9377	9451	9525	9599	9673	9746	9820	9894	9968	770042	74
67	9	770115	770189	770263	770336	770410	770484	770557	770631	770705	0778	74
	590	770852	770926	770999	771073	771146	771220	771293	771367	771440	771514	74
7	1	1587	1661	1734	1808	1881	1955	2028	2102	2175	2248	73
15	2	2322	2395	2468	2542	2615	2698	2762	2835	2908	2981	73
22	3	3055	3128	3201	3274	3348	3421	3494	3567	3640	3713	73
29	4	3786	3860	3933	4006	4079	4152	4225	4298	4371	4444	73
37	5	4517	4590	4663	4736	4809	4882	4955	5028	5100	5173	73
44	6	5246	5319	5392	5465	5538	5610	5683	5756	5829	5902	73
51	7	5974	6047	6120	6193	6265	6338	6411	6483	6556	6629	73
58	8	6701	6774	6846	6919	6992	7064	7137	7209	7282	7354	73
66	9	7427	7499	7572	7644	7717	7789	7862	7934	8006	8079	72
	600	778151	778224	778296	778368	778441	778513	778585	778658	778730	778802	72
7	1	8874	8947	9019	9091	9163	9236	9308	9380	9452	9524	72
14	2	9596	9669	9741	9813	9885	9957	780029	780101	780173	780245	72
22	3	780317	780389	780461	780533	780605	780677	0749	0821	0893	0965	72
29	4	1037	1109	1181	1253	1324	1396	1468	1540	1612	1684	72
36	- 5	1755	1827	1899	1971	2042	2114	2186	2258	2329	2401	72
43	6	2473	2544	2616	2688	2759	2831	2902	2974	3046	3117	72
50	7	3189	3260	3332	3403	3475	3546	3618	3689	3761	3832	71
58	8	3904	3975	4046	4118	4189	4261	4332	4403	4475	4546	71
65	9	4617	4689	4760	4831	4902	4974	5045	5116	5187	5259	71
	610	785330	785401	785472	785543	785615	785686	785757	785828	785899	785970	71
7	1	6041	6112	6183	6254	6325	6396	6467	6538	6609	6680	71
14	2	6751	6822	6893	6964	7035	7106	7177	7248	7319	7390	71
21	3	7460	7531	7602	7673	7744	7815	7885	7956	8027	8098	71
28	4	8168	8239	8310	8381	8451	8522	8593	8663	8734	8804	71
36	5	8875	8946	9016	9087	9157	9228	9299	9369	9440	9510	71
43	6	9581	9651	9722	9792	9863	9933	790004	790074	790144	790215	70
50	7	790285	790356	790426	790496	790567	790637	0707	0778	0848	0918	70
57	8	0988	1059	1129	1199	1269	1340	1410	1480	1550	1620	70
64	9	1691	1761	1831	1901	1971	2041	2111	2181	2252	2322	70
	620	792392	792462	792532	792602	792672	792742	792812	792882	792952	793022	70
7	1	3092	3162	3231	3301	3371	3441	3511	3581	3651	3721	70
14	2	3790	3860	3930	4000	4070	4139	4209	4279	4349	4418	70
21	3	4488	4558	4627	4697	4767	4836	4906	4976	5045	5115	70
28	4	5185	5254	5324	5393	5463	5532	5602	5672	5741	5811	70
35	5	5880	5949	6019	6088	6158	6227	6297	6366	6436	6505	69
42	6	6574	6644	6713	6782	6852	6921	6990	7060	7129	7198	69
49	7	7268	7337	7406	7475	7545	7614	7683	7752	7821	7890	69
56	8	7960	8029	8098	8167	8236	8305	8374	8443	8513	8582	69
63	9	8651	8720	8789	8858	8927	8996	9065	9134	9203	9272	69
	630	799341	799409	799478	799547	799616	799685	799754	799823	799892	799961	69
7	1	800029	800098	800167	800236	800305	800373	800442	800511	800580	800648	69
14	2	0717	0786	0854	0923	0992	1061	1129	1198	1266	1335	69
21	3	1404	1472	1541	1609	1678	1747	1815	1884	1952	2021	69
28	4	2089	2158	2226	2295	2363	2432	2500	2568	2637	2705	69
35	5	2774	2842	2910	2979	3047	3116	3184	3252	3321	3389	68
41	6	3457	3525	3594	3662	3730	3798	3867	3935	4003	4071	68
48	7	4139	4208	4276	4344	4412	4480	4548	4616	4685	4753	68
55	8	4821	4889	4957	5025	5093	5161	5229	5297	5365	5433	68
62	9	5501	5569	5637	5705	5773	5841	5908	5976	6044	6112	68

P P	N.	0	1	2	3	4	5	6	7	8	9	D.
	640	806180	806248	806316	806384	806451	806519	806587	806655	806723	806790	68
7	1	6858	6926	6994	7061	7129	7197	7264	7332	7400	7467	68
13	2	7535	7603	7670	7738	7806	7873	7941	8008	8076	8143	68
20	3	8211	8279	8346	8414	8481	8549	8616	8684	8751	8818	67
27	4	8886	8953	9021	9088	9156	9223	9290	9358	9425	9492	67
34	5	9560	9627	9694	9762	9829	9896	9964	810031	810098	810165	67
40	6	810233	810300	810367	810434	810501	810569	810636	0703	0770	0837	67
47	7	0904	0971	1039	1106	1173	1240	1307	1374	1441	1508	67
54	8	1575	1642	1709	1776	1843	1910	1977	2044	2111	2178	67
60	9	2245	2312	2379	2445	2512	2579	2646	2713	2780	2847	67
	650	812913	812980	813047	813114	813181	813247	813314	813381	813448	813514	67
7	1	3581	3648	3714	3781	3848	3914	3981	4048	4114	4181	67
13	2	4248	4314	4381	4447	4514	4581	4647	4714	4780	4847	67
20	3	4913	4980	5046	5113	5179	5246	5312	5378	5445	5511	66
26	4	5578	5644	5711	5777	5843	5910	5976	6042	6109	6175	66
33	5	6241	6308	6374	6440	6506	6573	6639	6705	6771	6838	66
40	6	6904	6970	7036	7102	7169	7235	7301	7367	7433	7499	66
46	7	7565	7631	7698	7764	7830	7896	7962	8028	8094	8160	66
53	8	8226	8292	8358	8424	8490	8556	8622	8688	8754	8820	66
59	9	8885	8951	9017	9083	9149	9215	9281	9346	9412	9478	66
	660	819544	819610	819676	819741	819807	819873	819939	820004	820070	820136	66
7	1	820201	820267	820333	820399	820464	820530	820595	0661	0727	0792	66
13	2	0858	0924	0989	1055	1120	1186	1251	1317	1382	1448	66
20	3	1514	1579	1645	1710	1775	1841	1906	1972	2037	2103	65
26	4	2168	2233	2299	2364	2430	2495	2560	2626	2691	2756	65
33	5	2822	2887	2952	3018	3083	3148	3213	3279	3344	3409	65
39	6	3474	3539	3605	3670	3735	3800	3865	3930	3996	4061	65
46	7	4126	4191	4256	4321	4386	4451	4516	4581	4646	4711	65
52	8	4776	4841	4906	4971	5036	5101	5166	5231	5296	5361	65
59	9	5426	5491	5556	5621	5686	5751	5815	5880	5945	6010	65
	670	826075	826140	826204	826269	826334	826399	826464	826528	826593	826658	65
6	1	6723	6787	6852	6917	6981	7046	7111	7175	7240	7305	65
13	2	7369	7434	7499	7563	7628	7692	7757	7821	7886	7951	65
19	3	8015	8080	8144	8209	8273	8338	8402	8467	8531	8595	64
26	4	8660	8724	8789	8853	8918	8982	9046	9111	9175	9239	64
32	5	9304	9368	9432	9497	9561	9625	9690	9754	9818	9882	64
38	6	9947	830011	830075	830139	830204	830268	830332	830396	830460	830525	64
45	7	830589	0653	0717	0781	0845	0909	0973	1037	1102	1166	64
51	8	1230	1294	1358	1422	1486	1550	1614	1678	1742	1806	64
58	9	1870	1934	1998	2062	2126	2189	2253	2317	2381	2445	64
	680	832509	832573	832637	832700	832764	832828	832892	832956	833020	833083	64
6	1	3147	3211	3275	3338	3402	3466	3530	3593	3657	3721	64
13	2	3784	3848	3912	3975	4039	4103	4166	4230	4294	4357	64
19	3	4421	4484	4548	4611	4675	4739	4802	4866	4929	4993	64
25	4	5056	5120	5183	5247	5310	5373	5437	5500	5564	5627	63
32	5	5691	5754	5817	5881	5944	6007	6071	6134	6197	6261	63
38	6	6324	6387	6451	6514	6577	6641	6704	6767	6830	6894	63
44	7	6957	7020	7083	7146	7210	7273	7336	7399	7462	7525	63
50	8	7588	7652	7715	7778	7841	7904	7967	8030	8093	8156	63
57	9	8219	8282	8345	8408	8471	8534	8597	8660	8723	8786	63
	690	838849	838912	838975	839038	839101	839164	839227	839289	839352	839415	63
6	1	9478	9541	9604	9667	9729	9792	9855	9918	9981	840043	63
13	2	840106	840169	840232	840294	840357	840420	840482	840545	840608	0671	63
19	3	0733	0796	0859	0921	0984	1046	1109	1172	1234	1297	63
25	4	1359	1422	1485	1547	1610	1672	1735	1797	1860	1922	63
32	5	1985	2047	2110	2172	2235	2297	2360	2422	2484	2547	62
38	6	2609	2672	2734	2796	2859	2921	2983	3046	3108	3170	62
44	7	3233	3295	3357	3420	3482	3544	3606	3669	3731	3793	62
50	8	3855	3918	3980	4042	4104	4166	4229	4291	4353	4415	62
57	9	4477	4539	4601	4664	4726	4788	4850	4912	4974	5036	62

PP	N.	0	1	2	3	4	5	6	7	8	9	D.
	700	845098	845160	845222	845284	845346	845408	845470	845532	845594	845656	62
6	1	5718	5780	5842	5904	5966	6028	6090	6151	6213	6275	62
12	2	6337	6399	6461	6523	6585	6646	6708	6770	6832	6894	62
19	3	6955	7017	7079	7141	7202	7264	7326	7388	7449	7511	62
25	4	7573	7634	7696	7758	7819	7881	7943	8004	8066	8128	62
31	5	8189	8251	8312	8374	8435	8497	8559	8620	8682	8743	62
37	6	8805	8866	8928	8989	9051	9112	9174	9235	9297	9358	6
43	7	9419	9481	9542	9604	9665	9726	9788	9849	9911	9972	61
50	8	850033	850095	850156	850217	850279	850340	850401	850462	850524	850585	61
56	9	0646	0707	0769	0830	0891	0952	1014	1075	1136	1197	61
	710	851258	851320	851381	851442	851503	851564	851625	851686	851747	851809	61
6	1	1870	1931	1992	2053	2114	2175	2236	2297	2358	2419	61
12	2	2480	2541	2602	2663	2724	2785	2846	2907	2968	3029	61
18	3	3090	3150	3211	3272	3333	3394	3455	3516	3577	3637	61
24	4	3698	3759	3820	3881	3941	4002	4063	4124	4185	4245	61
31	5	4306	4367	4428	4488	4549	4610	4670	4731	4792	4852	61
37	6	4913	4974	5034	5095	5156	5216	5277	5337	5398	5459	61
43	7	5519	5580	5640	5701	5761	5822	5882	5943	6003	6064	61
49	8	6124	6185	6245	6306	6366	6427	6487	6548	6608	6668	60
55	9	6729	6789	6850	6910	6970	7031	7091	7152	7212	7272	60
	720	857332	857393	857453	857513	857574	857634	857694	857755	857815	857875	60
6	1	7935	7995	8056	8116	8176	8236	8297	8357	8417	8477	60
12	2	8537	8597	8657	8718	8778	8838	8898	8958	9018	9078	60
18	3	9138	9198	9258	9318	9379	9439	9499	9559	9619	9679	60
24	4	9739	9799	9859	9918	9978	860038	860098	860158	860218	860278	60
30	5	860338	860398	860458	860518	860578	0637	0697	0757	0817	0877	60
36	6	0937	0996	1056	1116	1176	1236	1295	1355	1415	1475	60
42	7	1534	1594	1654	1714	1773	1833	1893	1952	2012	2072	60
48	8	2131	2191	2251	2310	2370	2430	2489	2549	2608	2668	60
54	9	2728	2787	2847	2906	2966	3025	3085	3144	3204	3263	60
	730	863323	863382	863442	863501	863561	863620	863680	863739	863799	863858	59
6	1	3917	3977	4036	4096	4155	4214	4274	4333	4392	4452	59
12	2	4511	4570	4630	4689	4748	4808	4867	4926	4985	5045	59
18	3	5104	5163	5222	5282	5341	5400	5459	5519	5578	5637	59
24	4	5696	5755	5814	5874	5933	5992	6051	6110	6169	6228	59
30	5	6287	6346	6405	6465	6524	6583	6642	6701	6760	6819	59
35	6	6878	6937	6996	7055	7114	7173	7232	7291	7350	7409	59
41	7	7467	7526	7585	7644	7703	7762	7821	7880	7939	7998	59
47	8	8056	8115	8174	8233	8292	8350	8409	8468	8527	8586	59
53	9	8644	8703	8762	8821	8879	8938	8997	9056	9114	9173	59
	740	869232	869290	869349	869408	869466	869525	869584	869642	869701	869760	59
6	1	9818	9877	9935	9994	870053	870111	870170	870228	870287	870345	59
12	2	870404	870462	870521	870579	0638	0696	0755	0813	0872	0930	58
17	3	0989	1047	1106	1164	1223	1281	1339	1398	1456	1515	58
23	4	1573	1631	1690	1748	1806	1865	1923	1981	2040	2098	58
29	5	2156	2215	2273	2331	2389	2448	2506	2564	2622	2681	58
35	6	2739	2797	2855	2913	2972	3030	3088	3146	3204	3262	58
41	7	3321	3379	3437	3495	3553	3611	3669	3727	3785	3844	58
46	8	3902	3960	4018	4076	4134	4192	4250	4308	4366	4424	58
52	9	4482	4540	4598	4656	4714	4772	4830	4888	4945	5003	58
	750	875061	875119	875177	875235	875293	875351	875409	875466	875524	875582	58
6	1	5640	5698	5756	5813	5871	5929	5987	6045	6102	6160	58
12	2	6218	6276	6333	6391	6449	6507	6564	6622	6680	6737	58
17	3	6795	6853	6910	6968	7026	7083	7141	7199	7256	7314	58
23	4	7371	7429	7487	7544	7602	7659	7717	7774	7832	7889	58
29	5	7947	8004	8062	8119	8177	8234	8292	8349	8407	8464	57
35	6	8522	8579	8637	8694	8752	8809	8866	8924	8981	9039	57
41	7	9096	9153	9211	9268	9325	9383	9440	9497	9555	9612	57
46	8	9669	9726	9784	9841	9898	9956	880013	880070	880127	880185	57
52	9	880242	880299	880356	880413	880471	880528	0585	0642	0699	0756	57

P P	N.	0	1	2	3	4	5	6	7	8	9	D.
	760	880814	880871	880928	880985	881042	881099	881156	881213	881271	881328	57
6	1	1385	1442	1499	1556	1613	1670	1727	1784	1841	1898	57
11	2	1955	2012	2069	2126	2183	2240	2297	2354	2411	2468	57
17	3	2525	2581	2638	2695	2752	2809	2866	2923	2980	3037	57
23	4	3093	3150	3207	3264	3321	3377	3434	3491	3548	3605	57
29	5	3661	3718	3775	3832	3888	3945	4002	4059	4115	4172	57
34	6	4229	4285	4342	4399	4455	4512	4569	4625	4682	4739	57
40	7	4795	4852	4909	4965	5022	5078	5135	5192	5248	5305	57
46	8	5361	5418	5474	5531	5587	5644	5700	5757	5813	5870	57
51	9	5926	5983	6039	6096	6152	6209	6265	6321	6378	6434	56
	770	886491	886547	886604	886660	886716	886773	886829	886885	886942	886998	56
6	1	7054	7111	7167	7223	7280	7336	7392	7449	7505	7561	56
11	2	7617	7674	7730	7786	7842	7898	7955	8011	8067	8123	56
17	3	8179	8236	8292	8348	8404	8460	8516	8573	8629	8685	56
22	4	8741	8797	8853	8909	8965	9021	9077	9134	9190	9246	56
28	5	9302	9358	9414	9470	9526	9582	9638	9694	9750	9806	56
34	6	9862	9918	9974	890030	890086	890141	890197	890253	890309	890365	56
39	7	890421	890477	890533	0589	0645	0700	0756	0812	0868	0924	56
45	8	0580	1035	1091	1147	1203	1259	1314	1370	1426	1482	56
50	9	1537	1593	1649	1705	1760	1816	1872	1928	1983	2039	56
	780	892095	892150	892206	892262	892317	892373	892429	892484	892540	892595	56
6	1	2651	2707	2762	2818	2873	2929	2985	3040	3096	3151	56
11	2	3207	3262	3318	3373	3429	3484	3540	3595	3651	3706	56
17	3	3762	3817	3873	3928	3984	4039	4094	4150	4205	4261	55
22	4	4316	4371	4427	4482	4538	4593	4648	4704	4759	4814	55
27	5	4870	4925	4980	5036	5091	5146	5201	5257	5312	5367	55
33	6	5423	5478	5533	5588	5644	5699	5754	5809	5864	5920	55
38	7	5975	6030	6085	6140	6195	6251	6306	6361	6416	6471	55
44	8	6525	6581	6636	6692	6747	6802	6857	6912	6967	7022	55
49	9	7077	7132	7187	7242	7297	7352	7407	7462	7517	7572	55
	790	897627	897682	897737	897792	897847	897902	897957	898012	898067	898122	55
5	1	8176	8231	8286	8341	8396	8451	8506	8561	8615	8670	55
11	2	8725	8780	8835	8890	8944	8999	9054	9109	9164	9218	55
17	3	9273	9328	9383	9437	9492	9547	9602	9656	9711	9766	55
22	4	9821	9875	9930	9985	900039	900094	900149	900203	900258	900312	55
27	5	900367	900422	900476	900531	0586	0640	0695	0749	0804	0859	55
33	6	0913	0968	1022	1077	1131	1186	1240	1295	1349	1404	55
38	7	1458	1513	1567	1622	1676	1731	1785	1840	1894	1948	54
44	8	2003	2057	2112	2166	2221	2275	2329	2384	2438	2492	54
49	9	2547	2601	2655	2710	2764	2818	2873	2927	2981	3036	54
	800	903090	903144	903199	903253	903307	903361	903416	903470	903524	903578	54
5	1	3633	3687	3741	3795	3849	3904	3958	4012	4066	4120	54
11	2	4174	4229	4283	4337	4391	4445	4499	4553	4607	4661	54
16	3	4716	4770	4824	4878	4932	4986	5040	5094	5148	5202	54
22	4	5256	5310	5364	5418	5472	5526	5580	5634	5688	5742	54
27	5	5796	5850	5904	5958	6012	6066	6119	6173	6227	6281	54
32	6	6335	6389	6443	6497	6551	6604	6658	6712	6766	6820	54
38	7	6874	6927	6981	7035	7089	7143	7196	7250	7304	7358	54
43	8	7411	7465	7519	7573	7626	7680	7734	7787	7841	7895	54
49	9	7949	8002	8056	8110	8163	8217	8270	8324	8378	8431	54
	810	908485	908539	908592	908646	908699	908753	908807	908860	908914	908967	54
5	1	9021	9074	9128	9181	9235	9289	9342	9396	9449	9503	54
11	2	9556	9610	9663	9716	9770	9823	9877	9930	9984	910037	53
16	3	910091	910144	910197	910251	910304	910358	910411	910464	910518	0571	53
21	4	0624	0678	0731	0784	0838	0891	0944	0998	1051	1104	53
27	5	1158	1211	1264	1317	1371	1424	1477	1530	1584	1637	53
32	6	1690	1743	1797	1850	1903	1956	2009	2063	2116	2169	53
37	7	2222	2275	2328	2381	2435	2488	2541	2594	2647	2700	53
42	8	2753	2806	2859	2913	2966	3019	3072	3125	3178	3231	53
48	9	3284	3337	3390	3443	3496	3549	3602	3655	3708	3761	53

P P	N.	0	1	2	3	4	5	6	7	8	9	D.
	820	913814	913867	913920	913973	914026	914079	914132	914184	914237	914290	53
5	1	4343	4396	4449	4502	4555	4608	4660	4713	4766	4819	53
11	2	4872	4925	4977	5030	5083	5136	5189	5241	5294	5347	53
16	3	5400	5453	5505	5558	5611	5664	5716	5769	5822	5875	53
21	4	5927	5980	6033	6085	6138	6191	6243	6296	6349	6401	53
27	5	6454	6507	6559	6612	6664	6717	6770	6822	6875	6927	53
32	6	6980	7033	7085	7138	7190	7243	7295	7348	7400	7453	53
37	7	7506	7558	7611	7663	7716	7768	7820	7873	7925	7978	52
42	8	8030	8083	8135	8188	8240	8293	8345	8397	8450	8502	52
48	9	8555	8607	8659	8712	8764	8816	8869	8921	8973	9026	52
	830	919078	919130	919183	919235	919287	919340	919392	919444	919496	919549	52
5	1	9601	9653	9706	9758	9810	9862	9914	9967	920019	920071	52
10	2	920123	920176	920228	920280	920332	920384	920436	920489	0541	0593	52
16	3	0645	0697	0749	0801	0853	0906	0958	1010	1062	1114	52
21	4	1166	1218	1270	1322	1374	1426	1478	1530	1582	1634	52
26	5	1686	1738	1790	1842	1894	1946	1998	2050	2102	2154	52
31	6	2206	2258	2310	2362	2414	2466	2518	2570	2622	2674	52
36	7	2725	2777	2829	2881	2933	2985	3037	3089	3140	3192	52
42	8	3244	3296	3348	3399	3451	3503	3555	3607	3658	3710	52
47	9	3762	3814	3865	3917	3969	4021	4072	4124	4176	4228	52
	840	924279	924331	924383	924434	924486	924538	924589	924641	924693	924744	52
5	1	4796	4848	4899	4951	5003	5054	5106	5157	5209	5261	52
10	2	5312	5364	5415	5467	5518	5570	5621	5673	5725	5776	52
15	3	5828	5879	5931	5982	6034	6085	6137	6188	6240	6291	51
20	4	6342	6394	6445	6497	6548	6600	6651	6702	6754	6805	51
26	5	6857	6908	6959	7011	7062	7114	7165	7216	7268	7319	51
31	6	7370	7422	7473	7524	7576	7627	7678	7730	7781	7832	51
36	7	7883	7935	7986	8037	8088	8140	8191	8242	8293	8345	51
41	8	8396	8447	8498	8549	8601	8652	8703	8754	8805	8857	51
46	9	8908	8959	9010	9061	9112	9163	9215	9266	9317	9368	51
	850	929419	929470	929521	929572	929623	929674	929725	929776	929827	929879	51
5	1	9930	9981	930032	930083	930134	930185	930236	930287	930338	930389	51
10	2	930440	930491	0542	0592	0643	0694	0745	0796	0847	0898	51
15	3	0949	1000	1051	1102	1153	1204	1254	1305	1356	1407	51
20	4	1458	1509	1560	1610	1661	1712	1763	1814	1865	1915	51
26	5	1966	2017	2068	2118	2169	2220	2271	2322	2372	2423	51
31	6	2474	2524	2575	2626	2677	2727	2778	2829	2879	2930	51
36	7	2981	3031	3082	3133	3183	3234	3285	3335	3386	3437	51
41	8	3487	3538	3589	3639	3690	3740	3791	3841	3892	3943	51
46	9	3993	4044	4094	4145	4195	4246	4296	4347	4397	4448	51
	860	934498	934549	934599	934650	934700	934751	934801	934852	934902	934953	50
5	1	5003	5054	5104	5154	5205	5255	5306	5356	5406	5457	50
10	2	5507	5558	5608	5658	5709	5759	5809	5860	5910	5960	50
15	3	6011	6061	6111	6162	6212	6262	6313	6363	6413	6463	50
20	4	6514	6564	6614	6665	6715	6765	6815	6865	6916	6966	50
25	5	7016	7066	7117	7167	7217	7267	7317	7367	7418	7468	50
30	6	7518	7568	7618	7668	7718	7769	7819	7869	7919	7969	50
35	7	8019	8069	8119	8169	8219	8269	8320	8370	8420	8470	50
40	8	8520	8570	8620	8670	8720	8770	8820	8870	8920	8970	50
45	9	9020	9070	9120	9170	9220	9270	9320	9369	9419	9469	50
	870	939519	939569	939619	939669	939719	939769	939819	939869	939918	939968	50
5	1	940018	940068	940118	940168	940218	940267	940317	940367	940417	940467	50
10	2	0516	0566	0616	0666	0716	0765	0815	0865	0915	0964	50
15	3	1014	1064	1114	1163	1213	1263	1313	1362	1412	1462	50
20	4	1511	1561	1611	1660	1710	1760	1809	1859	1909	1958	50
25	5	2008	2058	2107	2157	2207	2256	2306	2355	2405	2455	50
30	6	2504	2554	2603	2653	2702	2752	2801	2851	2901	2950	50
35	7	3000	3049	3099	3148	3198	3247	3297	3346	3396	3445	49
40	8	3495	3544	3593	3643	3692	3742	3792	3841	3890	3939	49
45	9	3989	4038	4088	4137	4186	4236	4285	4335	4384	4433	49

P P	N.	0	1	2	3	4	5	6	7	8	9	D.
	880	944483	944532	944581	944631	944680	944729	944779	944828	944877	944927	49
5	1	4976	5025	5074	5124	5173	5222	5272	5321	5370	5419	49
10	2	5469	5518	5567	5616	5665	5715	5764	5813	5862	5912	49
15	3	5961	6010	6059	6108	6157	6207	6256	6305	6354	6403	49
20	4	6452	6501	6551	6600	6649	6698	6747	6796	6845	6894	49
25	5	6943	6992	7041	7090	7140	7189	7238	7287	7336	7385	49
29	6	7434	7483	7532	7581	7630	7679	7728	7777	7826	7875	49
34	7	7924	7973	8022	8070	8119	8168	8217	8266	8315	8364	49
39	8	8413	8462	8511	8560	8609	8657	8706	8755	8804	8853	49
44	9	8902	8951	8999	9048	9097	9146	9195	9244	9292	9341	49
	890	949390	949439	949488	949536	949585	949634	949683	949731	949780	949829	49
5	1	9878	9926	9975	950024	950073	950121	950170	950219	950267	950316	49
10	2	950365	950414	950462	0511	0560	0608	0657	0706	0754	0803	49
15	3	0851	0900	0949	0997	1046	1095	1143	1192	1240	1289	49
20	4	1338	1386	1435	1483	1532	1580	1629	1677	1726	1775	49
24	5	1823	1872	1920	1969	2017	2066	2114	2163	2211	2260	48
29	6	2308	2356	2405	2453	2502	2550	2599	2647	2696	2744	48
34	7	2792	2841	2889	2938	2986	3034	3083	3131	3180	3228	48
39	8	3276	3325	3373	3421	3470	3518	3566	3615	3663	3711	48
44	9	3760	3808	3856	3905	3953	4001	4049	4098	4146	4194	48
	900	954243	054291	954339	954387	954435	954484	954532	954580	954628	954677	48
5	1	4725	4773	4821	4869	4918	4966	5014	5062	5110	5158	48
10	2	5207	5255	5303	5351	5399	5447	5495	5543	5592	5640	48
14	3	5688	5736	5784	5832	5880	5928	5976	6024	6072	6120	48
19	4	6168	6216	6265	6313	6361	6409	6457	6505	6553	6601	48
24	5	6649	6697	6745	6793	6840	6888	6936	6984	7032	7080	48
29	6	7128	7176	7224	7272	7320	7368	7416	7464	7512	7559	48
34	7	7607	7655	7703	7751	7799	7847	7894	7942	7990	8038	48
38	8	8086	8134	8181	8229	8277	8325	8373	8421	8468	8516	48
43	9	8564	8612	8659	8707	8755	8803	8850	8898	8946	8994	48
	910	959041	959089	959137	959185	959232	959280	959328	959375	959423	959471	48
5	1	9518	9566	9614	9661	9709	9757	9804	9852	9900	9947	48
9	2	9995	960042	960090	960138	960185	960233	960281	960328	960376	960423	48
14	3	960471	0518	0566	0613	0661	0709	0756	0804	0851	0899	48
19	4	0946	0994	1041	1089	1136	1184	1231	1279	1326	1374	47
24	5	1421	1469	1516	1563	1611	1658	1706	1753	1801	1848	47
28	6	1895	1943	1990	2038	2085	2132	2180	2227	2275	2322	47
33	7	2369	2417	2464	2511	2559	2606	2653	2701	2748	2795	47
38	8	2843	2890	2937	2985	3032	3079	3126	3174	3221	3268	47
42	9	3316	3363	3410	3457	3504	3552	3599	3646	3693	3741	47
	920	963788	963835	963882	963929	963977	964024	964071	964118	964165	964212	47
5	1	4260	4307	4354	4401	4448	4495	4542	4590	4637	4684	47
9	2	4731	4778	4825	4872	4919	4966	5013	5061	5108	5155	47
14	3	5202	5249	5296	5343	5390	5437	5484	5531	5578	5625	47
19	4	5672	5719	5766	5813	5860	5907	5954	6001	6048	6095	47
23	5	6142	6189	6236	6283	6329	6376	6423	6470	6517	6564	47
28	6	6611	6658	6705	6752	6799	6845	6892	6939	6986	7033	47
33	7	7080	7127	7173	7220	7267	7314	7361	7408	7454	7501	47
38	8	7548	7595	7642	7688	7735	7782	7829	7875	7922	7969	47
42	9	8016	8062	8109	8156	8203	8249	8296	8343	8390	8436	47
	930	968483	968530	968576	968623	968670	968716	968763	968810	968856	968903	47
5	1	8950	8996	9043	9090	9136	9183	9229	9276	9323	9369	47
9	2	9416	9463	9509	9556	9602	9649	9695	9742	9789	9835	47
14	3	9882	9928	9975	970021	970068	970114	970161	970207	970254	970300	47
18	4	970347	970393	970440	0486	0533	0579	0626	0672	0719	0765	46
23	5	0812	0858	0904	0951	0997	1044	1090	1137	1183	1229	46
28	6	1276	1322	1369	1415	1461	1508	1554	1601	1647	1693	46
32	7	1740	1786	1832	1879	1925	1971	2018	2064	2110	2157	46
37	8	2203	2249	2295	2342	2388	2434	2481	2527	2573	2619	46
41	9	2666	2712	2758	2804	2851	2897	2943	2989	3035	3082	46

P P	N.	0	1	2	3	4	5	6	7	8	9	D.
	940	973128	973174	973220	973266	973313	973359	973405	973451	973497	973543	46
5	1	3590	3636	3682	3728	3774	3820	3866	3913	3959	4005	46
9	2	4051	4097	4143	4189	4235	4281	4327	4374	4420	4466	46
14	3	4512	4558	4604	4650	4696	4742	4788	4834	4880	4926	46
18	4	4972	5018	5064	5110	5156	5202	5248	5294	5340	5386	46
23	5	5432	5478	5524	5570	5616	5662	5707	5753	5799	5845	46
28	6	5891	5937	5983	6029	6075	6121	6167	6212	6258	6304	46
32	7	6350	6396	6442	6488	6533	6579	6625	6671	6717	6763	46
37	8	6808	6854	6900	6946	6992	7037	7083	7129	7175	7220	46
41	9	7266	7312	7358	7403	7449	7495	7541	7586	7632	7678	46
	950	977724	977769	977815	977861	977906	977952	977998	978043	978089	978135	46
5	1	8181	8226	8272	8317	8363	8409	8454	8500	8546	8591	46
9	2	8637	8683	8728	8774	8819	8865	8911	8956	9002	9047	46
14	3	9093	9138	9184	9230	9275	9321	9366	9412	9457	9503	46
18	4	9548	9594	9639	9685	9730	9776	9821	9867	9912	9958	46
23	5	980003	980049	980094	980140	980185	980231	980276	980322	980367	980412	45
27	6	0458	0503	0549	0594	0640	0685	0730	0776	0821	0867	45
32	7	0912	0957	1003	1048	1093	1139	1184	1229	1275	1320	45
36	8	1366	1411	1456	1501	1547	1592	1637	1683	1728	1773	45
41	9	1819	1864	1909	1954	2000	2045	2090	2135	2181	2226	45
	960	982271	982316	982362	982407	982452	982497	982543	982588	982633	982678	45
5	1	2723	2769	2814	2859	2904	2949	2994	3040	3085	3130	45
9	2	3175	3220	3265	3310	3356	3401	3446	3491	3536	3581	45
14	3	3626	3671	3716	3762	3807	3852	3897	3942	3987	4032	45
18	4	4077	4122	4167	4212	4257	4302	4347	4392	4437	4482	45
23	5	4527	4572	4617	4662	4707	4752	4797	4842	4887	4932	45
27	6	4977	5022	5067	5112	5157	5202	5247	5292	5337	5382	45
32	7	5426	5471	5516	5561	5606	5651	5696	5741	5786	5830	45
36	8	5875	5920	5965	6010	6055	6100	6144	6189	6234	6279	45
41	9	6324	6369	6413	6458	6503	6548	6593	6637	6682	6727	45
	970	986772	986817	986861	986906	986951	986996	987040	987085	987130	987175	45
5	1	7219	7264	7309	7353	7398	7443	7488	7532	7577	7622	45
9	2	7666	7711	7756	7800	7845	7890	7934	7979	8024	8068	45
14	3	8113	8157	8202	8247	8291	8336	8381	8425	8470	8514	45
18	4	8559	8604	8648	8693	8737	8782	8826	8871	8916	8960	45
23	5	9005	9049	9094	9138	9183	9227	9272	9316	9361	9405	45
27	6	9450	9494	9539	9583	9628	9672	9717	9761	9806	9850	44
32	7	9895	9939	9983	990028	990072	990117	990161	990206	990250	990294	44
36	8	990339	990383	990428	0472	0516	0561	0605	0650	0694	0738	44
41	9	0783	0827	0871	0916	0960	1004	1049	1093	1137	1182	44
	980	991226	991270	991315	991359	991403	991448	991492	991536	991580	991625	44
4	1	1669	1713	1758	1802	1846	1890	1935	1979	2023	2067	44
9	2	2111	2156	2200	2244	2288	2333	2377	2421	2465	2509	44
13	3	2554	2598	2642	2686	2730	2774	2819	2863	2907	2951	44
18	4	2995	3039	3083	3127	3172	3216	3260	3304	3348	3392	44
22	5	3436	3480	3524	3568	3613	3657	3701	3745	3789	3833	44
26	6	3877	3921	3965	4009	4053	4097	4141	4185	4229	4273	44
31	7	4317	4361	4405	4449	4493	4537	4581	4625	4669	4713	44
35	8	4757	4801	4845	4889	4933	4977	5021	5065	5108	5152	44
40	9	5196	5240	5284	5328	5372	5416	5460	5504	5547	5591	44
	990	995635	995679	995723	995767	995811	995854	995898	995942	995986	996030	44
4	1	6074	6117	6161	6205	6249	6293	6337	6380	6424	6468	44
9	2	6512	6555	6599	6643	6687	6731	6774	6818	6862	6906	44
13	3	6949	6993	7037	7080	7124	7168	7212	7255	7299	7343	44
18	4	7386	7430	7474	7517	7561	7605	7648	7692	7736	7779	44
22	5	7823	7867	7910	7954	7998	8041	8085	8129	8172	8216	44
26	6	8259	8303	8347	8390	8434	8477	8521	8564	8608	8652	44
31	7	8695	8739	8782	8826	8869	8913	8956	9000	9043	9087	44
35	8	9131	9174	9218	9261	9305	9348	9392	9435	9479	9522	44
40	9	9565	9609	9652	9696	9739	9783	9826	9870	9913	9957	43

No.	Square.	Cube.	Sq. Root.	Cube Root	No.	Square.	Cube.	Sq. Root.	Cube Root
1	1	1	1·0000000	1·000000	64	4096	262144	8·0000000	4·000000
2	4	8	1·4142136	1·259921	65	4225	274625	8·0622577	4·020726
3	9	27	1·7320508	1·442250	66	4356	287496	8·1240384	4·041240
4	16	64	2·0000000	1·587401	67	4489	300763	8·1853528	4·061548
5	25	125	2·2360680	1·709976	68	4624	314432	8·2462113	4·081656
6	36	216	2·4494897	1·817121	69	4761	328509	8·3066239	4·101566
7	49	343	2·6457513	1·912931	70	4900	343000	8·3666003	4·121285
8	64	512	2·8284271	2·000000	71	5041	357911	8·4261498	4·140818
9	81	729	3·0000000	2·080084	72	5184	373248	8·4852814	4·160168
10	100	1000	3·1622777	2·154435	73	5329	389017	8·5440037	4·179339
11	121	1331	3·3166248	2·223980	74	5476	405224	8·6023253	4·198336
12	144	1728	3·4641016	2·289428	75	5625	421875	8·6602540	4·217163
13	169	2197	3·6055513	2·351335	76	5776	438976	8·7177979	4·235824
14	196	2744	3·7416574	2·410142	77	5929	456533	8·7749644	4·254321
15	225	3375	3·8729833	2·466212	78	6084	474552	8·8317609	4·272659
16	256	4096	4·0000000	2·519842	79	6241	493039	8·8881944	4·290841
17	289	4913	4·1231056	2·571282	80	6400	512000	8·9442719	4·308870
18	324	5832	4·2426407	2·620741	81	6561	531441	9·0000000	4·326749
19	361	6859	4·3588989	2·668402	82	6724	551368	9·0553851	4·344481
20	400	8000	4·4721360	2·714418	83	6889	571787	9·1104336	4·362071
21	441	9261	4·5825757	2·758924	84	7056	592704	9·1651514	4·379519
22	484	10648	4·6904158	2·802039	85	7225	614125	9·2195445	4·396830
23	529	12167	4·7958315	2·843867	86	7396	636056	9·2736185	4·414005
24	576	13824	4·8989795	2·884499	87	7569	658503	9·3273791	4·431047
25	625	15625	5·0000000	2·924018	88	7744	681472	9·3808315	4·447960
26	676	17576	5·0990195	2·962496	89	7921	704969	9·4339811	4·464745
27	729	19683	5·1961524	3·000000	90	8100	729000	9·4868330	4·481405
28	784	21952	5·2915026	3·036589	91	8281	753571	9·5393920	4·497941
29	841	24389	5·3851648	3·072317	92	8464	778688	9·5916630	4·514357
30	900	27000	5·4772256	3·107232	93	8649	804357	9·6436508	4·530655
31	961	29791	5·5677644	3·141381	94	8836	830584	9·6953597	4·546836
32	1024	32768	5·6568542	3·174802	95	9025	857375	9·7467943	4·562903
33	1089	35937	5·7445626	3·207534	96	9216	884736	9·7979590	4·578857
34	1156	39304	5·8309519	3·239612	97	9409	912673	9·8488578	4·594701
35	1225	42875	5·9160798	3·271066	98	9604	941192	9·8994949	4·610436
36	1296	46656	6·0000000	3·301927	99	9801	970299	9·9498744	4·626065
37	1369	50653	6·0827625	3·332222	100	10000	1000000	10·0000000	4·641589
38	1444	54872	6·1644140	3·361975	101	10201	1030301	10·0498756	4·657010
39	1521	59319	6·2449980	3·391211	102	10404	1061208	10·0995049	4·672329
40	1600	64000	6·3245553	3·419952	103	10609	1092727	10·1488916	4·687548
41	1681	68921	6·4031242	3·448217	104	10816	1124864	10·1980390	4·702669
42	1764	74088	6·4807407	3·476027	105	11025	1157625	10·2469508	4·717694
43	1849	79507	6·5574385	3·503398	106	11236	1191016	10·2956301	4·732624
44	1936	85184	6·6332496	3·530348	107	11449	1225043	10·3440804	4·747459
45	2025	91125	6·7082039	3·556893	108	11664	1259712	10·3923048	4·762203
46	2116	97336	6·7823300	3·583048	109	11881	1295029	10·4403065	4·776856
47	2209	103823	6·8556546	3·608826	110	12100	1331000	10·4880885	4·791420
48	2304	110592	6·9282032	3·634241	111	12321	1367631	10·5356538	4·805896
49	2401	117649	7·0000000	3·659306	112	12544	1404928	10·5830052	4·820284
50	2500	125000	7·0710678	3·684031	113	12769	1442897	10·6301458	4·834588
51	2601	132651	7·1414284	3·708430	114	12996	1481544	10·6770783	4·848808
52	2704	140608	7·2111026	3·732511	115	13225	1520875	10·7238053	4·862944
53	2809	148877	7·2801099	3·756286	116	13456	1560896	10·7703296	4·876999
54	2916	157464	7·3484692	3·779763	117	13689	1601613	10·8166538	4·890973
55	3025	166375	7·4161985	3·802953	118	13924	1643032	10·8627805	4·904868
56	3136	175616	7·4833148	3·825862	119	14161	1685159	10·9087121	4·918685
57	3249	185193	7·5498344	3·848501	120	14400	1728000	10·9544512	4·932424
58	3364	195112	7·6157731	3·870877	121	14641	1771561	11·0000000	4·946088
59	3481	205379	7·6811457	3·892996	122	14884	1815848	11·0453610	4·959675
60	3600	216000	7·7459667	3·914867	123	15129	1860867	11·0905365	4·973190
61	3721	226981	7·8102497	3·936497	124	15376	1906624	11·1355287	4·986631
62	3844	238328	7·8740079	3·957892	125	15625	1953125	11·1803399	5·000000
63	3969	250047	7·9372539	3·979057	126	15876	2000376	11·2249722	5·013298

No.	Square.	Cube.	Sq. Root.	Cube Root	No.	Square.	Cube.	Sq. Root.	Cube Root
127	16129	2048383	11·2694277	5 026526	190	36100	6859000	13 7840488	5 748897
128	16384	2097152	11·3137085	5·039684	191	36481	6967871	13 8202750	5·758965
129	16641	2146689	11·3578167	5·052774	192	36864	7077888	13 8564065	5·768998
130	16900	2197000	11·4017543	5 065797	193	37249	7189057	13·8924440	5 778996
131	17161	2248091	11 4455231	5 078753	194	37636	7301384	13 9283883	5 788960
132	17424	2299968	11·4891253	5 091643	195	38025	7414875	13 9642400	5 798890
133	17689	2352637	11·5325626	5 104469	196	38416	7529536	14 0000000	5 808786
134	17956	2406104	11 5758369	5 117230	197	38809	7645373	14 0356688	5 818648
135	18225	2460375	11 6189500	5·129928	198	39204	7762392	14 0712473	5·828476
136	18496	2515456	11 6619038	5 142563	199	39601	7880599	14 1067360	5 838272
137	18769	2571353	11 7046999	5 155137	200	40000	8000000	14 1421356	5 848035
138	19044	2628072	11·7473444	5 167649	201	40401	8120601	14 1774469	5 857766
139	19321	2685619	11 7898261	5 180101	202	40804	8242408	14 2126704	5·867464
140	19600	2744000	11 8321596	5 192494	203	41209	8365427	14·2478068	5·877130
141	19881	2803221	11 8743421	5 204828	204	41616	8489664	14 2828569	5 886765
142	20164	2863288	11·9163753	5·217103	205	42025	8615125	14 3178211	5·896368
143	20449	2924207	11 9582607	5 229321	206	42436	8741816	14 3527001	5 905943
144	20736	2985984	12 0000000	5 241483	207	42849	8869743	14·3874946	5·915483
145	21025	3048625	12 0415946	5·253588	208	43264	8998912	14 4222051	5·924993
146	21316	3112136	12 0830460	5 265637	209	43681	9123329	14 4568323	5·934473
147	21609	3176523	12 1243557	5·277632	210	44100	9261000	14 4913767	5 943921
148	21904	3241792	12 1655251	5 289572	211	44521	9393931	14 5258390	5·953341
149	22201	3307949	12 2065556	5 301459	212	44944	9528128	14 5602198	5·962731
150	22500	3375000	12 2474487	5 313293	213	45369	9663597	14 5945195	5 972091
151	22801	3442951	12 2882056	5·325074	214	45796	9800344	14 6287388	5·981426
152	23104	3511808	12·3288280	5·336803	215	46225	9938375	14·6628783	5·990727
153	23409	3581577	12 3693169	5 348481	216	46656	10077696	14 6969385	6·000000
154	23716	3652264	12·4096736	5·360108	217	47089	10218313	14 7309199	6·009244
155	24025	3723875	12 4498996	5 371685	218	47524	10360232	14 7648231	6·018463
156	24336	3796416	12 4899960	5 383213	219	47961	10503459	14 7986486	6·027650
157	24649	3869893	12 5299641	5 394691	220	48400	10648000	14 8323970	6·036811
158	24964	3944312	12·5698051	5·406120	221	48841	10793861	14 8660687	6 045943
159	25281	4019679	12 6095202	5 417501	222	49284	10941048	14·8996644	6·055048
160	25600	4096000	12 6491106	5 428835	223	49729	11089567	14 9331845	6 064126
161	25921	4173281	12 6885775	5 440122	224	50176	11239424	14·9666295	6 073178
162	26244	4251528	12 7279221	5 451362	225	50625	11390625	15·0000000	6 082201
163	26569	4330747	12 7671453	5·462556	226	51076	11543176	15 0332964	6 091199
164	26896	4410944	12 8062485	5 473704	227	51529	11697083	15 0665192	6·100170
165	27225	4492125	12 8452326	5 484806	228	51984	11852352	15 0996689	6·109115
166	27556	4574296	12 8840987	5 495865	229	52441	12008989	15 1327460	6 118033
167	27889	4657463	12 9228480	5 506879	230	52900	12167000	15·1657509	6·126925
168	28224	4741632	12·9614814	5·517848	231	53361	12326391	15 1986842	6·135792
169	28561	4826809	13 0000000	5 528775	232	53824	12487168	15 2315462	6·144634
170	28900	4913000	13 0384048	5 539658	233	54289	12649337	15 2643375	6 153449
171	29241	5000211	13 0766968	5·550499	234	54756	12812904	15 2970585	6·162239
172	29584	5088448	13·1148770	5 561298	235	55225	12977875	15 3297097	6·171005
173	29929	5177717	13 1529464	5 572055	236	55696	13144256	15·3622915	6·179747
174	30276	5268024	13 1909060	5 582770	237	56169	13312053	15 3948043	6·188463
175	30625	5359375	13 2287566	5 593445	238	56644	13481272	15 4272486	6·197154
176	30976	5451776	13 2664992	5 604079	239	57121	13651919	15·4596248	6·205821
177	31329	5545233	13 3041347	5 614673	240	57600	13824000	15·4919334	6·214464
178	31684	5639752	13 3416641	5 625226	241	58081	13997521	15 5241747	6 223084
179	32041	5735339	13 3790882	5 635741	242	58564	14172488	15 5563492	6·231679
180	32400	5832000	13 4164079	5 646216	243	59049	14348907	15 5884573	6·240251
181	32761	5929741	13 4536240	5 656651	244	59536	14526789	15 6204994	6 248800
182	33124	6028568	13 4907376	5 667051	245	60025	14706125	15 6524758	6·257324
183	33489	6128487	13 5277493	5 677411	246	60516	14886936	15 6843871	6·265826
184	33856	6229504	13 5646600	5 687734	247	61009	15069223	15 7162336	6·274305
185	34225	6331625	13 6014705	5·698019	248	61504	15252992	15 7480157	6·282760
186	34596	6434856	13 6381817	5 708267	249	62001	15438249	15 7797338	6 291194
187	34969	6539203	13 6747943	5 718479	250	62500	15625000	15 8113883	6·299604
188	35344	6644672	13 7113092	5·728654	251	63001	15813251	15 8429795	6·307993
189	35721	6751269	13 7477271	5·738794	252	63504	16003008	15 8745079	6·316359

No.	Square.	Cube.	Sq. Root.	Cube Root	No.	Square.	Cube.	Sq. Root.	Cube Root
253	64009	16194277	15·9059737	6·324704	316	99856	31554496	17·7763888	6·811284
254	64516	16387064	15 9373775	6 333026	317	100489	31855013	17 8044938	6 818462
255	65025	16581375	15 9687194	6·341326	318	101124	32157432	17·8325545	6·825624
256	65536	16777216	16 0000000	6 349604	319	101761	32461759	17 8605711	6 832771
257	66049	16974593	16·0312195	6 357861	320	102400	32768000	17 8885438	6 839904
258	66564	17173512	16 0623784	6·366095	321	103041	33076161	17 9164729	6 847021
259	67081	17373979	16·0934769	6 374311	322	103684	33386248	17·9443584	6 854124
260	67600	17576000	16 1245155	6 382504	323	104329	33698267	17 9722008	6·861212
261	68121	17779581	16 1554944	6 390676	324	104976	34012224	18·0000000	6·868285
262	68644	17984728	16 1864141	6 398828	325	105625	34328125	18 0277564	7·875344
263	69169	18191447	16 2172·47	6 406958	326	106276	34645976	18 0554701	6·882388
264	69696	18399744	16 2480768	6 415068	327	106929	34965783	18 0831413	6·889419
265	70225	18609625	16 2788206	6 423158	328	107584	35287552	18 1107703	6·896435
266	70756	18821096	16 3095064	6 431228	329	108241	35611289	18 1383571	6 903436
267	71289	19034163	16 3401346	6 439277	330	108900	35937000	18 1659021	6 910423
268	71824	19248832	16 3707055	6 447305	331	109561	36264691	18 1934054	6·917396
269	72361	19465109	16 4012195	6·455315	332	110224	36594368	18 2208672	6 924355
270	72900	19683000	16 4316767	6·463304	333	110889	36926037	18 2482876	6·931301
271	73441	19902511	16 4620776	6·471274	334	111556	37259704	18·2756669	6·938232
272	73984	20123648	16 4924225	6·479224	335	112225	37595375	18·3030052	6·945149
273	74529	20346417	16 5227116	6 487154	336	112896	37933056	18 3303028	6 952053
274	75076	20570824	16 5529454	6 495065	337	113569	38272753	18·3575598	6·958943
275	75625	20796875	16 5831240	6 502956	338	114244	38614472	18 3847763	6 965819
276	76176	21024576	16 6132477	6 510830	339	114921	38958219	18 4119526	6 972683
277	76729	21253933	16 6433170	6 518684	340	115600	39304000	18 4390889	6 979532
278	77284	21484952	16·6733320	6 526519	341	116281	39651821	18 4661853	6 986368
279	77841	21717639	16 7032931	6·534335	342	116964	40001688	18·4932420	6 993191
280	78400	21952000	16 7332005	6 542133	343	117649	40353607	18 5202592	7 000000
281	78961	22188041	16 7630546	6·549912	344	118336	40707584	18·5472370	7 006796
282	79524	22425768	16 7928556	6 557672	345	119025	41063625	18·5741756	7·013579
283	80089	22665187	16 8226038	6·565415	346	119716	41421736	18 6010752	7 020349
284	80656	22906304	16 8522995	6 573139	347	120409	41781923	18 6279360	7 027106
285	81225	23149125	16 8819439	6 580844	348	121104	42144192	19·6547581	7 033850
286	81796	23393656	16 9115345	6 588532	349	121801	42508549	18 6815417	7·040581
287	82369	23639903	16 9410743	6 596202	350	122500	42875000	18 7082869	7·047298
288	82944	23887872	16 9705627	6 603854	351	123201	43243551	18·7349940	7 054004
289	83521	24137569	17 0000000	6 611489	352	123904	43614208	18 7616630	7 060696
290	84100	24389000	17 0293864	6 619106	353	124609	43986977	18 7882942	7·067376
291	84681	24642171	17 0587221	6 626705	354	125316	44361864	18 8148877	7 074044
292	85264	24897088	17 0880075	6 634287	355	126025	44738875	18 8414437	7 080699
293	85849	25153757	17 1172428	6 641852	356	126736	45118016	18 8679623	7 087341
294	86436	25412184	17 1464282	6 649399	357	127449	45499293	18 8944436	7 093971
295	87025	25672375	17 1755640	6 656930	358	128164	45882712	18·9208879	7·100588
296	87616	25934336	17 2046505	6 664444	359	128881	46268279	18 9472953	7·107194
297	88209	26198073	17 2336879	6·671940	360	129600	46656000	18 9736660	7·113786
298	88804	26463592	17 2626762	6·679420	361	130321	47045881	19 0000000	7 120367
299	89401	26730899	17 2916165	6 686882	362	131044	47437928	19 0262976	7·126936
300	90000	27000000	17 3205081	6 694329	363	131769	47832147	19·0525589	7 133492
301	90601	27270901	17 3493516	6 701759	364	132496	48228544	19 0787840	7 140037
302	91204	27543608	17 3781472	6 709173	365	133225	48627125	19 1049732	7 146569
303	91809	27818127	17 4068952	6 716570	366	133956	49027896	19·1311265	7 153090
304	92416	28094464	17 4355958	6 723951	367	134689	49430863	19 1572441	7 159599
305	93025	28372625	17·4642492	6 731316	368	135424	49836032	19·1833261	7·166096
306	93636	28652616	17·4928557	6 738665	369	136161	50243409	19 2093727	7 172580
307	94249	28934443	17 5214155	6·745997	370	136900	50653000	19 2353841	7 179054
308	94864	29218112	17 5499288	6 753313	371	137641	51064811	19·2613603	7·185516
309	95481	29503629	17 5783958	6·760614	372	138384	51478848	19 2873015	7·191966
310	96100	29791000	17 6068169	6 767899	373	139129	51895117	19 3132079	7·198405
311	96721	30080231	17 6351921	6 775169	374	139876	52313624	19·3390796	7·204832
312	97344	30371328	17·6635217	6·782423	375	140625	52734375	19·3649167	7·211248
313	97969	30664297	17 6918060	6·789661	376	141376	53157376	19·3907194	7 217652
314	98596	30959144	17 7200451	6 796884	377	142129	53582633	19·4164878	7·224045
315	99225	31255875	17 7482393	6 804092	378	142884	54010152	19 4422221	7·230427

No.	Square.	Cube.	Sq. Root.	Cube Root.	No.	Square.	Cube.	Sq. Root.	Cube Root
379	143641	54439939	19·4679223	7·236797	442	195364	86350888	21·0237960	7 617412
380	144400	54872000	19 4936887	7·243156	443	196249	86938307	21·0475652	7·623152
381	145161	55306341	19·5192213	7·249504	444	197136	87528384	21·0713075	7·628884
382	145924	55742968	19·5448203	7·255841	445	198025	88121125	21·0950231	7·634607
383	146689	56181887	19·5703858	7·262167	446	198916	88716536	21·1187121	7 640321
384	147456	56623104	19·5959179	7 268482	447	199809	89314623	21 1423745	7 646027
385	148225	57066625	19·6214169	7 274786	448	200704	89915392	21·1660105	7·651725
386	148996	57512456	19·6468827	7· 281079	449	201601	90518844	21·1896201	7 657414
387	149769	57960603	19 6723156	7·287362	450	202500	91125000	21·2132034	7 663094
388	150544	58411072	19·6977156	7·293633	451	203401	91733851	21·2367500	7 668766
389	151321	58863869	19·7230829	7 299894	452	204304	92345408	21 2602916	7·674430
390	152100	59319000	19·7484177	7 306143	453	205209	92959677	21 2837967	7 680086
391	152881	59776471	19·7737199	7 312383	454	206116	93576664	21 3072758	7 685733
392	153664	60236288	19 7989899	7 318611	455	207025	94196375	21 3307290	7·691372
393	154449	60698457	19 8242276	7 324829	456	207936	94818816	21 3541565	7 697002
394	155236	61162984	19·8494332	7 331037	457	208849	95443993	21·3775583	7 702625
395	156025	61629875	19·8746069	7·337234	458	209764	96071912	21·4009346	7 708239
396	156816	62099136	19·8997487	7 343420	459	210681	96702579	21·4242853	7·713845
397	157609	62570773	19·9248588	7 349597	460	211600	97336000	21 4476106	7·719442
398	158404	63044792	19·9499373	7·355762	461	212521	97972181	21·4709106	7 725032
399	159201	63521199	19 9749844	7 361918	462	213444	98611128	21·4941853	7·730614
400	160000	64000000	20 0000000	7·368063	463	214369	99252847	21·5174348	7·736188
401	160801	64481201	20·0249844	7 374198	464	215296	99897344	21 5406592	7 741753
402	161604	64964808	20 0499377	7 380322	465	216225	100544625	21 5638587	7 747311
403	162409	65450827	20 0748599	7 386437	466	217156	101194696	21 5870331	7 752861
404	163216	65939264	20 0997512	7 392542	467	218089	101847563	21·6101828	7·758402
405	164025	66430125	20·1246118	7 398636	468	219024	102503232	21 6333077	7 763936
406	164836	66923416	20·1494417	7·404720	469	219961	103161709	21·6564078	7·769462
407	165649	67419143	20·1742410	7·410795	470	220900	103823000	21 6794834	7 774980
408	166464	67917312	20 1990099	7·416859	471	221841	104487111	21·7025344	7 780490
409	167281	68417929	20 2237484	7·422914	472	222784	105154048	21·7255610	7 785993
410	168100	68921000	20 2484567	7·428959	473	223729	105823817	21 7485632	7 791487
411	168921	69426531	20 2731349	7·434994	474	224676	106496424	21·7715411	7 796974
412	169744	69934528	20 2977831	7·441019	475	225625	107171875	21 7944947	7 802454
413	170569	70444997	20 3224014	7·447034	476	226576	107850176	21 8174242	7 807925
414	171396	70957944	20·3469899	7 453040	477	227529	108531333	21 8403297	7 813389
415	172225	71473375	20·3715488	7 459036	478	228484	109215352	21·8632111	7 818846
416	173056	71991296	20 3960781	7 465022	479	229441	109902239	21 8860686	7·824294
417	173889	72511713	20 4205779	7 470999	480	230400	110592000	21 9089023	7·829735
418	174724	73034632	20 4450183	7 476966	481	231361	111284641	21 9317122	7 835169
419	175561	73560059	20 4694895	7·482924	482	232324	111980168	21 9544984	7 840595
420	176400	74088000	20 4939015	7·488872	483	233289	112678587	21 9772610	7 846013
421	177241	74618461	20 5182845	7 494811	484	234256	113379904	22·0000000	7 851424
422	178084	75151448	20 5426386	7·500741	485	235225	114084125	22·0227155	7 856828
423	178929	75686967	20 5669638	7·506661	486	236196	114791256	22 0454077	7 862224
424	179776	76225024	20 5912603	7 512571	487	237169	115501303	22 0680765	7·867613
425	180625	76765625	20 6155281	7 518473	488	238144	116214272	22 0907220	7·872994
426	181476	77308776	20 6397674	7·524365	489	239121	116930169	22 1133444	7 878368
427	182329	77854483	20 6639783	7 530248	490	240100	117649000	22 1359436	7·883735
428	183184	78402752	20·6881609	7 536121	491	241081	118370771	22 1585198	7·889095
429	184041	78953589	20 7123152	7 541986	492	242064	119095488	22 1810730	7 894447
430	184900	79507000	20 7364414	7 547842	493	243049	119823157	22 2036033	7 899792
431	185761	80062991	20·7605395	7 553688	494	244036	120553784	22 2261108	7 905129
432	186624	80621568	20 7846097	7 559526	495	245025	121287375	22·2485955	7 910460
433	187489	81182737	20·8086520	7 565355	496	246016	122023936	22 2710575	7·915783
434	188356	81746504	20·8326667	7 571174	497	247009	122763473	22 2934968	7·921100
435	189225	82312875	20 8566536	7·576985	498	248004	123505992	22 3159136	7 926408
436	190096	82881856	20 8806130	7·582786	499	249001	124251499	22 3383079	7 931710
437	190969	83453453	20 9045450	7 588579	500	250000	125000000	22 3606798	7·937005
438	191844	84027672	20·9284495	7 594363	501	251001	125751501	22 3830293	7·942293
439	192721	84604519	20·9523268	7·600138	502	252004	126506008	22 4053565	7·947574
440	193600	85184000	20 9761770	7 605905	503	253009	127263527	22 4276615	7·952848
441	194481	85766121	21·0000000	7 611662	504½	254016	128024064	22 4499443	7·958114

No.	Square.	Cube.	Sq. Root.	Cube Root.	No.	Square.	Cube.	Sq. Root.	Cube Root
505	255025	128787625	22·4722051	7·963374	568	322624	183250432	23·8327506	8·218635
506	256036	129554216	22·4944438	7·968627	569	323761	184220009	23·8537209	8·286493
507	257049	130323843	22·5166605	7·973873	570	324900	185193000	23·8746728	8·291344
508	258064	131096512	22·5388553	7·979112	571	326041	186169411	23·8956063	8·296190
509	259081	131872229	22·5610283	7·984344	572	327184	187149248	23·9165215	8·301030
510	260100	132651000	22·5831796	7·989570	573	328329	188132517	23·9374184	8·305865
511	261121	133432831	22·6053091	7·994788	574	329476	189119224	23·9582971	8·310694
512	262144	134217728	22·6274170	8·000000	575	330625	190109375	23·9791576	8·315517
513	263169	135005697	22·6495033	8·005205	576	331776	191102976	24·0000000	8·320335
514	264196	135796744	22·6715681	8·010403	577	332929	192100033	24·0208243	8·325147
515	265225	136590875	22·6936114	8·015595	578	334084	193100552	24·0416306	8·329954
516	266256	137388096	22·7156334	8·020779	579	335241	194104539	24·0624188	8·334755
517	267289	138188413	22·7376340	8·025957	580	336400	195112000	24·0831892	8·339551
518	268324	138991832	22·7596134	8·031129	581	337561	196122941	24·1039416	8·344341
519	269361	139798359	22·7815715	8·036293	582	338724	197137368	24·1246762	8·349126
520	270400	140608000	22·8035085	8·041451	583	339889	198155287	24·1453929	8·353905
521	271441	141420761	22·8254244	8·046603	584	341056	199176704	24·1660919	8·358678
522	272484	142236648	22·8473193	8·051748	585	342225	200201625	24·1867732	8·363446
523	273529	143055667	22·8691933	8·056886	586	343396	201230056	24·2074369	8·368209
524	274576	143877824	22·8910463	8·062018	587	344569	202262003	24·2280829	8·372967
525	275625	144703125	22·9128785	8·067143	588	345744	203297472	24·2487113	8·377719
526	276676	145531576	22·9346899	8·072262	589	346921	204336469	24·2693222	8·382465
527	277729	146363183	22·9564806	8·077374	590	348100	205379000	24·2899156	8·387206
528	278784	147197952	22·9782500	8·082480	591	349281	206425071	24·3104916	8·391942
529	279841	148035889	23·0000000	8·087579	592	350464	207474688	24·3310501	8·396673
530	280900	148877000	23·0217289	8·092672	593	351649	208527857	24·3515913	8·401398
531	281961	149721291	23·0434372	8·097759	594	352836	209584584	24·3721152	8·406118
532	283024	150568768	23·0651252	8·102839	595	354025	210644875	24·3926218	8·410833
533	284089	151419437	23·0867928	8·107913	596	355216	211708736	24·4131112	8·415542
534	285156	152273304	23·1084400	8·112980	597	356409	212776173	24·4335834	8·420246
535	286225	153130375	23·1300670	8·118041	598	357604	213847192	24·4540385	8·424945
536	287296	153990656	23·1516738	8·123096	599	358801	214921799	24·4744765	8·429638
537	288369	154854153	23·1732605	8·128145	600	360000	216000000	24·4948974	8·434327
538	289444	155720872	23·1948270	8·133187	601	361201	217081801	24·5153013	8·439010
539	290521	156590819	23·2163735	8·138223	602	362404	218167208	24·5356883	8·443688
540	291600	157464000	23·2379001	8·143253	603	363609	219256227	24·5560583	8·448360
541	292681	158340421	23·2594067	8·148276	604	364816	220348864	24·5764115	8·453028
542	293764	159220088	23·2808935	8·153294	605	366025	221445125	24·5967478	8·457691
543	294849	160103007	23·3023604	8·158305	606	367236	222545016	24·6170673	8·462348
544	295936	160989184	23·3238076	8·163310	607	368449	223648543	24·6373700	8·467000
545	297025	161878625	23·3452351	8·168309	608	369664	224755712	24·6576560	8·471647
546	298116	162771336	23·3666429	8·173302	609	370881	225866529	24·6779254	8·476289
547	299209	163667323	23·3880311	8·178289	610	372100	226981000	24·6981781	8·480926
548	300304	164566592	23·4093998	8·183269	611	373321	228099131	24·7184142	8·485558
549	301401	165469149	23·4307490	8·188244	612	374544	229220928	24·7386338	8·490185
550	302500	166375000	23·4520788	8·193213	613	375769	230346397	24·7588368	8·494806
551	303601	167284151	23·4733892	8·198175	614	376996	231475544	24·7790234	8·499423
552	304704	168196608	23·4946802	8·203132	615	378225	232608375	24·7991935	8·504035
553	305809	169112377	23·5159520	8·208082	616	379456	233744896	24·8193473	8·508642
554	306916	170031464	23·5372046	8·213027	617	380689	234885113	24·8394847	8·513243
555	308025	170953875	23·5584380	8·217966	618	381924	236029032	24·8596058	8·517840
556	309136	171879616	23·5796522	8·222898	619	383161	237176659	24·8797106	8·522432
557	310249	172808693	23·6008474	8·227825	620	384400	238328000	24·8997992	8·527019
558	311364	173741112	23·6220236	8·232746	621	385641	239483061	23·9198716	8·531601
559	312481	174676879	23·6431808	8·237661	622	386884	240641848	24·9399278	8·536178
560	313600	175616000	23·6643191	8·242571	623	388129	241804367	24·9599679	8·540750
561	314721	176558481	23·6854386	8·247474	624	389376	242970624	24·9799920	8·545317
562	315844	177504328	23·7065392	8·252371	625	390625	244140625	25·0000000	8·549879
563	316969	178453547	23·7276210	8·257263	626	391876	245314376	25·0199920	8·554437
564	318096	179406144	23·7486842	8·262149	627	393129	246491883	25·0399681	8·558990
565	319225	180362125	23·7697286	8·267029	628	394384	247673152	25·0599282	8·563538
566	320356	181321496	23·7907545	8·271904	629	395641	248858189	25·0798724	8·568081
567	321489	182284263	23·8117618	8·286773	630	396900	250047000	25·0998008	8·572619

No.	Square.	Cube.	Sq. Root.	Cube Root.	No.	Square.	Cube.	Sq. Root.	Cube Root.
631	398161	251239591	25·1197134	8·577152	694	481636	334255384	26 3438797	8·853598
632	399424	252435968	25·1396102	8 581681	695	483025	335702375	26·3628527	8 857849
633	400689	253636137	25·1594913	8 586205	696	484416	337153536	26 3818119	8 862095
634	401956	254840104	25·1793566	8·590724	697	485809	338608873	26·4007576	8·866337
635	403225	256047875	25·1992063	8·595238	698	487204	340068392	26 4196896	8·870576
636	404496	257259456	25·2190404	8 599747	699	488601	341532099	26·4386081	8 874810
637	405769	258474853	25·2388589	8 604252	700	490000	343000000	26·4575131	8·879040
638	407044	259694072	25·2586619	8 608753	701	491401	344472101	26·4764046	8 883266
639	408321	260917119	25·2784493	8·613248	702	492804	345948408	26·4952826	8 887488
640	409600	262144000	25·2982213	8·617739	703	494209	347428927	26 5141472	8 891706
641	410881	263374721	24·3179778	8·622225	704	495616	348913664	26·5329983	8 895920
642	412164	264609288	25·3377189	8·626706	705	497025	350402625	26·5518361	8 900130
643	413449	265847707	25·3574447	8·631183	706	498436	351895816	26·5706605	8 904336
644	414736	267089984	25·3771551	8·635655	707	499849	353393243	26·5894716	8·908538
645	416025	268336125	25·3968502	8·640123	708	501264	354894912	26·6082694	8·912737
646	417316	269586136	25·4165301	8·644585	709	502681	356400829	26·6270539	8·916931
647	418609	270840023	25·4361947	8·649044	710	504100	357911000	26 6458252	8 921121
648	419904	272097792	25·4558441	8 653497	711	505521	359425431	26·6645833	8 925308
649	421201	273359449	25·4754784	8·657946	712	506944	360944128	26 6833281	8·929490
650	422500	274625000	25 4950976	8 662391	713	508369	362467097	26·7020598	8·933668
651	423801	275894451	25 5147016	8·666831	714	509796	363994344	26·7207784	8·937843
652	425104	277167808	25·5342907	8·671266	715	511225	365525875	26·7394839	8 942014
653	426409	278445077	25·5538647	8·675697	716	512656	367061696	26·7581763	8 946181
654	427716	279726264	25·5734237	8·680124	717	514089	368601813	26·7768557	8·950344
655	429025	281011375	25 5929678	8 684546	718	515524	370146232	26 7955220	8 954503
656	430336	282300416	25 6124969	8 688963	719	516961	371694959	26·8141754	8 958658
657	431649	283593393	25 6320112	8 693376	720	518400	373248000	26 8328157	8 962809
658	432964	284890312	25·6515107	8 697784	721	519841	374805361	26 8514432	8 966957
659	434281	286191179	25 6709953	8·702188	722	521284	376367048	26·8700577	8 971101
660	435600	287496000	25 6904652	8·706587	723	522729	377933067	26·8886593	8 975240
661	436921	288804781	25·7099203	8·710983	724	524176	379503424	26 9072481	8·979376
662	438244	290117528	25·7203607	8·715373	725	525625	381078125	26·9258240	8 983509
663	439569	291434247	25·7487864	8·719759	726	527076	382657176	26 9443872	8·987637
664	440896	292754994	25 7681975	8·724141	727	528529	384240583	26·9629375	8·991762
665	442225	294079625	25 7875939	8·728518	728	529984	385828352	26 9814751	8·995883
666	443556	295408296	25 8069758	8·732892	729	531441	387420489	27·0000000	9·000000
667	444889	296740963	25 8263431	8·737260	730	532900	389017000	27·0185122	9·004113
668	446224	298077632	25 8456960	8·741624	731	534361	390617891	27 0370117	9·008223
669	447561	299418309	25 8650343	8·745985	732	535824	392223168	27·0554985	9·012329
670	448900	300763000	25·8843582	8 750340	733	537289	393832837	27·0739727	9·016431
671	450241	302111711	25·9036677	8·754691	734	538756	395446904	27·0924344	9·020529
672	451584	303464448	25 8229628	8 759038	735	540225	397065375	27·1108834	9·024624
673	452929	304821217	25 9422435	8·763381	736	541696	398688256	27·1293199	9·028715
674	454276	306182024	25 9615100	8 767719	737	543169	400315553	27 1477439	9 032802
675	455625	307546875	25·9807621	8 772053	738	544644	401947272	27 1661554	9·036886
676	456976	308915776	26 0000000	8 776383	739	546121	403583419	27·1845544	9·040965
677	458329	310238733	26 0192237	8 780708	740	547600	405224000	27·2029410	9·045041
678	459684	311665752	26 0384331	8 785029	741	549081	406869021	27·2213152	9·049114
679	461041	313046839	26 0576284	8 789346	742	550564	408518488	27·2396769	9·053183
680	462400	314432000	26·0768096	8·793659	743	552049	410172407	27 2580263	9·057248
681	463761	315821241	26·0959767	8 797968	744	553536	411830784	27 2763634	9·061310
682	465124	317214568	26 1151297	8·802272	745	555025	413493625	27·2946881	9·065367
683	466189	318611987	26 1342687	8 806572	746	556516	415160936	27·3130006	9·069422
684	467856	320013504	26·1533937	8 810868	747	558009	416832723	27 3313007	9·073473
685	469225	321419125	26·1725047	8·815160	748	559504	418508992	27 3495887	9 077520
686	470596	322828856	26 1916017	8·819447	749	561001	420189749	27·3678644	9·081563
687	471969	324242703	26·2106848	8 823731	750	562500	421875000	27 3861279	9·085603
688	473344	325660672	26·2297541	8 828009	751	564001	423564751	27·4043792	9·089639
689	474721	327082769	26·2488095	8 832285	752	565504	425259008	27·4226184	9·093672
690	476100	328509000	26 2678511	8 836556	753	567009	426957777	27·4408455	9 097701
691	477481	329939371	26·2868789	8 840823	754	568516	428661064	27 4590604	9·101726
692	478864	331373888	26 3058929	8·845085	755	570025	430368875	27 4772633	9·105748
693	480249	332812557	26·3248932	8·849344	756	571536	432081216	27·4954542	9·109766

No.	Square.	Cube.	Sq. Root.	Cube Root.	No.	Square.	Cube.	Sq. Root.	Cube Root
757	573049	433798093	27 5136330	9·113781	820	672400	551368000	28·6356421	9·359902
758	574564	435519512	27·5317998	9 117793	821	674041	553387661	28·6530976	9·363705
759	576081	437245479	27·5499546	9 121801	822	675684	555412248	28 6705424	9·367505
760	577600	438976000	27·5680975	9·125805	823	677329	557441767	28·6879766	9 371302
761	579121	440711081	27·5862284	9·129806	824	678976	559476224	28 7054002	9·375096
762	580644	442450728	27 6043475	9·133803	825	680625	561515625	28·7228132	9 378887
763	582169	444194947	27 6224546	9·137797	826	682276	563359976	28 7402157	9·382675
764	583696	445943744	27·6105499	9·141788	827	683929	565609283	28·7576077	9·386460
765	585225	447697125	27·6586334	9·145774	828	685584	567663552	28·7749891	9·390242
766	586756	449455096	27·6767050	9·149757	829	687241	569722789	28·7923601	9·394020
767	588289	451217663	27·6947648	9 153737	830	688900	571787000	28 8097206	9 397796
768	589824	452984832	27·7128129	9·157714	831	690561	573856191	28 8270706	9·401569
769	591361	454756609	27·7308492	9 161686	832	692224	575930368	28·8444102	9·405339
770	592900	456533000	27·7488739	9·165656	833	693889	578009537	28·8617394	9·409105
771	594141	458314011	27·7668868	9·169622	834	695556	580093704	28 8790582	9·412869
772	595984	460099648	27·7848880	9·173585	835	697225	582182875	28·8963666	9·416630
773	597529	461889917	27·8028775	9·177544	836	698896	584277056	28 9236646	9·420387
774	599076	463684824	27·8208555	9·181500	837	700569	586376253	28 9309523	9·424142
775	600625	465484375	27·8388218	9·185453	838	702244	588480472	28 9482297	9·427894
776	602176	467288576	27 8567766	9·189402	839	703921	590589719	28 9654967	9·431642
777	603729	469097433	27·8747197	9 193347	840	705600	592704000	28 9827535	9·435388
778	605284	470910952	27·8926514	9·197289	841	707281	594823321	29 0000000	9·439131
779	606841	472729139	27·9105715	9 201229	842	708964	596947688	29 0172363	9·442870
780	608400	474552000	27·9284801	9 205164	843	710649	599077107	29 0344623	9 446607
781	609961	476379541	27·9463772	9·209096	844	712336	601211584	29 0516781	9·450341
782	611524	478211768	27 9642629	9·213025	845	714025	603351125	29·0688837	9·454072
783	613089	480048687	27·9821372	9·216950	846	715716	605495736	29·0860791	9·457800
784	614656	481890304	28·0000000	9·220873	847	717409	607645423	29·1032644	9·461525
785	616225	483736625	28·0178515	9·224791	848	719104	609800192	29·1204396	9·465247
786	617796	485587656	28·0356915	9·228707	849	720801	611960049	29 1376046	9·468966
787	619369	487443403	28·0535203	9 232619	850	722500	614125000	29·1547595	9·472682
788	620944	489303872	28 0713377	9·237528	851	724201	616295051	29·1719043	9·476395
789	622521	491169069	28 0891438	9·240433	852	725904	618470208	29·1890390	9·480106
790	624100	493039000	28 1069386	9·244335	853	727609	620650477	29·2061637	9 483813
791	625681	494913671	28·1247222	9·248234	854	729316	622835864	29·2232784	9·487518
792	627264	496793088	28·1424946	9·252130	855	731025	625026375	29 2403830	9·491220
793	628849	498677257	28·1602557	9 256022	856	732736	627222016	29·2574777	9·494919
794	630436	500566184	28·1780056	9·259911	857	734449	629422793	29·2745623	9·498615
795	632025	502459875	28·1957444	9 263797	858	736164	631628712	29·2916370	9·502308
796	633616	504358336	28·2134720	9 267680	859	737881	633839779	29·3087018	9·505998
797	635209	506261573	28·2311884	9 271559	860	739600	636056000	29 3257566	9·509685
798	636804	508169592	28 2488938	9·275435	861	741321	638277381	29 3428015	9 513370
799	638401	510082399	28 2665881	9·279308	862	743044	640503928	29·3598365	9·517051
800	640000	512000000	28·2842712	9 283178	863	744769	642735647	29·3768616	9 520730
801	641601	513922401	28·3019434	9·287044	864	746496	644972544	29·3938769	9 524406
802	643204	515849608	28·3196045	9 290907	865	748225	647214625	29·4108823	9 528079
803	644809	517781627	28·3372546	9·294767	866	749956	649461896	29 4278779	9·531749
804	646416	519718464	28 3548938	9·298624	867	751689	651714363	29·4448637	9 535417
805	648025	521660125	28 3725219	9 302477	868	753424	653972032	29·4618397	9·539082
806	649636	523606616	28·3901391	9·306328	869	755161	656234909	29·4788059	9 542744
807	651249	525557943	28 4077454	9·310175	870	756900	658503000	29 4957624	9 546403
808	652864	527514112	28 4253408	9·314019	871	758641	660776311	29 5127091	9 550059
809	654481	529475129	28 4429253	9 317860	872	760384	663054848	29·5296461	9·553712
810	656100	531441000	28 4604989	9 321697	873	762129	665338617	29 5465734	9·557363
811	657721	533411731	28 4780617	9·325532	874	763876	667627624	29 5634910	9 561011
812	659344	535387328	28·4956137	9·329363	875	765625	669921875	29 5803989	9·564656
813	660969	537367797	28 5131549	9 333192	876	767376	672221376	29·5972972	9·568298
814	662596	539353144	28 5306852	9·337017	877	769129	674526133	29 6141858	9·571938
815	664225	541343375	28 5482048	9·340838	878	770884	676836152	29 6310648	9 575574
816	665856	543338496	28·5657137	9 344657	879	772641	679151439	29 6479325	9 579208
817	667489	545338513	28 5832119	9 348473	880	774400	681472000	29·6647939	9·582840
818	669124	547343432	28 6006993	9·352286	881	776161	683797841	29 6816442	9 586468
819	670761	549353259	28·6181760	9 356095	882	777924	686128968	29·6984848	9·590094

No.	Square.	Cube.	Sq. Root.	Cube Root	No.	Square.	Cube.	Sq. Root.	Cube Root
883	779689	688465387	29·7153159	9·593716	942	887364	835896888	30 6920185	9·802804
884	781456	690807104	29·7321375	9·597337	943	889249	838561807	30·7083051	9·806271
885	783225	693154125	29 7489496	9 600955	944	891136	841232384	30·7245830	9·809736
886	784996	695550456	29 7657521	9·604570	945	893025	843908625	30 7408523	9·813199
887	786769	697864103	29 7825452	9 608182	946	894916	846590536	30 7571130	9·816659
888	788544	700227072	29·7993289	9·611791	947	896809	849278123	30 7733651	9 820117
889	790321	702595369	29·8161030	9·615398	948	898704	851971392	30·7896086	9·823572
890	792100	704969000	29·8328678	9 619002	949	900601	854670349	30·8058436	9·827025
891	793881	707347971	29·8496231	9·622603	950	902500	857375000	30 8220700	9·830476
892	795664	709732288	29·8663690	9 626201	951	904401	860085351	30 8382879	9·833924
893	797449	712121957	29·8831056	9·629797	952	906304	862801408	30 8544972	9 837369
894	799236	714516984	29 8998328	9 633390	953	908209	865523177	30 8706981	9·840813
895	801025	716917375	29 9165506	9 636981	954	910116	868250664	30 8868904	9·844254
896	802816	719323136	29·9332591	9 640569	955	912025	870983875	30 9030743	9·847692
897	804609	721734273	29·9499583	9 644154	956	913936	873722816	30·9192497	9·851128
898	806404	724150792	29 9666481	9 647737	957	915849	876467493	30 9354166	9 854562
899	808201	726572699	29·9833287	9 651317	958	917764	879217912	30 9515751	9·857993
900	810000	729000000	30·0000000	9 654894	959	919681	881974079	30·9677251	9 861422
901	811801	731431701	30 0166620	9·658468	960	921600	884736000	30 9838668	9·864848
902	813604	733870808	30 0333148	9 662040	961	923521	887503681	31·0000000	9·868272
903	815409	736314327	30 0499584	9 665609	962	925444	890277128	31·0161248	9 871694
904	817216	738763264	30 0665928	9 669176	963	927369	893056347	31·0322413	9·875113
905	819025	741217625	30 0832179	9 672740	964	929296	895841344	31 0483494	9·878530
906	820836	743677416	30·0998339	9 676302	965	931225	898632125	31·0644491	9·881945
907	822649	746142643	30·1164407	9 679860	966	933156	901428696	31·0805405	9·885357
908	824464	748613312	30 1330383	9·683416	967	935089	904231063	31·0966230	9·888767
909	826281	751089429	30 1496269	9 686970	968	937024	907039232	31 1126984	9·892175
910	828100	753571000	30·1662063	9·690521	969	938961	909853209	31 1287648	9 895580
911	829921	756058031	30 1827765	9·694069	970	940900	912673000	31·1448230	9 898983
912	831744	758550528	30·1993377	9·697615	971	942841	915498611	31 1608729	9·902383
913	833569	761048497	30·2158999	9 701158	972	944784	918330048	31·1769145	9·905782
914	835396	763551944	30·2324329	9·704699	973	946729	921167317	31·1929479	9·909178
915	837225	766060875	30·2489669	9 708237	974	948676	924010424	31 2089731	9·912571
916	839056	768575296	30 2654919	9 711772	975	950625	926859375	31·2249900	9 915962
917	840889	771095213	30·2820079	9·715305	976	952576	929714176	31 2409537	9 919351
918	842724	773620632	30 2985148	9 718835	977	954529	932574833	31·2569992	9 922738
919	844561	776151559	30·3150128	9 722363	978	956484	935441352	31 2729915	9 926122
920	846400	778688000	30·3315018	9·725888	979	958441	938313739	31·2889757	9·929504
921	848241	781229961	30 3479813	9 729411	980	960400	941192000	31 3049517	9·932884
922	850084	783777448	30 3644529	9 732931	981	962361	944076141	31·3209195	9 936261
923	851929	786330467	30 3809151	9·736448	982	964324	946966168	31 3368792	9 939636
924	853776	788889024	30 3973683	9·739963	983	966289	949862087	31 3528308	9·943009
925	855625	791453125	30·4138127	9 743476	984	968256	952763904	31·3687743	9·946380
926	857476	794022776	30·4302481	9·746986	985	970225	955671625	31·3847097	9·949748
927	859329	796597983	30 4466747	9 750493	986	972196	958585256	31 4006369	9 953114
928	861184	799178752	30 4630924	9 753998	987	974169	961504803	31·4165561	9 956477
929	863041	801765089	30 4795013	9·757500	988	976144	964430272	31·4324673	9·959839
930	864900	804357000	30 4959014	9·761000	989	978121	967361669	31·4483704	9 963198
931	866761	806954491	30·5122926	9 764497	990	980100	970299000	31 4642654	9 966555
932	868624	809557568	30 5286750	9 767992	991	982081	973242271	31·4801525	9·969909
933	870489	812166237	30 5450487	9 771484	992	984064	976191488	31·4960315	9·973262
934	872356	814780504	30 5614136	9 774974	993	986049	979146657	31·5119025	9 976612
935	874225	817400375	30·5777697	9·778462	994	988036	982107784	31·5277655	9·979960
936	876096	820025856	30·5941171	9·782946	995	990025	985074875	31 5436206	9 983305
937	877969	822656953	30·6104557	9 785429	996	992016	988047936	31·5594677	9 986649
938	879844	825293672	30 6267857	9 788909	997	994009	991026973	31·5753068	9·989990
939	881721	827936019	30 6431069	9·792386	998	996004	994011992	31·5911380	9 993329
940	883600	830584000	30 6594194	9·795861	999	998001	997002999	31 6069613	9·99666f
941	885481	833237621	30·6757233	9 799334	1000	1000000	1000000000	31·6227766	10 000000

TABLE OF THE AMOUNTS OF £1 AT COMPOUND INTEREST.

No. of Pay-ments	3 per cent.	4 per cent.	5 per cent.	6 per cent.	No. of Pay-ments	3 per cent.	4 per cent.	5 per cent.	6 per cent.
1	1·03000	1·04000	1·05000	1·06000	26	2·15659	2·77247	3·55567	4·54933
2	1·06090	1·08160	1·10250	1·12360	27	2·22129	2·88337	3·73346	4·82235
3	1·09273	1·12486	1·15762	1·19102	28	2·28793	2·99870	2·92013	5·11169
4	1·12551	1·16986	1·21551	1·26248	29	2·35657	3·11865	4·11614	5·41839
5	1·15927	1·21665	1·27628	1·33823	30	2·42726	3·24340	4·32194	5·74349
6	1·19405	1·26532	1·34010	1·41852	31	2·50008	3·37313	4·53804	6·08810
7	1·22987	1·31593	1·40710	1·50363	32	2·57508	3·50806	4 76494	6·45339
8	1·26677	1·36857	1·47745	1·59385	33	2·65233	3·64838	5·00319	6·84059
9	1·30477	1·42331	1·55133	1·68948	34	2·73190	3·79432	5·25335	7·25102
10	1·34392	1·48024	1·62889	1·79085	35	2·81386	3·94609	5·51601	7·68609
11	1·38423	1·53945	1·71034	1·89830	36	2·89828	4·10393	5·79182	8·14725
12	1·42576	1·60103	1 79586	2·01220	37	2·98523	4·26809	6·08141	8·63609
13	1·46853	1·66507	1·88565	2·13293	38	3·07478	4·43881	6·38548	9·15425
14	1·51259	1·73168	1·97993	2·26090	39	3·16703	4·61637	6·70475	9·70351
15	1·55779	1·80094	2·07893	2·39656	40	3·26204	4·80102	7·03999	10·28572
16	1·60471	1·87298	2·18287	2·54035	41	3·35990	4 99306	7·39199	10·90286
17	1·65285	1·94790	2·29202	2·69277	42	3·46070	5·19278	7·76159	11·55703
18	1·70243	2·02582	2·40662	2 85434	43	3·56452	5·40049	8·14967	12·25045
19	1·75351	2·10685	2·52695	3·02560	44	3·67145	5·61651	8 55715	12·93548
20	1·80611	2·19112	2·65330	3·20713	45	3·78160	5·84118	8·98501	13 76461
21	1·86029	2·27877	2·78596	3·39956	46	3 89504	6·07482	9 43426	14 59049
22	1·91610	2·36992	2·92526	3 60354	47	4·01190	6·31782	9·90597	15·46592
23	1·97359	2·46472	2·07152	3·81975	48	4·13225	6·57053	10·40127	16·39387
24	2·03279	2·56330	3·22510	4·04893	49	4·25622	6·83335	10·92133	17·37750
25	2·09378	2·66584	3·38635	4·29187	50	4·38391	7·10668	11·46740	18·42015

TABLE OF THE AMOUNTS OF AN ANNUITY OF £1.

No. of Pay-ments	3 per cent.	4 per cent.	5 per cent.	6 per cent.	No. of Pay-ments	3 per cent.	4 per cent.	5 per cent.	6 per cent
1	1·00000	1·00000	1·00000	1 00000	26	38 55304	44 31174	51 11345	59·15638
2	2·03000	2 04000	2 05000	2 06000	27	40 70963	47 08421	54·66913	63·70576
3	3 09090	3·12160	3 15250	3 18360	28	42·93092	49 96758	58 40258	68 52811
4	4·18363	4 24646	4 31012	4 37462	29	45 21885	52 96629	62 32271	73·63980
5	5 30913	5·41632	4·52563	5 63709	30	47 57541	56 08494	66 43885	79·05819
6	6 46841	6 63297	6·80191	6 97532	31	50·00268	59 32833	70 76079	84 80168
7	7·66246	7 89829	8 14201	8 39334	32	52 50276	62·70147	75 29829	90·88978
8	8·89234	9·21423	9 54911	9 89747	33	55 07784	66 20953	80·06377	97 34316
9	10·15911	10·58279	11 02656	11 49131	34	57·73018	69 85791	85 06696	104·18375
10	11·46388	12 00611	12 57789	13 18079	35	60 46208	73 65222	90 32031	111·43478
11	12·80779	13·48635	14 20679	14 97164	36	63·27594	77 59831	95 83632	119 12087
12	14·19203	15·02580	15 91713	16 86994	37	66·17422	81 70225	101 62814	127 26812
13	15·61779	16 62684	17 71298	18 88214	38	69·15945	85 97034	107 70954	135 90420
14	17·08632	18 29191	19 59863	21 01506	39	72 23423	90·40915	114 09502	145 05846
15	18·59891	20 02359	21 57856	23 27597	40	75·40126	95 02551	120·79977	154 76196
16	20·15688	21·82453	23 65749	25 67253	41	78·66330	99 82654	127·83976	165 04768
17	21·76159	23 69751	25·84037	28 21288	42	82 02320	104·81960	135 23175	175 95054
18	23·41443	25·64541	28·13238	30·90565	43	85 48389	110 01238	142 99334	187 50758
19	25 11687	27·67123	30·53900	33 75999	44	89 04841	115·41288	151 14300	199 75803
20	26·87037	29 77808	33·06595	36 78559	45	92 71986	121·02939	159·70015	212·74351
21	28·67648	31·96920	35 71925	39 99273	46	96 50146	126 87057	168·68516	226·50812
22	30·53678	34 24797	58·50521	43 39229	47	100 39650	132 94539	178 11942	241·09861
23	32·45288	36·61789	41·43047	46 99583	48	104 40839	139·26321	188·02539	256·56453
24	34·42647	39·08260	44·50200	50 81558	49	108 54065	145 83373	198 42666	272·95840
25	36·45926	41·64591	47·72710	54 86451	50	112 79687	152 66708	209 34799	290·33590

TABLE OF THE PRESENT VALUES OF AN ANNUITY OF £1.

No. of Payments	3 per cent.	4 per cent.	5 per cent.	6 per cent.	No. of Payments	3 per cent.	4 per cent.	5 per cent.	6 per cent.
1	0 97087	0 96154	0 95238	0·94340	26	17·87684	15 98277	14 37518	13·00316
2	1 91347	1 88619	1 85941	1·83339	27	18 32703	16·32958	14 64303	13·21053
3	2 82861	2·77519	2 72325	2·67301	28	18 76411	16 66306	14·89812	13·40616
4	3 71710	3 62999	3·54595	3 46510	29	19 18846	16·98371	15 14107	13·59072
5	4 57971	4 45182	4 32948	4 21236	30	19 60044	17 29203	15 37245	13·76483
6	5·41719	5 24214	5 07569	4 91732	31	20·00043	17 58849	15 59281	13·92908
7	6 23028	6 00205	5 78637	5 58238	32	20 38877	17 87355	15 80267	14·08404
8	7·01969	6 73274	6·46321	6 20979	33	20·76579	18 14764	16 00255	14·23023
9	7·78611	7·43533	7 10782	6 80169	34	21 13184	18·41119	16·19290	14·36814
10	8·53020	8·11089	7·72173	7 36009	35	21·48722	18 66461	16 37419	14·49824
11	9·25262	8·76058	8 30641	7·88687	36	21 83225	18 90828	16 54685	14·62099
12	9·95400	9·38507	8 80325	8 38384	37	22 16724	19·14258	16 71128	14·73678
13	10·63496	9·98565	9 39357	8 85268	38	22 49246	19 36786	16 86789	14·84602
14	11 29607	10·56312	9 89864	9 29498	39	22 80822	19 58448	17 01704	14·94907
15	11·93794	11 11849	10 37965	9·71225	40	23·11477	19 79277	17 15908	15·04630
16	12 56110	11·65230	10·83777	10·10589	41	23 41240	19 99305	17 29436	15·13801
17	13 16612	12 16567	11·27406	10 47726	42	23·70136	20 18562	17 42320	15·22454
18	13 75351	12 65940	11·68958	10 82760	43	23 98190	20 37079	17 54591	15 30617
19	14 32380	13 13394	12·08532	11 15811	44	24 25428	20 54884	17 66277	15 38318
20	14 87748	13 59032	12 46221	11·46992	45	24 51871	20 72004	17 77407	15·45583
21	15 41502	14 02916	12 82115	11 76407	46	24·77545	20 88465	17 88006	15·52437
22	15 93692	14 45111	13 16300	12 04158	47	25·02471	21 04293	17 98101	15·58903
23	16·44361	14·85684	13·48857	12 30338	48	25 26671	21 19513	18 07715	15·65002
24	16 93554	15·24696	13 79864	12 55036	49	25·50166	21 34147	18·16872	15·70757
25	17·41315	15 62208	14·09394	12·78335	50	25·72977	21 48218	18 25592	15·76186

IRISH CONVERTED INTO STATUTE ACRES.

Irish.		Statute.				Irish.	Statute.				Irish.	Statute.			
R.	P.	A.	R.	P.	Y.	A.	A.	R.	P.	Y.	A.	A.	R.	P.	Y.
0	1	0	0	1	18¾	1	1	2	19	5¼	20	32	1	23	14¼
0	2	0	0	3	7¼	2	3	0	38	10½	30	48	2	15	6¼
0	3	0	0	4	26	3	4	3	17	15¾	40	64	3	6	28½
0	4	0	0	6	14½	4	6	1	36	21	50	80	3	38	20½
0	5	0	0	8	3	5	8	0	15	26¼	100	161	3	37	10¾
0	10	0	0	16	6	6	9	2	35	1¼	200	323	3	34	21¼
0	20	0	0	32	12	7	11	1	14	6½	300	485	3	32	2
1	0	0	1	24	24	8	12	3	33	11¾	400	647	3	29	12¾
2	0	0	3	9	17¾	9	14	2	12	17	500	809	3	26	23
3	0	1	0	34	11½	10	16	0	31	22¼	1000	1619	3	13	16¾

VALUE OF FOREIGN MONEY IN BRITISH,

Silver being 5*s*. per ounce.

	s.	*d.*		*s.*	*d.*
1 Florin is worth	1	8	1 Dollar (New York) .	4	2
16 Schillings (Hamburg)	1	5½	96 Skillings (Copenhagen)	2	2⅛
1 Mark (Frankfort) .	1	7⅞	1 Lira (Venice) . .	0	8⅕
1 Franc . . .	0	9½	1 Lira (Genoa) . .	0	9 1/7
1 Milree (Lisbon) . .	4	8	1 Lira (Leghorn) . .	0	7¾
8 Reals . . .	3	1¾	1 Ruble . . .	3	1½

DUBLIN: Printed by ALEXANDER THOM, 87, Abbey-street.

PUBLISHED BY THE

Commissioners of National Education in Ireland.

	Price to National Schools.		Price to the Public.	
	s.	*d.*	*s.*	*d*
First Book of Lessons,	0	0½	0	2
Second Book of Lessons,	0	2	0	7
Sequel to Second Book of Lessons,	0	3	0	9
Third Book of Lessons,	0	4	1	2
Fourth Book of Lessons,	0	5	1	4
Fifth Book of Lessons,	0	6	1	8
Reading Book for Girls,	0	6	1	8
Introduction to the Art of Reading,	0	4	1	6
English Grammar,	0	2	0	8
Key to English Grammar,	0	1	0	3
First Arithmetic,	0	2	0	8
Key to First Arithmetic,	0	2	0	8
Treatise on Arithmetic; Theory and Practice,	0	8	2	6
Book-keeping,	0	3	0	10
Key to Book-keeping,	0	3	0	10
Epitome of Geographical Knowledge,	0	10	2	6
Compendium of Geography,	0	3	1	0
Elements of Geometry,	0	2	0	9
Mensuration,	0	4	1	6
Appendix to Mensuration,	0	3	1	0
Scripture Lessons (Old Testament), No. 1,	0	3	0	9
———— No. 2,	0	3	0	9
Scripture Lessons (New Testament), No. 1,	0	3	0	9
———— No. 2,	0	3	0	9
Sacred Poetry,	0	2	0	6
Lessons on the Truth of Christianity,	0	2	0	7
Directions for Needlework, with Specimens,	4	6	12	0
———— Large,	5	3	14	0
Set Tablet Lessons, Arithmetic, 60 Sheets,	0	7	2	0
Set Tablet Lessons, Spelling and Reading, 33 Sheets,	0	4	1	0
Set Copy Lines,	0	6	1	3
Map of the World,	6	0	21	0
Maps of the Ancient World, Europe, Asia, Africa, America, England, Scotland, Ireland, Palestine, each	4	6	16	0

BOOKS NOT PUBLISHED, BUT SANCTIONED BY THE COMMISSIONERS OF EDUCATION.

By whom arrangements have been made for supplying National Schools with the following Works at the reduced prices affixed.

Lectures on Natural Philosophy, by Professor M'GAULEY,	3	6	7	6
The Spelling Book Superseded, by Professor SULLIVAN,	0	3	1	4
Geography Generalized, by Professor SULLIVAN,	0	9	2	0
Introduction to Geography, &c , by Professor SULLIVAN,	0	3	1	0
Dower's Atlas, 12 Maps, coloured,	1	6	5	0
Kirkwood's Atlas, 12 Maps, plain,	0	6	2	6
Easy Lessons on Christian Evidences,	0	2	0	6
Easy Lessons on Money Matters,	0	3	1	0
Easy Lessons on Reasoning,	0	9	1	6
Treatise on Arithmetic, by JAMES THOMSON, LL D.,	1	0	3	6
Elements of Euclid, Part I., by JAMES THOMSON, LL.D.,	0	10	3	0
———— Part II , by JAMES THOMSON, LL.D.,	0	10	2	6
Introduction to Algebra, by JAMES THOMSON, LL D.,	1	6	5	0
Arithmetical Table Books, *per* 100,	2	0	8	4

Lightning Source UK Ltd.
Milton Keynes UK
UKHW021258171218
334146UK00012B/725/P